AF630818

Climate Change and Food Security

NIPA® GENX ELECTRONIC RESOURCES & SOLUTIONS P. LTD.
New Delhi-110 034

ABOUT THE EDITORS

Dr. Mrinmoy Datta (b. 1954) completed Ph.D. in Soil Science and Agricultural Chemistry from University of Calcutta in 1989 and is now in the position of Principal Scientist (Soil Science) in Tripura Centre of ICAR Research Complex for NEH Region. He has been closely associated with the functioning of Agro-Advisory Services as Nodal Officer since the time (1996) of inception of the service under National Centre for Medium Range Weather Forecasting (NCMRWF) presently in the Ministry of Earth Sciences. Govt of India. Over the years, Agro advisory services has become very popular and are in a position to cater the need of the farming community in modulating the farming operation in accordance with weather forecast as received from NCMRWF/ IMD. To his credit, he has 51 full length paper in reputed national/ international scientific journals, 13 popular articles, 24 abstract papers presented in conference/ symposium, 3 books, 6 book chapters, 1 research bulletin and 6 leaflets/ extension bulletins.

Dr. Narendra Pratap Singh (b. 1957) completed his M.Sc and Ph.D. in Agronomy from the Institute of Agricultural Sciences, Banaras Hindu University, Varanasi, UP in the year 1980 and 1984, respectively. He has maintained good academic record throughout his studies and recipient of District Merit Fellowship, BHU Fellowship and Pesticides India LTd. He has joined in 1984 as a lecturer in the Department of Agronomy, College of Agriculture, North Eastern Hill University, Medziphema, Nagaland and in 1988 he became Reader & Head in Department of Agronomy, School of Agricultural Sciences & Rural Development, North Eastern Hill University, Medziphema, Nagaland. After the establishment of Nagaland University (A Central University), he was on deputation to Nagaland University up to 1996 and he joined later on as a Professor & Head, Department of Agronomy, School of Agriculture Sciences & Rural Development, Nagaland University, Medziphema, Nagaland. In 1999, he joined ICAR Research Complex for NEH Region, Umiam as Principal Scientist and Head in the Department of Agronomy. He is life member of different academic societies in India and abroad. He has visited The Netherlands, France, Thailand, Indonesia and Laos for different academic programmes. Presently, he is working as Joint Director in ICAR Research Complex for NEH Region, Tripura Centre, Lembucherra. He has 45 full length papers in many reputed national/ international scientific journals, 104 popular articles, 46 book chapters, 4 books and 32 research/ extension bulletins in his credit. He is also associated with the functioning of Agro Advisory Services at ICAR Research Complex for NEH Region, Tripura Centre, Lembucherra. He has successfully completed many research projects and also coordinated many training, symposium and seminar program.

Er. Dhiman Daschaudhuri (b. 1972) has been working as Technical Officer in Agromet Advisory Service in ICAR Research Complex for NEH Region, Tripura Centre, Lembucherra. Being an Agricultural Engineer from College of Agricultural Enginnering, MAU, Parbhani, he has started his career in 1997 as Research Associate under Agromet Advisory Service at Division of Agricultural Engineering, ICAR Research Complex for NEH Region, Umiam (Meghalaya) and thereafter joined in the Agromet Advisory Services at Tripura in 2001. He has 5 full length papers in reputed national scientific journals, 2 short communications, 6 book chapters, 1 research bulletin in his credit.

Climate Change
&
Food Security

Edited by

Mrinmoy Datta
(Principal Scientist, Soil Science)

Narendra Pratap Singh
(Joint Director)

Er. Dhiman Daschaudhuri
(Technical Officer, Agromet Advisory Service)

ICAR Research Complex for NEH Region
Tripura Centre, Lembucherra-799 210, Tripura, India

NIPA® GENX ELECTRONIC RESOURCES & SOLUTIONS P. LTD.
New Delhi-110 034

NIPA® GENX ELECTRONIC
RESOURCES & SOLUTIONS P. LTD.

101,103, Vikas Surya Plaza, CU Block
L.S.C. Market, Pitam Pura, New Delhi-110 034
Ph : +91 11 27341616, 27341717, 27341718
E-mail: newindiapublishingagency@gmail.com
www: www.nipabooks.com

For customer assistance, please contact
Phone: + 91-11-27 34 17 17
Fax: + 91-11- 27 34 16 16
E-Mail: feedbacks@nipabooks.com

© 2023, Publisher

ISBN: 978-81-19072-40-8

All rights reserved. No part of this publication may be reproduced, stored in a retrieval system or transmitted in any form or by any means, including electronic, mechanical, photocopying recording or otherwise without the prior written permission of the publisher or the copyright holder.

This book contains information obtained from authentic and highly reliable sources. Reasonable efforts have been made to publish reliable data and information, but the author/s, editor/s and publisher cannot assume responsibility for the validity, accuracy or completeness of all materials or information published herein or the consequences of their use. The work is published with the understanding that the publisher and author/s are not attempting to render any professional services. The author/s, editor/s and publisher have attempted to trace and acknowledge the copyright holders of all material reproduced in this publication and apologize to copyright holders if permission and/or acknowledgements to publish in this form have not been taken. If any copyrighted material has not been acknowledged, please write to us and let us know so that we may rectify the error, in subsequent reprints.

Trademark Notice: NIPA®, the NIPA® logos and their presentations (the way they are written/presented) in this book are the trademarks of the publisher and hence may not be used without written permission, if copied or used without authorization, the infringer will be prosecuted as per law.

NIPA® also publishes books in a variety of electronic formats. Some content that appears in print may not be available in electronic books, and vice versa.

Composed and Designed by NIPA®.

भाकृ अनुप
ICAR

Indian Council of Agricultural Research
Office of the Secretary, DARE & Director General, ICAR
Krishi Bhavan, New Delhi - 110 001

FOREWORD

The science led developments in agriculture enabled us to increase the food production nearly 4 times since independence, but we can not afford to become complacent in the area of food and nutritional security in view of our ever increasing population. Moreover, the globalization and the new world trade scenario demands that every country must bring about a paradigm change in agriculture so as to make their farm produce competitive in terms of cost and quality.

The climate change is a natural process but the rate of change in the last century is certainly alarming. Meteorologists have recorded an increase in torrential rains in the north- west part of the country in recent decades during the monsoon and alternatively a drop in the number of rainy days along the east coast. Devastating floods and cyclones have also struck with great intensity and frequency. Meanwhile, droughts too, have become far more common. In the east, consecutive droughts between 2000 and 2002 saw over 11 million people affected in Orissa alone. While there was a marked decline in the frequency of depression and cyclone formations over the Bay of Bengal and Arabian Sea during the monsoon, their intensity was sometimes higher, resulting in greater loss of lives and property.

Moreover, inter-government panel on climate change (IPCC) has come out its assessment of global warming that strips 'Mother Earth' of her aura of invincibility. The predictions are that even a 0.5 degree centigrade rise in winter temperature could reduce wheat yields by 0.45 tonnes per hectare - 17 per cent drop in productivity. There is likely to be an identical impact on rice cultivation. The international community has also responded to the threat of climate change with the 1997 'Kyoto Protocol' that sets the target of sizeable reduction in global GHG in the commitment period of 2008- 2012.

The edited book "Climate Change and Food Security" has amply covered the various aspects of the impact of climate on productivity of agriculture and other allied sectors. In order to arrest the loss in the agrarian sectors due to weather vagaries, there is a need of timely and accurate weather forecast to be made available at the farmers' level. The prospects of agro-meteorological advisory services which were initiated by India Meteorological Department (IMD) in 1945, have been dealt in details in the book so as to produce location specific useful farm advisories. The publication could prove to be a useful reference for students, researchers, planning and development agencies.

September 12, 2007

(MANGALA RAI)
Secretary
Department of Agricultural Research &
Education and Director General
Indian Council of Agricultural Research
Dr. Rajendra Prasad Road, Krishi Bhawan
New Delhi 110 001, India

PREFACE

Climate of a particular place is dynamic in nature. Climatic changes which have recently been occurring are causing a great concern. Over the years, the anthropogenic activities have resulted in an increasing emission of greenhouse gases and the global mean annual temperature has been increased by 0.4 to 0.7°C above that recorded at the end of the 19th Century.

Due to sudden spurts in temperature before ripening of wheat, India has already witnessed a damage to this crop. Perennial rivers are gradually drying up and aberrations in rainfall are being noticed at regional levels. A shift in monsoon rainfall patterns has occurred since the last 3 years with heavy rainfall in drought prone areas and scanty rain in flood prone areas.

The farm productivity, as it is predicted by US- based Centre for Global Development and Peterson Institute for International Economics, would decline in the range of 3 to 16% by 2080 in India. Considering the issue of climate change and its impact on food supply, a National Symposium on Agrometeorological Advisory Service to Ensure Food Security in North East India was held on the 7th February 2006. The symposium was inaugurated by his Excellency, the Governor of Tripura Shri D.N. Sahay and was attended by many experts in the discipline of Agrometeorology. We do express our heartful thanks for the logistic support provided by the National Centre for Medium Range Weather Forecast (NCMRWF), National Horticultural Board (NHB), National Bank for Agriculture and Rural Development (NABARD), State Bank of India (SBI), Tripura, Pelican Equipments, Chennai, Metos Instrument (India) Pvt. Ltd, Delhi in organizing the symposium.

The recommendations as emerged out of the deliberations made in the National Symposium are as follows. There is a necessity to create a meteorological data bank in North Eastern Region and ICAR Research Complex for NEH Region with its headquarter at Umiam (Barapani), Meghalaya should take the lead to set up such data bank. A project proposal on the creation of data bank is to be submitted to NCMRWF, Noida. Forewarning models on pest and disease incidence in major crops need to be developed in NE Region and location specific dissemination may be targeted preferably at the Panchayat level. Major crops include rice, maize, oilseeds, pulses, vegetables and fruits. Attempts may also be made to develop models to forewarn disease incidence in animals and fish components. Keeping in view the wide variability in weather elements in NE region particularly hilly terrains, planning to set up meteorological observatory at the block level may be thought upon.

The book on "Climate Change and Food Security" is an outcome of the deliberations made in the symposium and contains 22 chapters on various aspects such as, Climate change, past and future strategy of agro advisory services in India, crop weather out look, effect of climate on productivity of food crops, pulses, oilseeds, and horticultural crops, relationship of weather with pest/ disease incidence in crops, impact of weather on livestock, fish and mushroom productivity *etc*. Thus the book which takes into account of the impact of variable weather elements on various aspects of agriculture and allied activities, is, as it is hoped, would be an useful one.

Thanks are due to NABARD for their financial support.

Editors

LIST OF CONTRIBUTORS

A.DHANDAPANI
Senior Scientist
National Centre for Integrated Pest Management, IARI Campus
New Delhi-110 012

A.K.VISHWAKARMA
Scientist (Agronomy)
ICAR Research Complex for NEH Region, Umiam - 793 103
Meghalaya

A.K.SINGH
Principal Scientist (Plant Pathology)
Division of Plant Pathology
ICAR Research Complex for NEH Region, Umiam - 793 103
Meghalaya

A.V.R.KESAVARAO
Visiting Scientist
ICRISAT, Hyderabad

SUDIP DEY
Asstt. Professor
Department of Geography
Tripura University, Suryamaninagar
Agartala - 799 006, Tripura

ASHUTOSH D.DEO
Asstt. Professor
College of Fisheries
Central Agricultural University
Lembucherra - 799 210, Tripura

A.V.M. S.RAO
Scientist (Sr. Scale)
Central Research Institute for Dryland Agriculture
Santosh Nagar, Hyderabad - 500 059
Andhra Pradesh

B.SANTHOSH
Scientist (Fisheries)
ICAR Research Complex for NEH Region, Tripura Centre
Lembucherra - 799 210, Tripura

BRAJENDRA
Scientist (Soil Science)
ICAR Research Complex for NEH Region, Umiam - 793 103
Meghalaya

CHANDER DATT
Scientist, SS (Animal Nutrition)
ICAR Research Complex for NEH Region, Tripura Centre
Lembucherra 799 210, Tripura

D.DASCHAUDHURI
Technical Officer
(Agromet Advisory Service)
ICAR Research Complex for NEH Region, Tripura Centre
Lembucherra - 799 210, Tripura

D.K.DAS
Senior Scientist
National Centre for Integrated Pest Management, IARI Campus
New Delhi - 110 012

G.G.S.N.RAO
Project Coordinator (Ag. Met)
Central Research Institute for Dryland Agriculture
Santosh Nagar, Hyderabad - 500 059
Andhra Pradesh

H.P.DAS
Deputy Director General
Agricultural Meteorology Division
India Meteorological Department
Pune, Maharastra

JAI PRAKASH
Scientist (Horticulture)
ICAR Research Complex for NEH Region, Tripura Centre
Lembucherra - 799 210, Tripura

K.K.SATAPATHY
Principal Scientist & Head
(Soil & Water Cons)
Division of Agricultural Engineering
ICAR Research Complex for NEH Region, Umiam - 793 103, Meghalaya

K.A.PATHAK
Joint Director
ICAR Research Complex for NEH Region, Kolasib - 796 081, Mizoram

K.K.SINGH
Scientist E
National Centre for Medium Range Weather Forecasting (NCMRWF)
Ministry of Earth Science, Govt. of India
A-50 Institutional Area
Phase–II, Sector–62
Noida - 201 301, Uttar Pradesh

L.S.RATHORE
Scientist G/ Advisor & Head
Application Division
National Centre for Medium Range Weather Forecasting (NCMRWF)
Ministry of Earth Science
Govt. of India
A-50 Institutional Area
Phase –II, Sector –62
Noida - 201 301, Uttar Pradesh

M.DATTA
Principal Scientist (Soil Science)
ICAR Research Complex for NEH Region, Tripura Centre
Lembucherra - 799 210
Tripura

M.SANKARAN
Scientist (Horticulture)
ICAR Research Complex for NEH Region, Tripura Centre
Lembucherra - 799 210, Tripura

M.K.SINGH
Reader
Department of Agronomy
Institute of Agricultural Sciences
Banaras Hindu University
Varanasi - 221 005, Uttar Pradesh

N.P.SINGH
Joint Director
ICAR Research Complex for NEH Region, Tripura Centre
Lembucherra - 799 210, Tripura

P.T.SHARMA
Technical Officer
(Agromet Advisory Service)
ICAR Research Complex for NEH Region, Lamphelpat
Imphal - 795 004
Manipur

PANKAJ BAISWAR
Scientist (Plant Pathology)
Division of Plant Pathology
ICAR Research Complex for NEH
Region, Umiam - 793 103
Meghalaya

P.CHOUDHURI
Technical Officer
Agromet Advisory Service
Division of Agril Engineering, ICAR
Research Complex for NEH Region
Umroi Road - 793 103
Meghalaya

S.BANDYOPADHYAY
Senior Scientist
(Animal Parasitology)
Division of Veterinary Parasitology
ICAR Research Complex for NEH
Region, Umroi Road
Umiam - 793 103 Meghalaya

S.BISWAS
Senior Scientist (Plant Pathology)
ICAR Research Complex for NEH
Region, Tripura Centre
Lembucherra - 799 210
Tripura

S.K.DEY
Deputy Director
Rubber Research Institute of India
Regional Research Station
Agartala - 799 006 Tripura

S.R.SINGH
Director
Institute of Agricultural Sciences
Banaras Hindu University
Varanasi - 221 005 Uttar Pradesh

S.V.NGACHAN
Director
ICAR, Research Complex for NEH
Region, Umiam - 793 103
Meghalaya

SHAMMI RAJ
Agrometeorologist
Rubber Research Institute of India
Regional Research Station
Agartala - 799 006, Tripura

T.P.TRIVEDI
Principal Scientist,
National Centre for Integrated Pest
Management, IARI Campus
New Delhi - 110 012

TARKESWAR SINGH
Reader
Department of Agronomy
Institute of Agricultural Sciences
Banaras Hindu University
Varanasi - 221 005
Uttar Pradesh

U.S.SAIKIA
Scientist (Agromet)
Division of Agricultural Engineering
ICAR Research Complex for NEH
Region, Umiam - 793 103, Meghalaya

V.U.M.RAO
Principal Scientist (Ag Met)
Central Research Institute for Dryland
Agriculture, Santosh Nagar
Hyderabad - 500 059, Andhra Pradesh

Y.S.RAMAKRISHNA
Director
Central Research Institute for Dryland
Agriculture, Santosh Nagar
Hyderabad - 500 059 Andhra Pradesh

❑❑❑

CONTENTS

Foreword – Dr. Mangala Rai *v*
Preface *vii*
List of Contributors *ix*

Chapter

1 Climate Change 1-19
K.K. Singh

2 Agrometeorological Advisory Service : Past
Present and Future 21-36
L.S. Rathore

3 Impact of Natural Disasters on Agriculture with
Reference to Strategies Adopted to Cope with it 37-51
H.P. Das

4 Agro Advisory Services - A National Perspective 53-64
Y.S. Ramakrishna, G.G.S.N. Rao, A.V.M.S. Rao
and V.U.M. Rao

5 National Facility of Agromet Data Bank
A Boon to Sustainable Agriculture 65-78
G.G.S.N. Rao, Y.S. Ramakrishna, A.V.R. Kesavarao
A.V.M.S. Rao and V.U.M. Rao

6	Agroclimatic Variability: A Thought on Sustainable Agriculture *N.P. Singh, M. Datta and D. Daschaudhuri*	79-101
7	Effect of Climatic Variability on Crop Productivity in North East India *M Datta, D. Daschaudhuri and N.P. Singh*	103-117
8	Agrometeorological Scenario of North East India *K.K. Satapathy, U. S. Saikia and P. Choudhuri*	119-127
9	Agroclimatic Characterization: A Necessity for Scientific Crop Production and Crop Planning in Dryland Agriculture - A Case Study of Varanasi Region *Tarkeswar Singh, M.K. Singh and S.R. Singh*	129-147
10	Rubber Cultivation in North East India An Agrometeorological Perspective *Shammi Raj and S.K. Dey*	149-162
11	Influence of Weather Parameters on Plant Disease *A.K. Singh and Pankaj Baiswar*	163-170
12	Weather and Climatic Information for Integrated Pest Management *D.K. Das, A. Dhandapani and T.P. Trivedi*	171-189
13	Role of Agrometeorological Advisory Services in Mizoram *K.A. Pathak, Brajendra and A.K. Biswakarma*	191-198
14	Analysis of Rainfall Characteristics at Lembucherra (Tripura) in North East India *D. Daschaudhuri, M. Datta and N.P. Singh*	199-213
15	Scope of Agrometeorological Advisory Services in Manipur *S.V. Ngachan and P.T. Sharma*	215-227

16 Late Quaternary Environmental Changes and their Impact upon Coastal and Paleo-Coastal Bengal Basin in India: A Climato-Geomorphic Study 229-241
Sudip Dey

17 Climate and Horticultural Crops 243-265
M. Sankaran, Jai Prakash, N.P. Singh and M. Datta

18 Influence of Environment in Cultivating Edible Mushroom 267-274
S. Biswas and N.P. Singh

19 Climate and Livestock Production 275-291
Chander Datt and N.P. Singh

20 Incidence of Strongyle Infection in Cattle and Pig with Relevance to Weather Parameters 293-298
S. Bandyopadhyay

21 Climate and Freshwater Fisheries 299-317
B. Santhosh, M. Datta and N.P. Singh

22 Impact of Climate on Fish Diseases 319-325
Ashutosh D Deo

The colour plates can be found on pages

.... 11-12
.... 29-30
.... 61-62
.... 73-76
.... 83
.... 125-126
.... 195-198
.... 217-222

Index 327

Climate Change and Food Security, 2008
Edited by: M.Datta, N.P.Singh & Er.D.Daschaudhuri, pp.1-19
New India Publishing Agency, New Delhi (India)
E-mail : newindiapublishingagency@gmail.com
Web: www.bookfactoryindia.com

1

Climate Change

K.K. SINGH

INTRODUCTION

The debate on climate change is going on. From the report of the Inter Governmental Panel on climate change (IPCC), a premier organization set up in 1998 by the World Meteorological Organisation and United Nations Environment Programme, it is known that the average surface temperature has increased over the 20th century by about 0.6°C. The number may seem insignificant but it is an unprecedented rise. In the past 150 years, the 11 hottest years were witnessed. It says that 1990s was the warmest decade and 1998 the warmest year since 1861. Worse, the Artic is warming twice as rapidly as the rest of the world and there is a distinct possibility of the Greenland Ice Sheet collapsing altogether and pushing up sea levels by several metres. It is predicted that 2035 is the year when Himalayan glaciers may totally disappear causing catastrophic disruptions. Unfortunately, India is among the countries that will suffer the most serious consequences as a result of global warming. Following is the probable effect of global warming.

- 40 per cent of world population now faces chronic shortage of fresh water for daily needs.
- Two-thirds of the world's farm lands suffer from soil degradation.
- Half of the world's wetlands have been lost.
- Since 1990, 24% of the world's forests have been destroyed. The rate of loss is now 90,000 sq. km every year.

- Air pollution has now become major killer accounting for death of 3 million people every year.
- 800 wildlife species have become extinct and 11,000 more are threatened.
- Of the 9,964 known bird species, 70 per cent have declined in numbers.
- Almost 75 per cent of the world's marine captures is over fished or fully utilized. In North America, 10 fish species went extinct in the 1990s.
- One-fifth of the 10,000 fresh water species have become extinct.

GREENHOUSE GASES

Global warming is caused by greenhouse gases, of which the increase in atmospheric CO_2 is playing a vital role. In India, the per capita emission of greenhouse gases is 25 times less than that of average American and 15 times less than that of European emission.The US is responsible for almost one -third of the world's total greenhouse gas emissions. The 1992 Earth Summit in Rio de Janeiro hammered out a framework convention on climate change to bring down the carbon dioxide emissions worldwide. The Kyoto Protocol had called upon the developed countries to reduce the greenhouse gases to 1990 levels by 2012 - 5.2 % reduction. The EU responded well and is also actively engaged in carbon trading. But with US and Australia, not part of the protocol, its overall implementation has been tardy. The account of greenhouse gases and per capita emissions of carbon dioxide in the globe are presented in Figs.1 and 2.

Presently, 379 parts per million is the concentration of carbon dioxide in the atmosphere being highest in 6.5 lakh years.CO_2 emissions across the globe are due to burning of fossil / liquid / solid / gaseous fuels. In India, the fossil fuel consumption is exploding and efficiency in power generation, its distribution / consumption can also be improved upon with an avenue to search sustainable energy alternatives such as solar and wind power etc. Another source of GHGs is methane and nitrous oxide. Rice cultivation amply contributes to emissions of methane and nitrous oxide. Ruminants also produce methane.

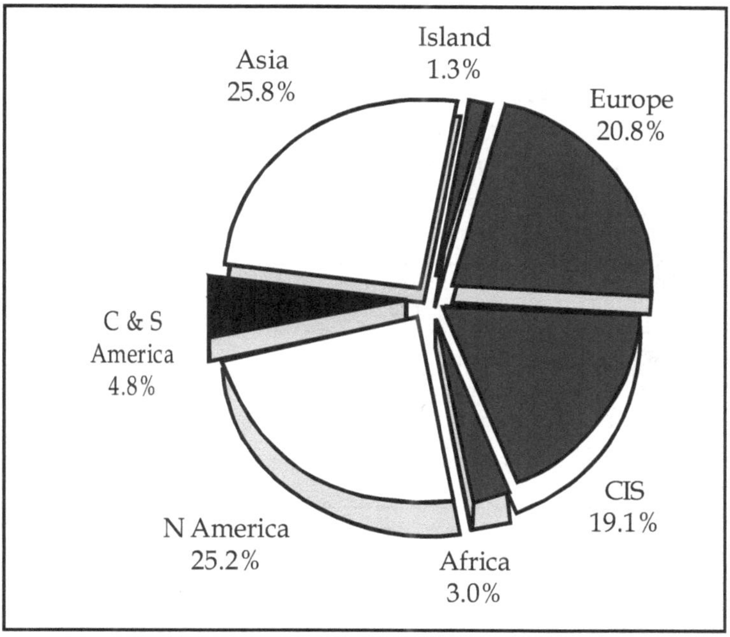

Fig. 1. Greenhouse gases in the globe

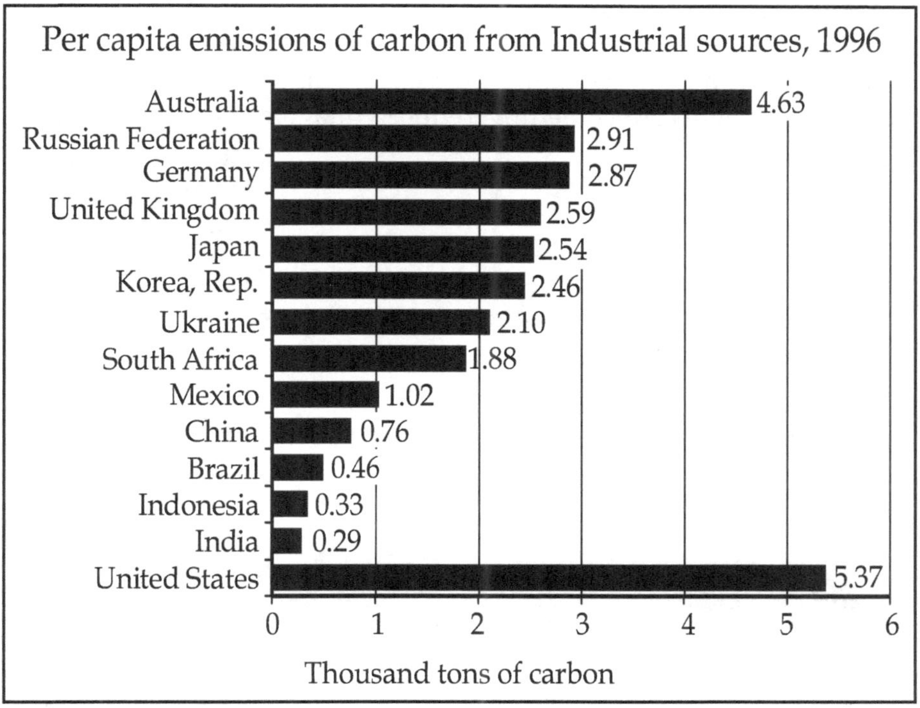

Fig. 2. Per capita emissions of CO_2 in countries

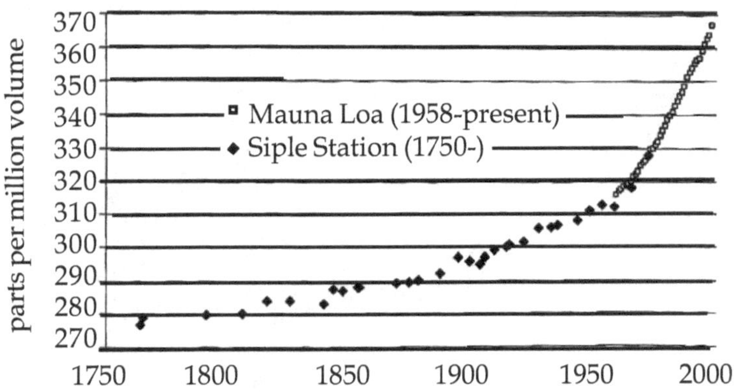

Data Source : C.D.Keeling and T.P. Whorf, Atmospheric CO2 Concentrations (ppmv) derived from in situ air samples collected at mauna Loa Observatory, Hawaii, Scripps Institute of Oceanography, August 1998. A Neftel et al., Historical Co2 Record from the Siple Station Ice Core, Physics Institute, University of Bern, Switzerland, September 1994. See http://cdiac.esd.ornl.gov/trends/co2/contents.htm

Fig. 3. Atmospheric carbon dioxide concentration

For a period of 150 years, there is a rise in methane concentration in the atmosphere from 800 to 1800 parts per million. Field measurements under rice cultivation in several Asian countries have identified possible technical options to mitigate methane emissions through modified water regimes (mid season drainage, alternate flooding), modified residue management (sequestration of straw), use of additives (phosphogypsum, nitrification inhibitors) and modified land management (direct seeding, reduced tillage, site- specific nutrient

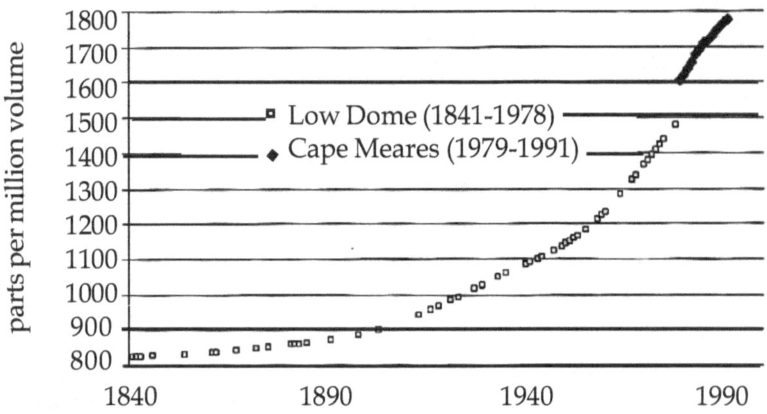

Data Source : D.M. Etherldge et al. Concentrations of CH4 from the Law Dome (East Side, "DE08" Site) Ice Core(a), Commonwealth Scientific and Industrial Research Organisation, Aspendate, Victoria, Australia. September 1994. Available: http://cdiac.esd.ornl.gov/ftp/trends/methane/lawdome.259. M.A.K. Khalli, R.A. Rasmussen, and F. Moraes. "Atmospheric methane at Cape Meares: Analysis of a high resolution data base and its environmental implications. "Journal of Geophysical Research 98:14,753-14,770. 1993. Available: http://cdiac.esd.ornl.gov/ftp/db1007/cmeares.mon

Fig. 4. Atmospheric methane concentration

management). Similarly novel approaches of demand driven N supply using leaf colour charts and other site- specific N management (placement of fertilizer in reduced soil layer under puddle condition) could minimize the pool of excessive nitrogen in the soil and thus reduce nitrous oxide emissions. Among the three greenhouse gases, carbon dioxide has contributed most to global warming over the past century.

Evidence of Global Warming

From the balloon and satellite data, it is observed that the troposphere which is a layer of atmospheric air extending about seven miles upwards from the earth's surface, is showing a warming trend since 1958. Normally , temperature falls in troposphere with increase in height.

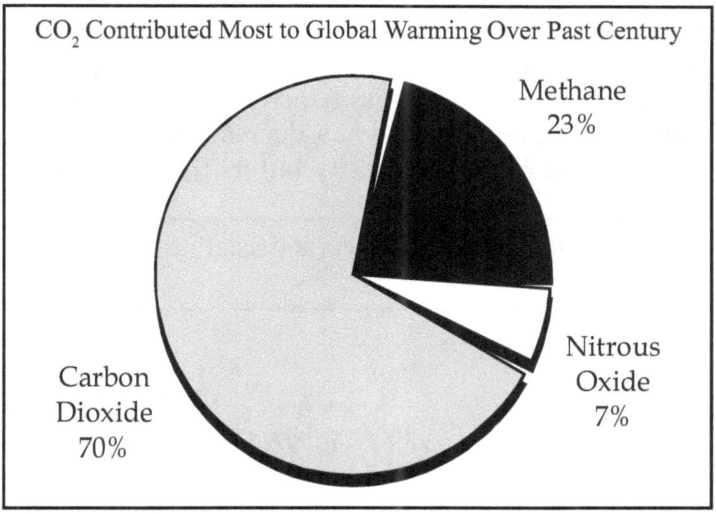

Fig. 5. Contribution of greenhouse gases for global warming

But in stratosphere, which is a layer of atmospheric air above troposphere, temperature is showing a cooling trend. Normally in this layer, temperature ceases to fall with increase in height but remains constant.

OBSERVED CLIMATE CHANGE (IPCC, 2001)

Temperature

Global average	
Surface temperature	+ 0.6 ± 0.2°C
Warmest decade	1990
Warmest year	1998

Rainfall	
Mid and high latitudes	+0.5-1%/ decade
Tropical land areas	(10°N-10°S) + 0.2 - 0.3% / decade
Sub-tropical regions	-0.3% / decade (10°N-30°N & 10°S-10°S)
Cloud Cover	
Mid to high latitudes	+ 2%
Snow Cover	-10%
Glaciers	Alpine and Continental mountain Retreat
Global sea level	+10-20cm

Moreover, warm episodes are more frequent, persistent and intense than cold episodes. There is an increase in extreme precipitation event but the total precipitation has decreased. There is no strong evidence for tropical cyclone intensity but its frequency has changed.

Evidence from Balloon and Satellite Data

- The troposphere is showing a warming trend
- The stratosphere is showing a cooling trend

Analysis of data since 1958

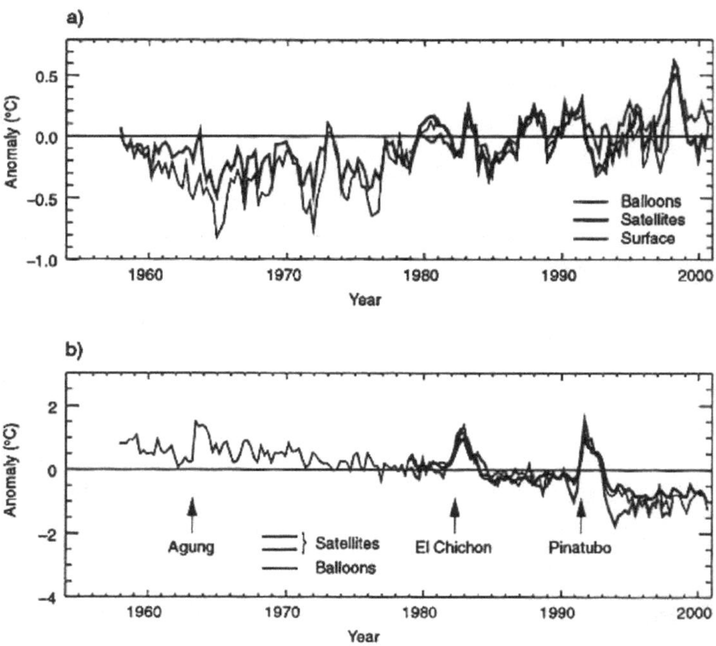

Fig. 6. Temperature trend in troposphere and stratosphere

EXTREME EVENTS

Asia

- *The 2002 drought in India :* The month of July, turned out to be the driest in the recorded history since 1877.
- *A 20 –day heat wave during May 2003 in Andhra Pradesh, India :* Maximum temperatures reached as high as 45-50^0 C, causing 1500 deaths.
- Conversely, extremely cold winter temperatures occurred across Asia in January, 2003.
- Thousands of deaths were attributed to extremely cold conditions in India and Bangladesh during January 2003.
- On 20 May, 2003, Sri Lanka witnessed one of the worst flooding in more than half a century which claimed about 150 lives and left about 3,00,000 people homeless.
- Drought like situation in India in July 2004. All India rainfall was 19% below LPA.
- Dubai for the first time received snowfall in the first week of the January, 2005.
- Kashmir Valley received one of the heaviest snowfall events in February, 2005.
- Very severe heat wave conditions over North and East India in June, 2005.
- Exceptionally heavy rains in Mumbai (94.4cm) on 26 July, 2005.
- *March 2005, Thailand:* The worst drought in decades strikes 70 out of Thailand's 76 provinces, affecting 8.3 million people. The agriculture sector, normally accounting for around 9 per cent of Thai GDP, is set to register a negative growth for 2005.

Rest of the World

- Moscow received snowfall in June, 2003 for the first time since 1963. Temperatures in northwestern Russia were as low as -45^{0}C.
- An all-time maximum temperature record in the United Kingdom was broken on Aug.10, when the mercury reached 38.1^{0}C in Kent.

- France had its warmest summer on record in July-August 2003. More than 14,000 people died.
- Normal-to-above normal rainfall in the Sahel led to ideal growing conditions in much of this region of Africa.
- Severe to extreme drought condition in western United States together with very heavy rain and snow during late December 2003 and upto early January 2004.
- Continuance of long term drought across areas of Africa, including the Greater Horn and parts of Southern Africa, in 2004.
- Excessive heavy rains during January-February 2004 in Brazil.
- Heavy rains in the United Kingdom in February, 2004 causing floods.
- *August 2005, United States:* Hurricane 'Katrina' devastates large areas along the coastline in Louisiana, Mississippi and Alabama, killing hundreds of people and causing upto $ 30 billion property loss. Scientists have noted that such hurricanes have grown in number and intensity by around 30 % over the last decade.
- *July 2005, Romania:* New severe flooding after continued heavy downpour damages properties with an estimated value of €650 million.
- *April2005, Romania :* The worst flooding in 40 years with continued downpours, strikes Romania, threatening 7000 farms near the Hungarian border, cutting off roads, rail and airports. Damage is said to be in millions of Euro.
- *February 2005, Greenland:* The Southern coast of Greenland experiences record breaking temperatures, with 16^0C even warmer than Greece.
- *January 2005, Northern Europe :* Countries across northern Europe face storm of high wind intensity in 40 years, leaving hundreds of thousands of homes in Scandinavia, Latvia, Estonia and Lithuania left without power. In Sweden, two nuclear power stations have to be shut down for safety reasons. Northern Britain is also hit by the storm and sees worst flooding in decades.

ANNUAL AND MONSOON RAINFALL

In India, the annual and monsoon rainfall for a period of 100 years from 1871 were recorded and presented in Fig.7 and both rise and fall compared to normal is recorded.

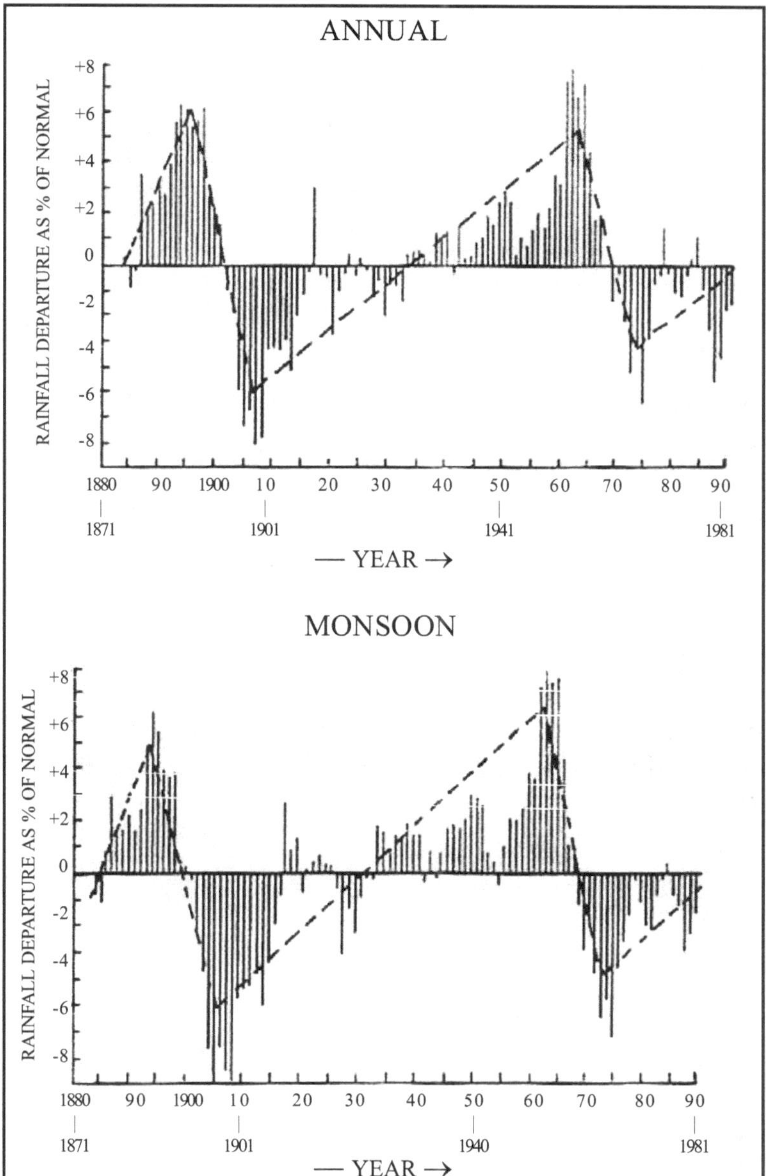

Epocal Trends in Annual and Monsoon Rainfall of India

Fig. 7. Trends in annual and monsoon rainfall in India for 100 years

The trends in the frequency of rainfall in India for a period of 90 years during winter, pre-monsoon, monsoon and post- monsoon are presented in Fig.8.

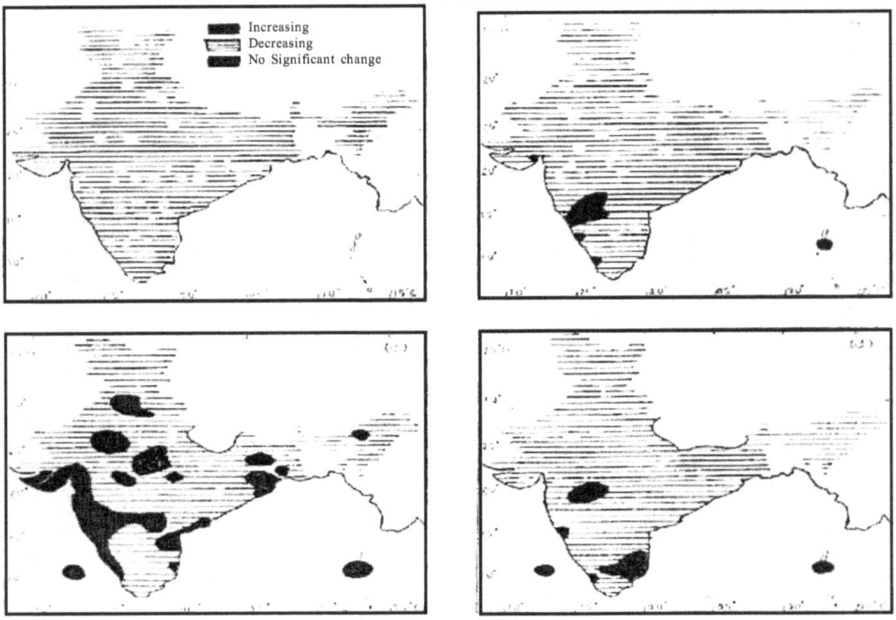

Trends in the frequency of rainfall > 7 cm in 24 hours (1901-1990)
(a) Winter (b) Pre-monsoon (c) Monsoon (d) Post-monsoon (De, U.S. 2001)

Fig. 8. Rainfall frequency in India for a period of 90 years (1901-1990)

The behaviour of monsoon rainfall recorded for a period of 50 years is presented in Table 1. It is noted that the number of break days is more in the month of July as compared to August.

Table 1. Behaviour of Indian Monsoon Rainfall: Multidecadal changes in Break Days

Period	Number of Break Days During					
	July			August		
	01-10	11-20	21-31	01-10	11-20	21-31
1888-1917	46	49	53	43	84	26
1918-1947	14	36	21	55	54	25
1948-1977	22	44	64	21	33	41
1978-2003	23	32	39	6	14	37

AIR TEMPERATURE

The variation in air temperature for a period of 100 years in India is presented in Fig.9. It is indicated that there are both rising and declining trend in air temperature. In Fig.10, seasonal variation in air temperature recorded for a period of 100 years in 31 meteorological stations is presented . It is found that there were fluctuations in winter, pre-monsoon and post monsoon season but monsoon showed low fluctuations form the base line.

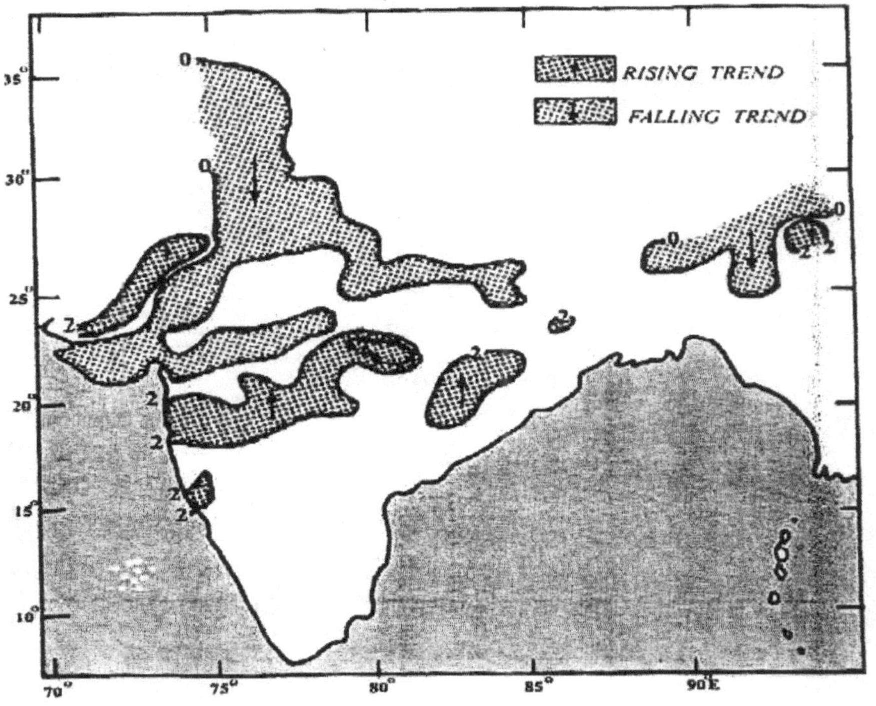

Spatial Variation in the Linear Trend of Air Temperature (in °C/100 Years) Over the Country

Fig. 9. Temperature variability in India for a period of 90 years (1901-1990)

CYCLONIC DISTURBANCES

For a period of 100 years, cyclonic disturbances/storms as recorded across the country are presented in Table11.

It is noted that cyclonic disturbances as compared to cyclonic storms were of wide variation over a period of 100 years thus indicating rising trend of cyclones over the globe in recent years.

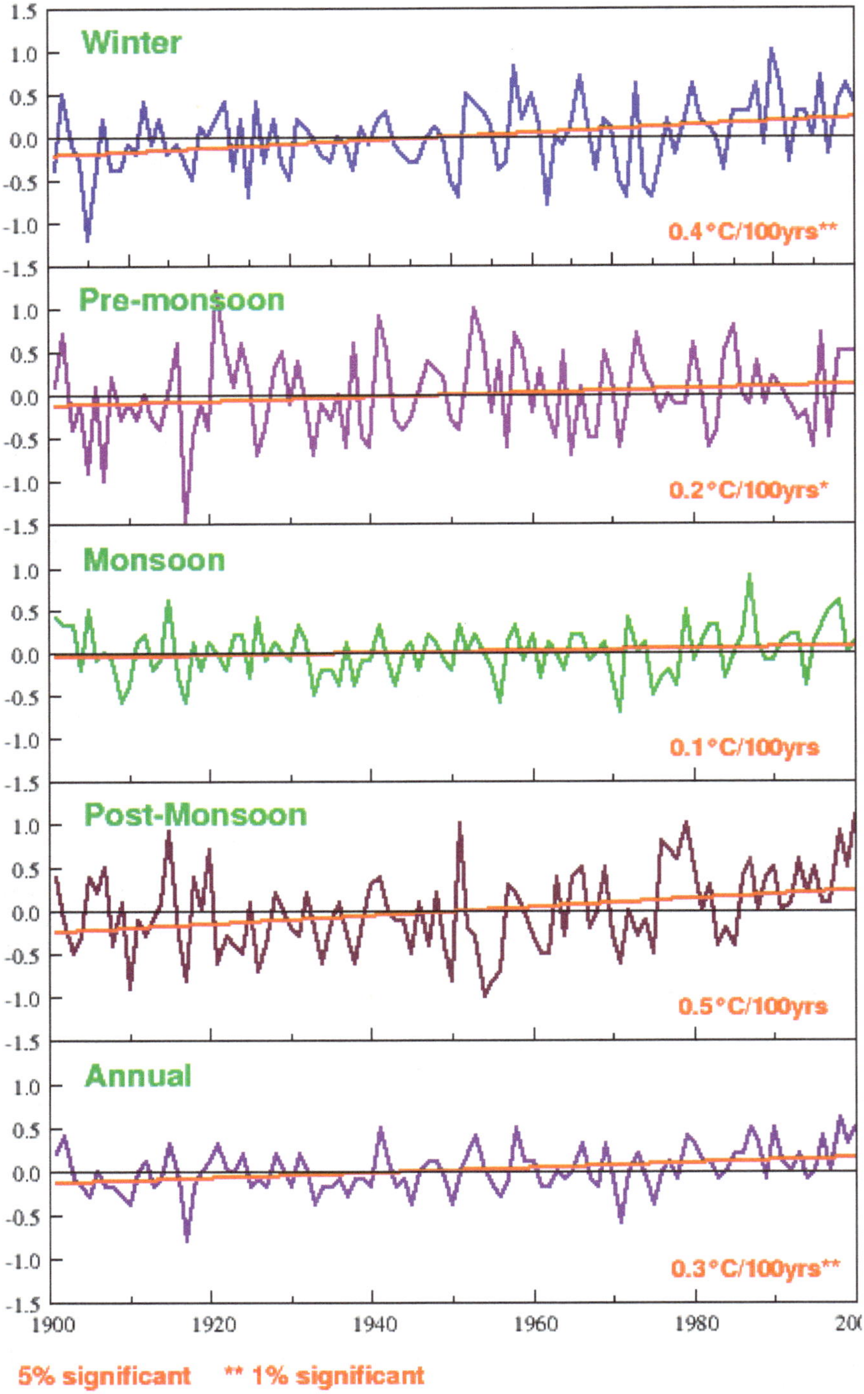

Fig. 10. Seasonal variation in air temperature for 100 years (1901-2000)

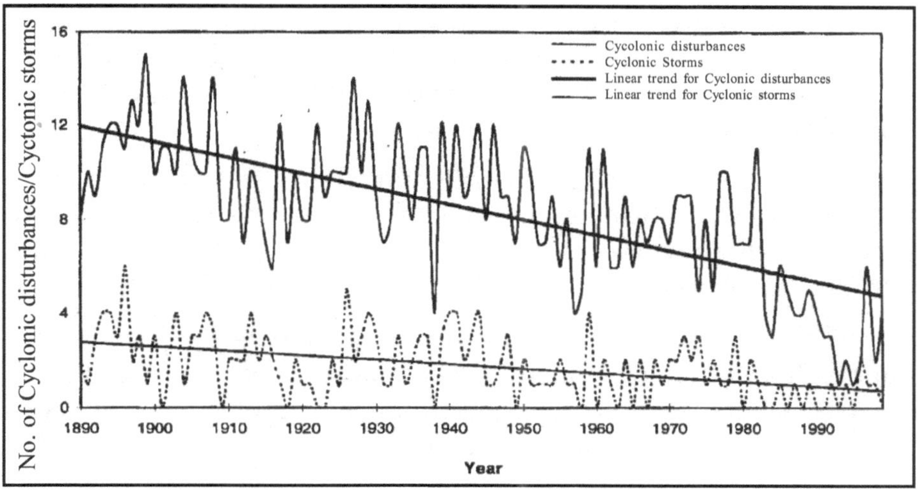

Fig. 11. Cyclonic disturbances/storms

FUTURE PROJECTIONS IN EMISSIONS AND CLIMATIC PARAMETERS

On the basis of special report on emission scenarios (SRES model simulation), following are the situations being derived globally.

A1: Rapid economic growth; Global population peaks in mid-century; Rapid introduction of new and more efficient technologies; Convergence among regions; increased cultural/social interaction; reduction in income disparities.

IPCC 2001 climate change scenarious

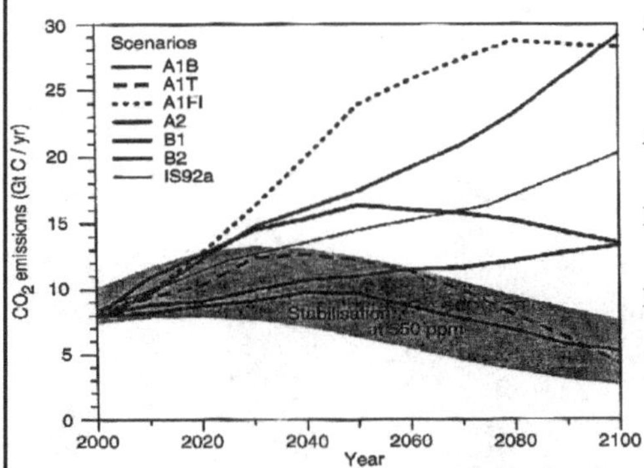

A1: A world of rapid economic growth and rapid introductions of new and more efficien technologies

A2: A very heterogenous world with an emphasis on family values and local traditions

B1: A world of "dematerialization" and introduction of clean technologies

B2: A world with an emphasis on local solutions to economic and environmental sustainability

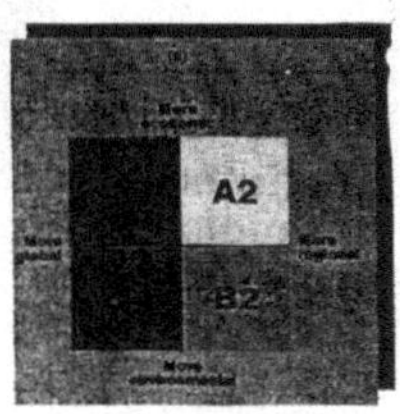

Fig. 12. Carbon dioxide emissions in future

A2 : Heterogeneous world; Self-reliance; Preservation of local identities; Continuously increasing population; Regionally oriented economic development; Fragmented per capita economic growth and technology change.

B1: Convergent world; Global population peaks in mid-century and declines thereafter; Rapid change in economic structures toward service and information; Reduction in material intensity and introduction of clean/resource-efficient technologies; Global solutions

B2 : Local solutions; Continuously increasing population; Intermediate levels of economic development; Less rapid and more diverse technological change; Environmental protection and social equity with focus on local/regional levels.

CO_2 Emissions

It is noted from the perusal of Fig.12 that carbon dioxide emissions may show an increasing trend under different global situations but may be stabilized at 500 parts per million around 2040-2060.

Monsoon Rainfall

On the basis of PRECIS (Providing Regional Climates for Impact Studies) Runs, monsoon rainfall scenarios for the states in India under two different scenarios (A2 and B2) are developed.The monsoon rainfall scenario for a period of 2071-2100 is presented in Fig.13 and the perusal

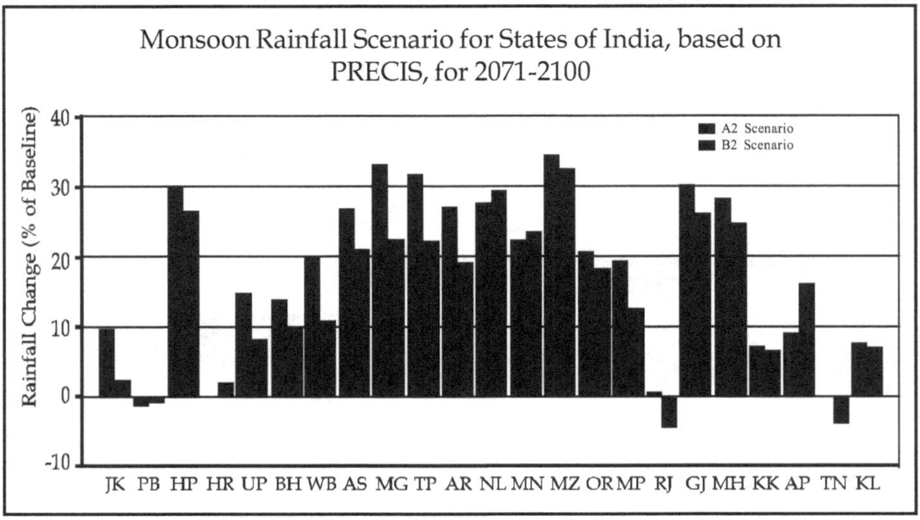

Fig. 13. Monsoon rainfall scenario in India

of projected increase/decrease in annual rainfall (Table2) indicates that besides Punjab,Rajasthan and Tamilnadu, all other states showed an increase (1-33%) in rainfall in the forthcoming years of 2071 -2100.

Table 2. Projected Increase or Decrease in Annual Rainfall for 2071-2100

State	% Rainfall Increase/ Decrease	State	% Rainfall Increase/ Decrease
J & k	+ 3-10	Manipur	+22-23
Punjab	-1	Mizorum	+32-33
Himachal	+26-30	Orissa	+19-21
Haryana	+1-3	Rajasthan	+1 to -4
UP + Uttranchal	+ 8-15	MP + Chattisgarh	+12-19
Bihar + Jharkhand	+10-12	Gujarat	+18-21
West Bengal	+10-20	Maharastra	+ 17-19
Assam	+21-25	Karnataka	+ 8
Meghalaya	+22-33	Andhra	+ 9-15
Tripura	+ 22-32	Tamilnadu	1 to -5
Nagaland	+ 28-29	Kerala	+7-8

Air Temperature

Changes in temperature in India over a period from 1800 to 2100 are presented in Fig.14 and it is predicted from SRES model that there would be change in temperature around 1-5^{0}C probably after 2020.

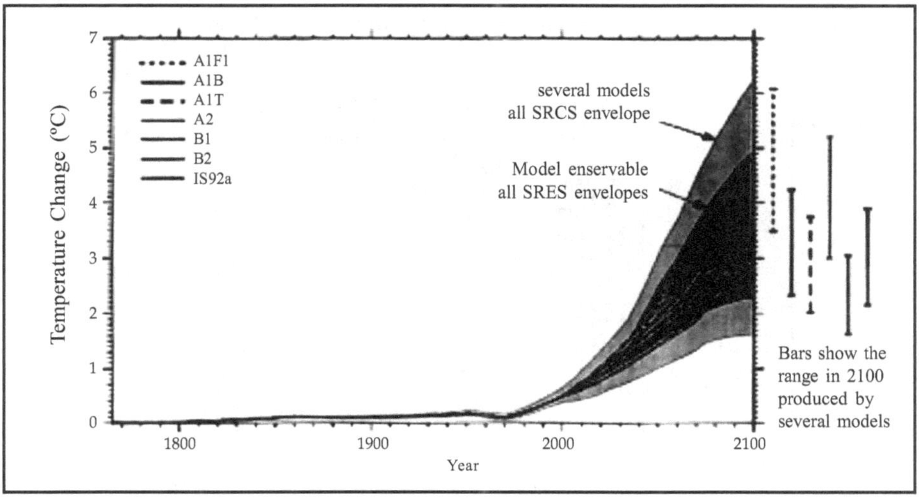

Fig. 14. Changes in temperature

On the basis of PRECIS model, annual temperature scenarios in states of India are presented in Fig. 15. Under A2 scenario, the temperature change would be around 3 to 5°C but under B2 scenario, the temperature change would be around 2.8 to 3.5°C.

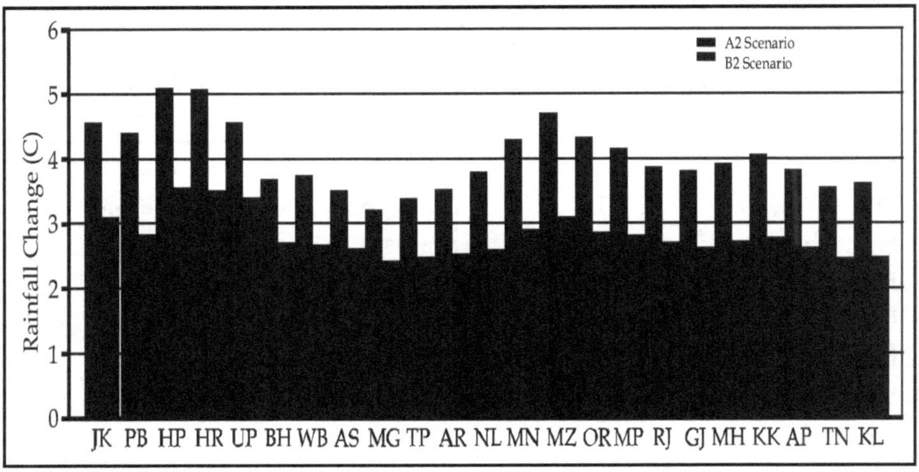

Fig. 15. Annual Temperature Scenario in States of India (2071-2100)

Under the impact of rise in temperature, the pod setting in groundnut under A2 emissions scenario may be drastically reduced(50%) and other crops would show 4-10% reduction in yield (Fig.16). Some adaptation may be made through use of tolerant varieties.

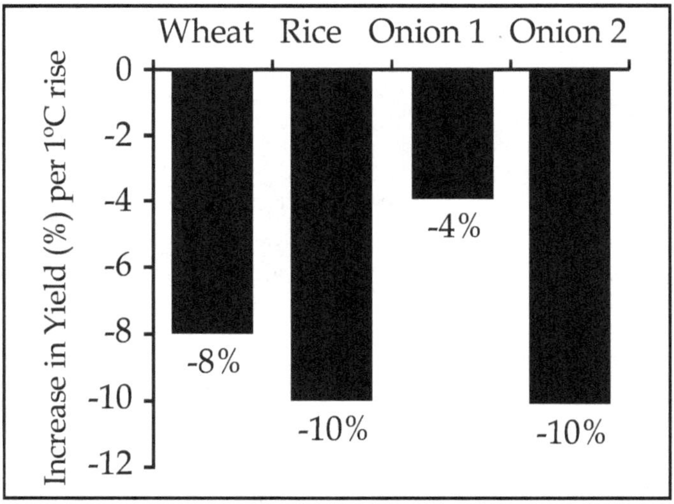

Fig 16. Effect of warmer seasonal mean temperature on crop yield

The increase in wheat yield with rise in carbon dioxide concentration (Fig.17) is offset by probable rise in temperature.

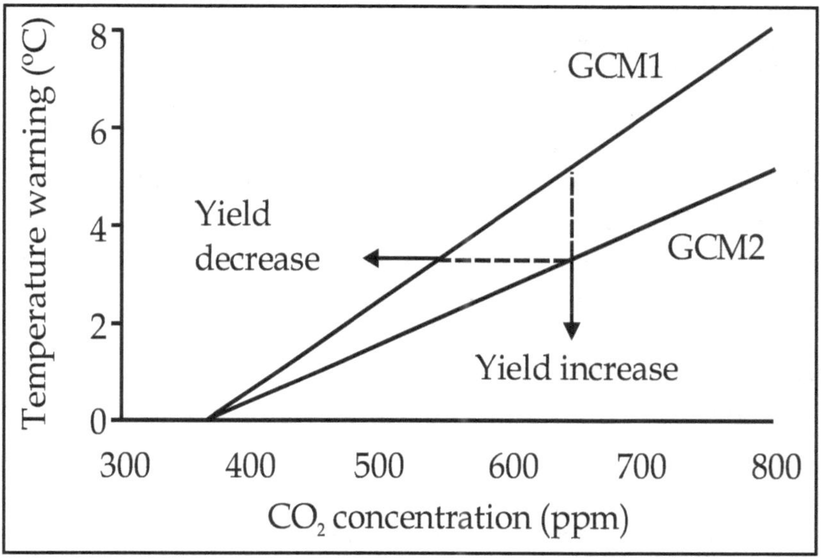

Fig.17. Wheat increase/decrease scenario under high CO_2 concentration and temperature.

Future Projections in Extreme Events

- Higher maximum temperature and more hot days.
- Higher minimum temperature and fewer cold/frost days.
- Reduced diurnal temperature.
- Increase of heat index areas.
- More intense precipitation.
- Increase in summer continental drying and associated risk of drought.
- Increase in tropical cyclone intensity.
- Retreat in glaciers and ice caps during 21st century.
- Decrease in snow cover and sea ice in northern hemisphere.
- Increase in mass of Antarctic ice sheet.
- Decrease in mass of artic ice sheet.
- Monsoon variability.

- Heavy precipitation events and floods.
- Snowmelt.
- Droughts and desertification.
- Heat and cold waves.
- Sea level rise and coastal impacts.

IMPLICATIONS OF EXTREME EVENTS IN INDIA

Tropical Cyclones

- An increasing trend in the frequency of cyclones/storms in the Bay of Bengal in May and November.

Monsoon Variability

- Strong inter-annual and decadal scale variability in Indian monsoon.
- Distortion in long range prediction of Indian monsoon rainfall due to the impact of climate change on Eurasian snow cover.

Floods /Droughts

- Probability in the occurrence of stronger monsoons due to increase in land-sea temperature contrast.
- Higher frequency in flood events.
- A strong correlation between mean precipitation and its inter-annual variability affecting floods/droughts.

Heat and Cold Waves

- Likely presence of both heat /cold waves due to increase in variability.

Sea Level Rise and Coastal Impacts

- Necessity to have coastal zone management in 7500 km long coastline and big/small islands.
- Proper planning of development activities in new ports and industries.
- Protection from increase in floods and coastal erosion.
- Conservation of natural ecosystems.

REASONS FOR CONCERN

Unique and threatened systems

- Tropical glaciers, coral reefs, mangroves.
- Communities highly dependent on natural resources.

Global aggregate impacts

- Difficult to estimate because gains for some may offset losses for others.

Distribution of impacts

- Impact of climate change will not be evenly distributed across the world.
- Developing countries may be more socially and economically vulnerable.

Extreme weather events

- Effects are more localized.
- Critical design thresholds may be crossed.

Large scale singular events

- Melting of west Antarctic ice sheet.
- Not properly understood.

CONCLUSION

The projections of future climate change must be interpreted very carefully and prudently. It must be remembered that regional impacts could differ substantially from global projections. The likely impacts of climate change have to be conceptualized in our long term planning for disaster preparedness, mitigation and its management. There may be areas not vulnerable to disasters now but could have been vulnerable in future. Such areas may be identified. The possibility that disasters could strike with increased fury as well as frequency than at present, should go into the process of our planning and design considerations for the future.

Climate Change and Food Security, 2008
Edited by: M.Datta, N.P.Singh & Er.D.Daschaudhuri, pp.21-36
New India Publishing Agency, New Delhi (India)
E-mail : newindiapublishingagency@gmail.com
Web: www.bookfactoryindia.com

2

Agro-Meteorological Advisory Service Past, Present and Future

L.S. RATHORE

INTRODUCTION

Research in agriculture and allied activities and the subsequent technological innovations are important requisites for enhanced productivity and generating income for the farmers. In India, farming is a way of livelihood to 65 per cent of nearly 1.1 billion population. Our farm population is increasing annually by 1.84 per cent. The average farm size is becoming smaller each year and the cost-risk-return structure of farming is becoming adverse with the result that farmers are getting increasingly indebted. There is no option except to produce more food and other commodities under conditions of diminishing per capita arable land and irrigation water resource. Hence, we must harness the best in frontier technologies, thus integrate them with traditional wisdom and thereby launch an eco-technology movement.

The productivity in agricultural sectors is dependent on several bio physical and socio-economic factors. The significant biophysical factors are weather parameters, soil health, water resources, crops/ varieties and pest environments. In order to monitor the impact of weather parameters, we need to have weather forecasts and the accuracy of which to harvest the economic benefits should be given top priority. Only fifty years ago, weather forecasting was an art which was derived after interpretation of data received from land based observing stations, balloons and aircraft etc. Now, modern technology particularly computers / weather satellites / radars has resulted in enormous improvements in the accuracy of weather forecasting with the help of various coordinated meteorological observing networks.

In India, agro-meteorological advisory service has gained immense popularity. In order to render effective service in this sector, all players such as National Centre for Medium Range Weather Forecasting, India Metrological Department, India Council of Agricultural Research, State Agricultural Universities, Agricultural Departments (Union/State) and the like have joined hands to run 107 Agro-Meteorological Advisory Service (AAS) units in different agro-climatic zones of the country.

The transmission of crop and weather data from agro meteorological service units to the rural areas is now much easier than in the past. The challenge for the meteorological community is to transmit weather warnings and forecasts to all those who are responsible to tailor weather based advisories for use by specific groups and mitigate risk by developing more effective means to integrate weather, water and climatic information into agri-management plans, strategies and decision. Weather service and Agricultural Extension sectors each have distinct roles in effective dissemination process of weather information. They must work strategically as partners for the public good and economic benefit of our country as a whole.

AAS-PAST

Agromet advisory service in the country was initiated through the interventions made by India Metrological Department in 1976. There was an attempt to publish 'Farmers Weather Bulletin' in 1945. The weather forecast issued by IMD was of short range and qualitative in nature. Crop weather calendar was also made and weather forecast was mainly disseminated through AIR.

AAS-PRESENT

National Centre for Medium Range Weather Forecast under the framework of Department of Science and Technology, Govt. of India has taken a lead to bring about a radical change in Agromet advisory Service across the country. There is a multi-institutional approach involving IMD/ICAR/SAUs/ Media in generating the modern agromet advisory service. Overwhelming increases in computing power along with sweeping acceleration in world wide communications are able to run sophisticated Numerical Weather Prediction (NWP) models simulating the atmosphere on a three dimensional grid of fine resolution representing earth-ocean atmosphere systems and finally to predict weather a week in advance. NWP is realized by integrating primitive equation modes. The dynamical quantities such as pressure and velocity gradients are evaluated in spectral space, while other parameters such as radiation, moisture conversion, turbulence are calculated in grid point space.

AGROMET ADVISORY SERVICE (AAS) NETWORK

The total units working for agromet advisory service network in India are 107 and out of which, 83 units are displayed in the map.

Fig. 1. Map of India

SALIENT FEATURES OF AGROMET ADVISORY SERVICE (AAS)

Agromet advisory service run by NCMRWF, Department of Science & Technology, Govt. of India has the following salient features enumerated below.

- At present there are 107 AAS units but there are plans to cover all 127 agro-climatic zones.
- Units are opened in State Agricultural Universities and ICAR Institutes.
- NCMRWF issues location specific quantitative weather forecasts upto one week in advance and twice a week.
- AAS Units translate these forecasts into Agro-advisories and disseminate them to Media and also directly to farmers.
- Farmers' feedback is an integral part of AAS.
- NCMRWF runs Crop Weather Models through inclusion of output in the day to day advisories.

INFRASTRUCTURE

NCMRWF has the following types of computers working in different models.

Computers

- Cary SV1 High end (One of the fastest computers in the country)
- Dec-Alpha (Indigenous)
- ORIGIN-200 (Parallel and Single CPU)
- PARAM PADMA (Indigenous)-Being procured

Models

- Global Models-Operationally used for day today weather forecast

  T-80 (150km x 150km)

  T-170 (75km x 75km)

- Mesoscale Models- Operationally used for weather forecast

  MM5 (Nested 90, 30 and 10 km)

  Eta (48 km resolution)

- Ocean Wave Model – for ocean state forecast
- Crop Weather Models (Implemented at AAS Units)

AGROMETEOROLOGICAL ADVISORY SERVICE OF NCMRWF

Agromet advisory bulletins published once or twice in a week by the AAS units are destined to reach the farmers through various dissemination modes and farmers' feedback is also taken into consideration to prepare the bulletins in the light of weather related agricultural operations.

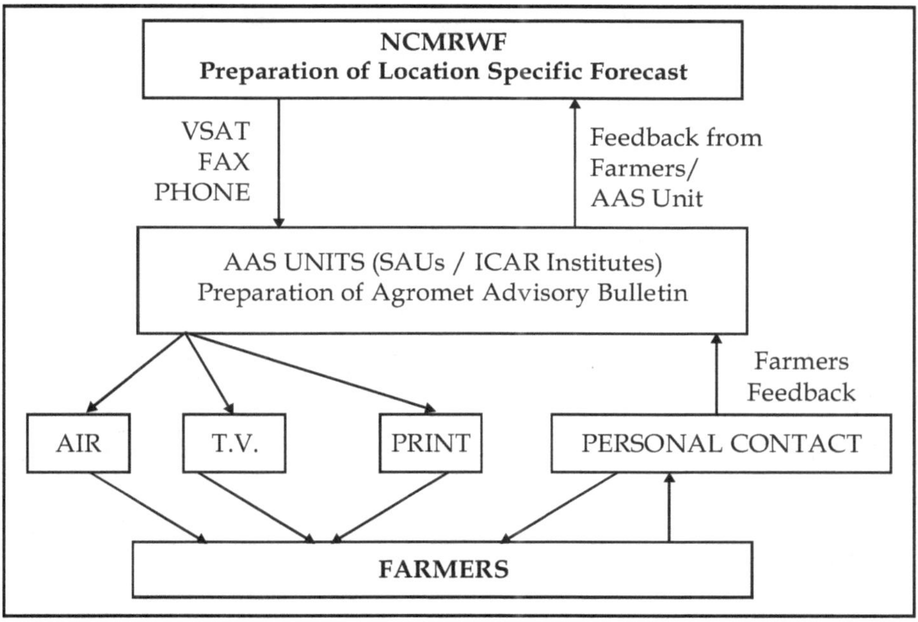

WEATHER FORECAST SYSTEM

Weather forecast system is mainly consisting of 3 types of forecasts, namely long range, medium range and short range. The operational modalities of weather forecast system are as follows.

Forecast Products

Following are the forecast products disseminated to AAS units by NCMRWF and on this basis agromet advisory bulletins are prepared.

- 24hr precipitation (mm)
- Average cloudiness (Okta)

- Average wind speed (kmph) at 10 ft Height
- Predominant wind Direction (Deg.) at 10 ft Height
- Maximum temperature trend (Deg. C) at 4.5 ft Height
- Minimum temperature trend (Deg. C) at 4.5 ft Height

Frequency of Forecast : Twice-a-week
Dissemination : On Tuesday and Friday
Period covered : 4 days

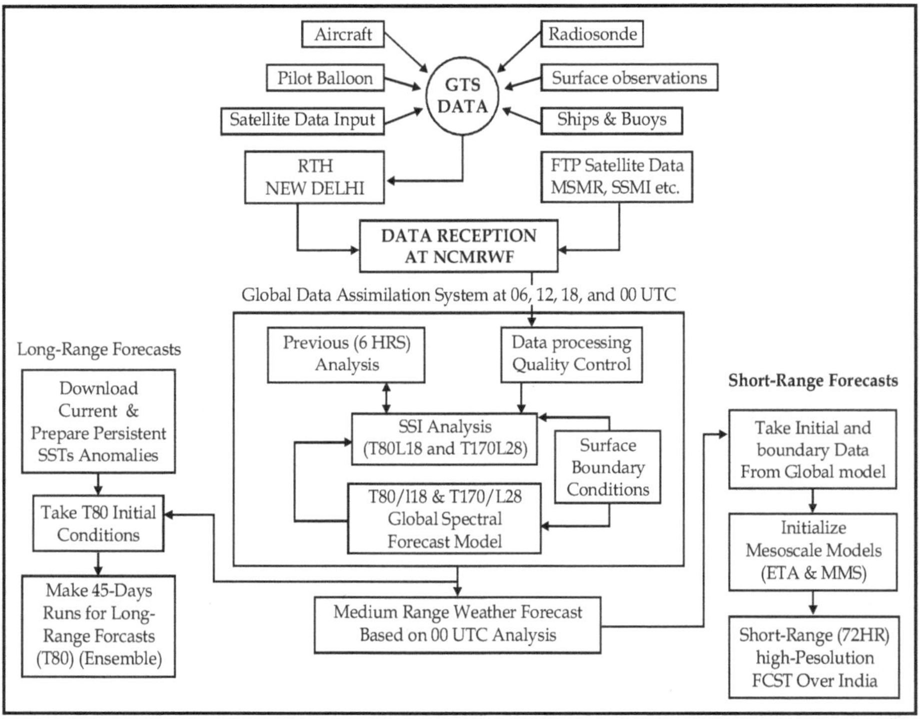

FORMAT FOR AGROMET ADVISORY BULLETIN

All AAS units are to prepare agromet advisory bulletin according to the following formats and as such bulletins should have crop wise farm/livestock management schedule to avoid weather related vagaries.

Weather Information

- Weather summary of preceding week
- Climatic normal for the week

- Weather forecast and
- Crop moisture index, Drought severity index etc.

Crop Information

- Type, state and phenological stages of the crops
- Information on pest and disease and
- Information on crop stresses

Advisory Bulletin

- Crop-wise farm management information tailored to weather sensitive agricultural practices like sowing, irrigation scheduling, p & d control operation, fertilizer use etc.
- Spraying condition for insect, weed and their products
- Wildfire rating forecasts in wildfire prone areas
- Livestock management information for housing, health and nutrition etc.

Weather Events

Following are the weather events covered in AAS for dissemination through bulletins.

Frost / Cold Wave

- Heat Wave
- Heavy Rain / Snowfall
- Dry Spell
- Strong Winds
- Hail
- Cyclone

Crops/ Livestock

Generally the following field crops, vegetables, horticultural crops and livestock components are to be covered by the AAS units considering the local conditions.

Field crops	Vegetables	Horticultural	Livestock
Rice	Tomato	Apple	Avian Birds
Wheat	Cauliflower	Mango	Sheep/ Goat (wool, meat)
Sorghum	Potato	Grapes	Cow/Buffalo (milk)
Millet	Toria	Orange	
Maize	Onion	Coconut	
Barley	Beans	Guava	
Gram	Ginger	Papaya	
Soybean	Peas		
Groundnut			
Sesame			
Cowpea			
Mustard			

Major Agricultural Operations

Agromet advisory bulletins are normally prepared to focus the major agricultural operations narrated below.

- Sowing / Transplanting
  Rainfed
  Irrigated
- Pest & Disease
  Spraying
- Water Management
  Irrigation
- Fertiliser Application
- Harvesting
- Post Harvesting / Storage

ECONOMIC IMPACT ASSESSMENT

The economic impact as assessed by the AAS units showing on an average 2 –20 % savings is enumerated below.

Based on Farmer's feedback, Field visits by scientists, Objective techniques (Crop cutting and saving due to reduction in farm inputs) and Notional savings

Field Unit	Crops/ Livestock	Weather Events/ Management Practices
Kovilpatti	Sorghum, Cotton	Early Sowing
Ludhiana	Potato, Tomato	Frost
Coimbatore	Sorghum, Maize, Banana Onion	Early Sowing, Wind speed, Cloudiness
Madras	Poultry	Thermal stress, Pest and Disease management
Namakkal	Poultry	Heat stroke, Disease management
Pune	Sorghum	Dry sowing
Raipur	Chilli, Potato	Irrigation, Wind speed, Cloudiness
Anand	Pigeonpea, Cotton, Potato	Pesticide application

AGRO-ADVISORIES FOR DOORDARSHAN

Doordarshan plays a vital role in the telecast of advisories to 83 locations on every Tuesday for the benefit of the farming community.

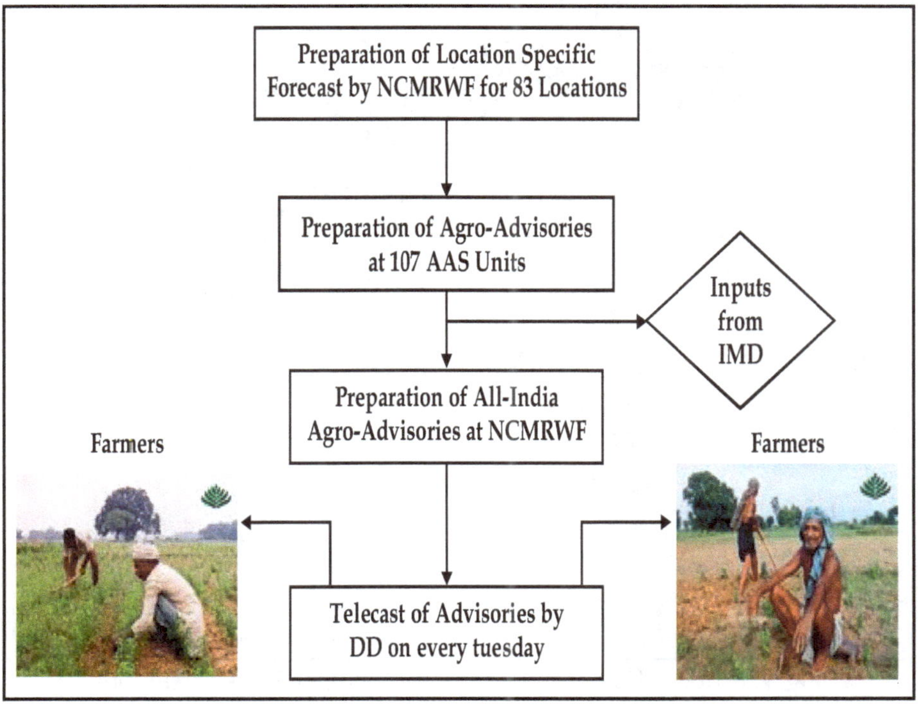

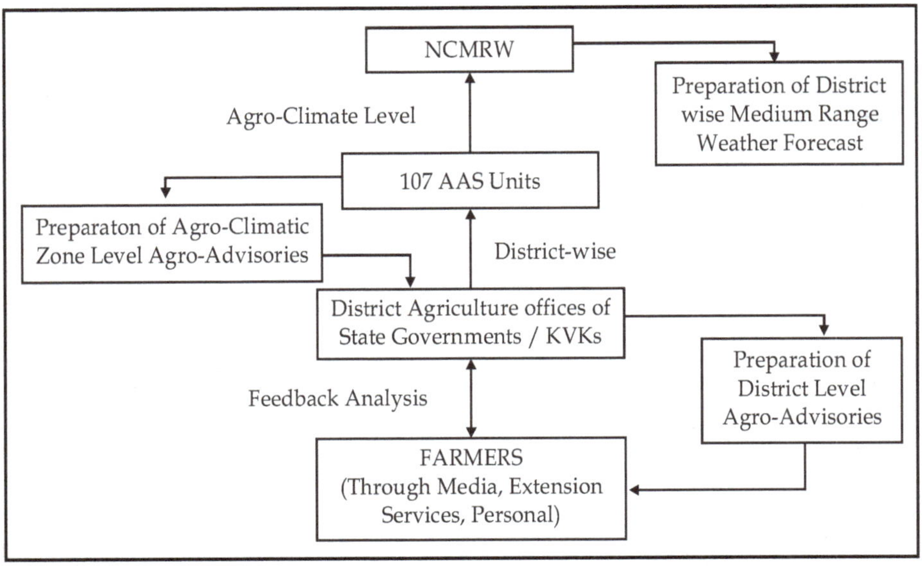

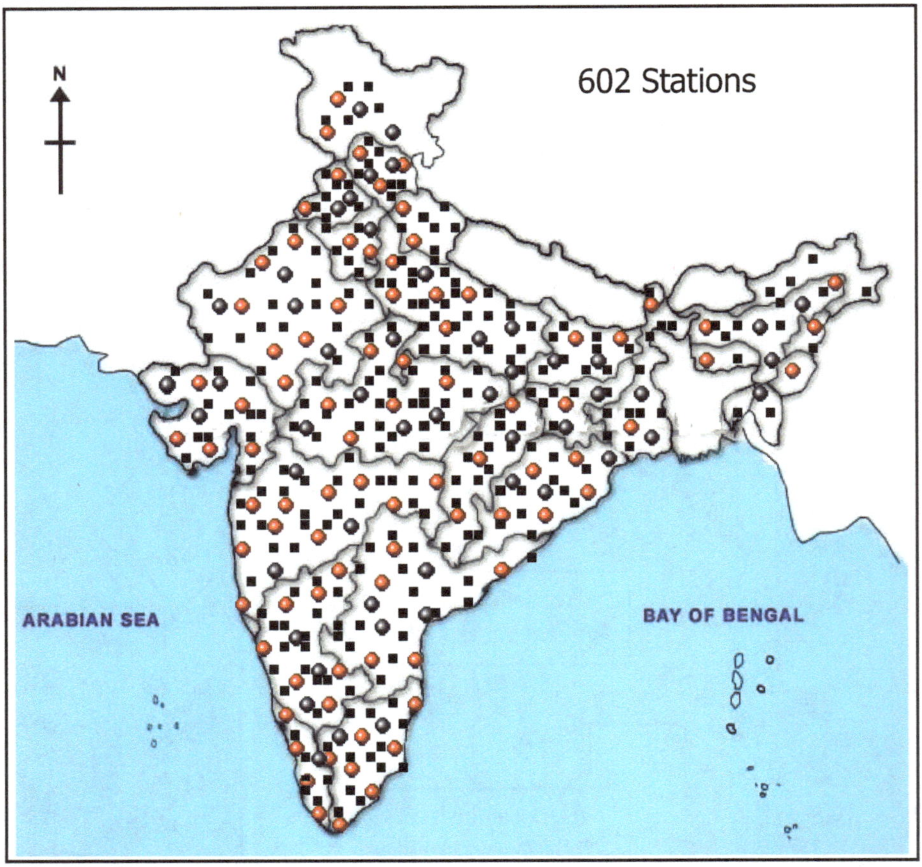

Fig. 2. Proposed District-Level Medium Range Forecast stations

AAS- FUTURE

The future of AAS revolves on some critical questions.

i) Assessment of information needed by diverse groups of end-users

- Identify Weather Sensitivity
- Of Crops/Animals
- Of Farm Operations
- For Input Management (e.g. Irrigation/ Water management)
- Mechanism for Backflow of Information
- Periodic Survey for Crop Assessment
- Use of Remote sensing and GIS Tools

ii) Generation of seamless weather forecast

- Improved observational System (Satellite/Radar/buoys/Argos)
- Date/Product Sharing
- Coupled NWP Models of Fine Resolution
- Wide Spectrum of Temporal Domain
- Multi-Model Ensemble

iii) Development of tools to integrate agro-meteorological data into useful information

- Improved Crop Models
- Crops Model Output Based Expert System
- Forecast for Derived Products (Soil Moisture/Temperature, ET, Aridity, Leaf Wetness Etc.)
- Improved P & D Models
- Vulnerability and Risk Assessment Tools

iv) Generation of farm level advisories

- Demand Driven Advisory
- Personalised Weather Forecast
- Cluster Approach
- Central Agromet-Clinics

- Village Observatories
- Direct Communication (Cellphone, Telephone, Internet)

v) Effective communication of agro-meteorological information

- Realtime Assimilation of Observations
- Three Tier Dissemination
  - Mass Media
  - User Groups
  - Individual User
- Innovative Tools for Easy Understanding of User
- AAS Popularisation Campaigns
- Develop Public-Private Partnership for Dissemination.

vi) Training needs of end-users

- Farm Level Demonstration
- Training Workshop
- Kisan Goshthi (Experience Sharing)
- Teaching Weather Sensitivity of Crop/Animal Production
- Teaching Methods of Crop Production from Adverse Weather

District-level Agro-advisory

There is a planning to introduce district level agro -advisory by NCMRWF and medium range forecast stations as proposed (Fig.2)across the country would be around 602.

Following are the States/Union Territories in which "District level Forecast of NCMRWF" may be initiated on experimental basis.

Table 1. District level forecast in different states of India under testing

Andhra Pradesh	Karnataka	Pondichery
Arunachal Pradesh	Kerala	Punjab
Assam	Lakshdweep	Rajasthan
Bihar	Madhya Pradesh	Sikkim
Goa	Maharashtra	Tamil Nadu
Gujrat	Manipur	Tripura
Haryana	Meghalaya	Uttar Pradesh
Himachal Pradesh	Mizoram	Uttaranchal
Jammu & Kashmir	Nagaland	West Bengal
Jharkhand	Orissa	

COMPONENTS OF PROPOSED DAAS

Following are the proposed components of district level agro - advosory systems.

Observational Network

- NCMRWF's proposed AWS at 127 AAS
- IMD's proposed network

District-wise MRWF

- Existing T-170 products
- Proposed higher resolution model products

DAAS Unit

- District Agriculture Office/ICAR's KVKs
- Proposed District Information System of IMD

Outreach

- Through NGOs/ Agri-Portal
- District authorities
- Media

Dissemination System

- Internet
- NIC network
- Web based dissemination system
- Agri-portal

Implementation Strategy

- Multi-institutional, Multi-state effort
- Linkage with NGOs/ Gram Panchayats

VILLAGE INFORMATION SYSTEM

A new program under pilot initiative for village information system is launched by DST in collaboration with AP Govt. Following are the modalities of such pilot initiatives.

1. Study Area

10 Villages of Nellore Districts of AP for Cyclone & flood vulnerability assessment and 10 Villages of Cuddapa District of AP for drought vulnerability assessment

2. Information to be bundled in addition to existing resources and other data at Village level-

a. Risk & vulnerability assessment
b. Mapping of Village scale risk & vulnerability
c. Details of best practices in respect of effective community level response to the disaster scenario.
d. Risk Mapping of infrastructure, houses, crops & other common facilities
e. Mitigation plans to reduce risk & vulnerability for evolving suitable developmental plans.

3. Computing & Network infrastructure establishment

4. Topographic Layers

- Habitation
- Revenue Village
- Settlements
- Roads
- Rail Network
- Canal
- Drainage Lines
- Water Bodies
- Wells
- Forest
- Contour
- Elevation
- Symbol

5. Thematic Layers from Satellite Images

- Hydrogeomorphology
- Slope
- Soil
- Land Use / Land Cover
- Land and Water Resources Development Plans

6. Non-Spatial Data for every Habitation

- Census
- Land Information
- Infrastructure

7. Census

- No. of Households
- Population
- Male/Female
- Age
- Caste
- Literacy
- Work Force

8. Land Use

- Total Geographical Area
- Forest Area
- Barren and Unculturable Waste Land
- Land put to Non-Agricultural Uses
- Culturable Waste Land
- Permanent Pastures / Grazing Lands
- Land put to Miscellaneous use (Trees/Groves)
- Fallow land
- Net Area Sown
- Irrigated Land(Canals, Tanks, Tube wells, Wells, Check dams, L.I. Schemes)

9. Infrastructure

- Schools
- Colleges
- Hostels
- Health Institutions
- Veterinay Institutions
- Electricity
- Police Station
- Post Office
- Telephone
- Fire Station
- Banks
- Drinking Water Sources
- Irrigation Sources
- Industries
- Public Distribution System
- Bus Services
- Women & Child Development Centres
- Community Latrines
- Garbage Yard
- Burial Ground
- Cable T.V.
- Rural Connectivity
- Self Help Groups
- Cooperative Institutions
- Market Yards
- Rail Services
- Village Library
- Non-formal Education Centres

NCMRWF's Plans for next 5 years

- District Level Weather Forecast and Agro-advisory Service
- Spatial Weather Information System for Mega-Cities
- Forecast for High Impact Weather Phenomena (Cyclone, Very Heavy Rainfall, Floods, etc)
- Disaster Mitigation Modelling, Mapping and Quantification of Impact for Hydrological disasters *(Targeting Village level community)*

Climate Change and Food Security, 2008
Edited by: M.Datta, N.P.Singh & Er.D.Daschaudhuri, pp.37-51
New India Publishing Agency, New Delhi (India)
E-mail : newindiapublishingagency@gmail.com
Web: www.bookfactoryindia.com

3

Impact of Natural Disasters on Agriculture with Reference to Strategies Adopted to Cope with It

H.P. DAS

INTRODUCTION

Inspite of recent technological developments which have helped to increase agricultural production in many countries, growth of plants and animals continue to depend to a large extent on the weather conditions. Each plant has its own climatic requirements for growth and development and any large scale deviation from it exert negative influence.

Growth of plants is most sensitive to temperature just above a threshold value and near the maximum value, where growth normally stops. Therefore, periods of extreme temperature i.e. low temperatures below the threshold value and high temperatures above the maxima are hazardous to plant development and growth. Extreme temperature conditions during cold spells cause stress and frost; high temperatures lead to heat stress, and both affect agricultural production. Snow and ice storms in late spring or early autumn are very hazardous to many temperate crops.

Similarly extremes of moisture conditions namely dry desiccating winds, drought episodes and low moisture conditions as well as very humid atmospheric conditions including wet spells tend to affect agriculture. High soil moisture situations of water logging and flooding associated with heavy rainfall and tropical storms have adverse effect on plant growth and development since they influence the rate of transpiration, leaf area expansion and ultimately plant productivity. Drastic changes in rainfall variability can have very significant impact, particularly in climatically marginal zones such as arid, semi-arid and sub-humid areas where incidence of widespread drought is frequent. Dry desiccating and strong winds reduce agricultural production as a result of very high evapotranspiration rates. It also causes mechanical damage to plant with weak stems by lodging such as the sugarcane and the banana.

AGROMETEOROLOGICAL IMPACT OF EXTREME EVENTS

The CAgM Working Group of WMO on "Agrometeorology related to extreme events", recently developed a survey of countries regarding extreme weather events and their agricultural impacts. The survey identified a number of extreme events including drought, desertification, cold/frost, floods and heavy rainfall, high winds and severe storms, tropical storms, forest and range fires, etc.

Positive Effect of Extreme Events on Agriculture

Impacts of extreme events on the agricultural sector can be positive or negative. While it is easier to contemplate negative impacts of extreme events such as droughts, tropical cyclones and floods, there are several positive impacts or benefits of extreme events. Increased rainfall in coastal areas from tropical cyclones, fixing of atmospheric nitrogen by thunderstorms, the germination of many native plant species resulting from bushfires and the maintenance of the fertility of the basin soils due to river flooding are some of the positive impact of extreme meteorological events. There are some advantages of drought at certain times in the development of some crops such as the sugarcane where a brief dry spell is essential during the pre-harvest stage which helps to increase the sucrose content of the cane. Also there is often a lower incidence of pests and diseases in periods of drought. Transport firms involved in shifting fodder and livestock and tourist organizations do derive benefit from drought.

Negative Effect of Extreme Events on Agriculture

It is, however, the negative or adverse impacts of extreme events referred to as damages or losses which are more pronounced and do affect agricultural operations significantly. Impacts of extreme events can be direct or indirect. Direct impacts arise from direct physical contact of the events with people, their animals and property. Indirect impacts of extreme agro meteorological events are those induced by the events. Indirect impacts often occur away from the scene of the extreme event or after the occurrence of the event. Indirect impacts include evacuation of people in the event of cyclone, landslide, and disruption to household and leisure activities, stress induced sickness and apprehension and anxiety of future extreme events like floods or bush fires.

In order to assess the impact of weather disaster on agriculture, one must link two fundamental aspects, first, the disaster proper i.e. the destructive power of the event and secondly, the characteristics of the agricultural system which has been hit. This is illustrated below in the example of Bangladesh (Rahman, 1985). Between 1970 and 1984, three major weather disasters struck Bangladesh. In mid-November 1970, one of the worst cyclones in history; in 1978-1979 a severe drought could affect the country, followed by extended heavy floods lashing from May to September in 1984. Each of these disasters resulted in a sharp decrease of the rice production, the staple food of the country (Gommes, 1997).

Weather disaster interactions with agriculture are complex and they are likely to involve non-agricultural factors as well.

Table 1. Weather factors which affect agriculture negatively

Weather factor	Negative effects on agriculture
Rainfall	Direct damage to fragile plant organs, like flowers; soil erosion; water logging; drought and floods; land slides; impeded drying of produce conditions favourable to crop and livestock conditions favourable to crop and livestock pest development; negative effect on pollination and pollinators. Pest development; negative effect on pollination and pollinators.
Wind	Physical damage to plant organs or whole plants (e.g. defoliation, particularly of shrubs and trees); soil erosion; excessive evaporation. Wind is an aggravating factor in the event of bush or forest fires.

Contd...

Table 1. Contd...

Air moisture	High values create conditions favourable to pest development; low values associated with high evaporation and often one of the most determinant factors in fire outbreaks.
High temperatures	Increased evapotranspiration; induced sterility in certain crops; poor vernalization; survival of pests during winter. High temperatures at night are associated with increased respiration loss. "Heat waves", lengthy spells of abnormally high temperatures are particularly harmful.
Low temperatures High cloudiness	Destruction of cell structure (frost); desiccation; slow growth, particularly during cold waves; cold dews.
Hail Lightning Snow Volcanic eruptions, avalanches and earthquakes Air and water pollution	Increased incidence of diseases; poor growth. Hail impact is usually rather localized, but the damage to crops- particularly at critical phonological stages- and infrastructure may be significant. Even light hail tends to be followed by pest and disease attacks. Lightning causes damage to buildings and the loss of farm animals. It is also one of the causes of wildfire. Heavy snowfall damages woody plants. Unseasonable occurrence particularly affects reproductive organs of plants.
	The events listed may disrupt infrastructure and cause the loss of crops and farmland, sometimes permanently. A recent example of carbon dioxide and hydrogen sulphide emissions from a volcanic lake in Cameroon caused significant loss of human life and farm animals.
	Air pollutants affect life in the immediate surrounding of point sources. Some pollutants, like ozone, are however known to have significant effects on crop yields over wide areas. In combination with fog, some pollutants have a more marked effect on plants and animals. Occurrences of irrigation water pollution have been reported.

As stated earlier, two broad categories of effects on the agricultural sector can be identified: direct and indirect effects. Direct effects to a farmer could be, for example, the loss of his current crop and damage to his irrigation facilities. Indirect effects on agriculture appear progressively, as a result of low income, decrease in production and other factors related to the weather disaster. The farmer may well have to pay high prices for seeds because of increased demand and disruption of the transportation system. He might also lose a portion of his future harvest because of storm surge-related salinization of soil or the destruction of perennial plantation crops which sometimes take 5–10 years to re-establish. Indirect effects are thus difficult to quantify and therefore are often termed "invisible effects. Conditions conducive to the development of pests and diseases are to be regarded as indirect effects. In fact, the conditions that trigger pest and disease development, are rarely directly harmful per se. They are usually a combination of moisture or temperature conditions which do not directly affect crops. Typical examples are desert locust outbreaks or increased disease incidence in sugar cane after hurricanes. Plants weakened by adverse weather are much more susceptible to cryptogamic diseases or pest attacks, such as the explosion of coconut black beetle on wind-damaged coconut trees. A perennial crop needs a number of years before it is restored to its normal situation. For example, it is taking 30 years for certain timber varieties to re establish after hurricane Allen hit Jamaica in August 1980 (FAO, 1982). It is frequently from 6–12 months for tea and banana plants and 4–5 years for coffee and sugar cane. Even for crops that regenerate easily after partial damage, harvesting is usually made difficult by the 'abnormal" morphology, thus further reducing yield. In agriculture, there can be large losses in primary production on account of delays in the recovery of arable land that has been inundated. For an annual crop, the office of the United Nations Disaster Relief Coordinator estimated a global loss (direct and indirect) at 1.5 to 2 times the direct loss (the lost harvest value). For a perennial crop, the global loss was evaluated at 5 to 7 times the value of the lost harvest. The following paragraphs starts with an attempt to list extreme events which can potentially interfere with agricultural productions.

Tropical Cyclone and Storm Surge

For the most developing countries, loss to life, property and the crop represents a bigger part of the damage caused by a climatic disaster. In monetary terms, the losses incurred by livestock rising, forestry and fisheries mostly remain below those suffered by agriculture. In Madagascar, for instance, according to FAO, a cyclone in 1983-84 caused crop losses of 85% of the total damage to the agricultural sector, whereas the damage to the infrastructure and equipment (drainage, irrigation channels, etc) barely reach 15%. The traditional small scale fisheries were also hit by the cyclone. The losses incurred by livestock raising, forestry and fisheries mostly remain below those suffered by crops (Osro, 1984).A rather different impact pattern occurs in small islands like Antigua and Barbuda where fisheries constitute the backbone of the economy. In eastern Caribbean countries affected by hurricane Hugo in 1989, 47 per cent of the losses occurred in fisheries, but crop losses still represented almost 40 per cent of the total damage (Osro, 1989).

It is worth mentioning that the losses affecting cash crops, which constitute a major source of export earnings in a number of developing countries, is rather high. In Nicaragua, for example, it is reported that direct losses of export crops due to Hurricane " Juana " (Joan) in late 1988 amounted to 21% of the total losses in the agricultural sector (Midinra, 1988). Coffee and bananas suffered a direct loss of their fruits and mechanical damage to their plants. Nonetheless, food crop losses were estimated to be higher (35%), while the livestock sector was less affected (8%, of which one fifth was poultry).

The effects of strong winds in coastal areas are seen in stunted and often much sculpted trees providing unmistakable evidence of the direction of the strong winds. In addition to the battering effects of winds, there is an additional damage caused by airborne sea salt which occurs within a few hundred meters of the coast. Winds which blow from coastal seas spray a lot of salt on coastal areas, making it impossible to grow crops sensitive to excessive salt.

Fields inundated by the storm surge suffer a loss of fertility due to salt deposition, even after the sea water has receded. The affected land takes a few years to regain its original fertility. The period of high water can last from 6 hours to several days, if the drainage is poor, and may leave the soil saline and unfit for crops. Saline soils are predominantly observed in the coastal areas of India. A rise in the sea level would adversely affect the 7000-km coastal belt of India, comprising 20 million ha of coastal ecosystem, increasing coastal salinity and reducing crop productivity.

The effects of cyclonic storms are not wholly bad; benefits principally are on the precipitation having considerable value to agriculture. The heavy rain associated with cyclonic storms guarantees a longer period of water availability and provides possibilities for off-site extra storage in rivers, lakes and artificial reservoirs (on farms or at the sub-catchment level) giving an improved rural water supply and expanded or more intensive irrigated agriculture and inland fisheries. The extra precipitation due to cyclonic storms on land helps to increase plant growth improving the protection of the land surface and increasing rainfed agricultural production.

Drought

Droughts have an immediate effect on the recharge of soil moisture resulting in reductions of stream flow reservoir levels and irrigation potential and even the availability of drinking water from wells.

In the event of a prolonged drought, farmers, either alone or with the entire family, may abandon their land in search of work and food in nearby cities. Fewer and weaker family members remain to till the land, affecting the area under cultivation (McCann, 1986). The acreage planted to food crops is also affected by land quality. Due to the uncertainty of rains during the drought, farmers sometimes make several attempts at sowing of seeds leading to a drastic reduction in seed reserves, which in due course are neither sufficient for planting nor for consumption. The farmer is then obliged to borrow offering labour or perhaps portion of the future harvest as payment for the loan.

Drought not only exposes and accelerates existing land quality problems, it also initiates new ones. The cultivation of lands subject to a high degree of rainfall variability makes them extremely susceptible to wind erosion (and desertification) during prolonged drought episodes, as the bare soil lacks the dense vegetative cover necessary to minimize the effects of aeolian processes. As the fertility of the land and crop yields decline farmers search for new land to cultivate. Farmers sometimes forced to cultivate lands considered marginal from the viewpoint of quality, terrain slope and rainfall (Glantz:, *et al.*, 1986). These newly cultivated lands are high risk areas in the long run for rainfed agriculture.

Flooding and Heavy Rainfall

Two of the principal hydro-meteorological events that often have deleterious effects on agriculture, are flooding and heavy rainfall. It is recognized that in some cases these events, can have positive effects. Examples include heavy rainfall from a tropical storm ending a long agricultural drought, or recurring, annual flooding that replenishes topsoil and soil nutrients, as was the historic situation along the lower stretches of both the Nile and Ganges rivers. Soil erosion, disruption to critical agricultural activities, the logging of crops, increased moisture leading to increased problems with diseases and insects, soil moisture saturation and runoff, soil temperature reduction, grain and fruit spoilage and interruption transportation are among the more significant agricultural impacts from heavy rainfall.

Direct damage to growing plants from flood is most often caused by depletion of oxygen available to the plant root zones. Flooding creates anaerobic soil conditions that can have significant impacts on vegetation. Root and shoot asphyxia, if prolonged, typically leads to plant death. Chemical reactions in anaerobic soils lead to a reduction in nitrate and the formation of nitrogen gas. The denitrification can be a significant cause of loss of plant vigour and growth (Foth, 1978).

Looking only at temporal characteristics of rain storms and their consequent impact on agriculture, the concept of intensity-duration-frequency (idf) becomes important. If any of these three rain storm characteristics increases, the potential for agricultural impacts increases. Heavy rainfall at a location can thus be defined using the idf convention and is framed by the agroclimatology of a location or region in question. Very intense (extreme) rainfall can result in catastrophic flood damage even though it occurred for a relatively short period of time, and/or at the proper location, at the proper time of year, etc. In general, greater damage to agriculture occurs from rain storms having higher intensities with sufficient duration, compared with low intensity, long duration storms.

Sand Storm / Dust Storm

The arid regions are characterised by frequent and strong winds which are due partly to considerable convection during the day. The usually sparse vegetation is not capable of slowing down air movement, so that dust storms and sand storm are frequent concomitant of wind movement. Winds in dry climatic zone also affect growth of the plant mechanically and physiologically. The sand and dust particles carried out by wind damage plant tissues. Emerging seedlings may be completely covered or alternatively, the roots of young plants may be exposed by strong winds. Winds also cause considerable losses by inducing lodging, breaking the stalks and shedding of grains and ultimately decreasing the yield. Many agricultural lands have been lost through wind action by the encroachment of sand dunes, dust storms and sand storms which carry away the top humus soil, leading to deterioration and degradation of the landscape and desertification.

Extreme Cold Weather / Frost Injury

Long periods of extreme cold weather combined with other meteorological phenomena are detrimental to many sectors of the economy, especially agriculture and can result in the loss of winter crops, fruit crops and vineyards due to frost injury. Frost injury is the most widespread type of cold injury to affect crops. Low soil temperature at the depth of plant roots can cause frost injury. Such reductions in soil temperature occur with strong frosts, in the absence of snow cover and with deep freezing of the soil.

Most frost injury to winter crops takes place in the first half of winter before sufficient snow cover which would afford protection ,has formed. In the second half of winter, frost injury happens in regions with unstable snow cover.

There have been many studies of plant injury caused by extreme cold weather. Under low temperatures, basically a plant dries out and the protoplasm (the living part of cells) dies. This happens because the concentration of cell sap rises, the distance between macromolecules is altered and the process of energy interchange are disrupted. The toxic products of metabolism accumulates within the plant's tissues and it is this which causes the protoplasm to die. The degree of damage depends on the intensity and duration of dangerous frosts as well as the stage of the plant's development. Damage to the part of a plant does not always result in damage or destruction of the whole plant. A determining factor is the degree of frost injury to a tillering node, if it is heavy the whole plant will perish. The winter crops most frequently destroyed by frost are those grown on uplands, where snow cover is less and the depth of soil freezing is greater.

The main agrometeorological factor influencing frost damage in winter crops is low soil temperature at the depth of the tillering node. Cooling to the critical temperature of frost injury, even for one day, and especially after a thaw, results in thinning out of crops. Long (three days or more) and intensive cooling causes complete devastation.

STRATEGIES ADOPTED IN AREAS WITH HIGH WEATHER RISK

The agrometeorological impacts of natural disasters has a pervasive societal ramification particularly in developing countries.

Associated human misery underscores our vulnerability to these natural hazards. A disastrous event does not pose much of the threat and ceases to be a disaster, if suitable and adequate mitigation measures are adopted well in advance. Prevention of the formation of tropical cyclones is not in the realm of possibility, but much of their disastrous potential can be reduced, restricting thereby the loss of human life and loss of property, by adopting appropriate strategies and taking timely precautions on the receipt of weather warnings. Climatological data helps in the advance preparation of long-range policies and programmes for disaster prevention. The following paragraphs list the agricultural strategies to be adopted in the events of some major natural disasters:

Drought Management

In arid, semi-arid and marginal areas with a probability of drought incidence of say, at least once in ten years, it is important for those responsible for planning of land-use, including agricultural programmes, to seek expert climatological advice regarding rainfall expectations. Drought is the result of the interaction of human pattern of landuse and the rainfall regimes. There is thus urgent need for a detailed examination of rainfall records of these regions. In this regards, the development of methods of predicting many weeks/months in advance the occurrence of rainfall deserves high priority.

Since technological inputs quickly reach an optimum level, more emphasis should be placed on drought management policies, especially in dryland farming areas. Agricultural planning and practices need to be worked out with consideration to the overall water requirements within an individual agroclimatic zone. Crops which need a short duration to mature and required relatively little water need to be encouraged in drought prone areas. Irrigation, through canals and groundwater resources, need to be monitored with optimum utilization avoiding soil salinity and excessive evaporation loss. A food reserve is needed to meet the emergency requirements of up to two consecutive droughts. A variety of policy decisions on farming, human migration, population dynamics, livestock survival, ecology, etc. must be formulated (Das, 1999).

Sustainable strategies must be developed to alleviate the impact of drought on crop productivity. In areas of recurring drought, one of the best strategies for alleviating drought is varietal manipulation, through which drought can be avoided or its effects can be minimised by adopting varieties that are drought-resistant at different growth stages.

If drought occurs during the middle of a growing season, corrective measures can be adopted; these vary from reducing plant population to fertilisation or weed management. In high rainfall areas where there are a series of wet and dry spells, rainfall can be harvested in either farm ponds or in village tanks and can be recycled as lifesaving irrigation during a prolonged dry spell. The remaining water can also be used to provide irrigation for a second crop with a lower water requirement, such as chickpea. However, no one strategy can be adopted universally. In fact, all such strategies are location, time, crop stage and (to some extent) and socio-economic condition specific. Developing such strategies for each specific factor can help make agriculture sustainable.

Cyclone Preparedness in Agriculture System

Disaster preparedness for impending cyclones, as we know, refers to the plan of action needed to minimize loss to human lives, damage to property and agriculture. The effectiveness of disaster preparedness ultimately depends on the effectiveness of planning and response at a district or local government level (Oakley, 1993).

Preparedness for cyclone in the agriculture system can include early harvesting of crops, if matured, safe storage of the harvest, etc. Irrigation canals and embankment of rivers in the risk zone should be repaired to avoid breaching. Beyond this, as the storm approaches the area, nothing can be done.

Damage Mitigation

Preconditioning of an area is very important for determining how significant and damaging a flood will be. Important considerations here are soil, vegetation and water supply factors. Soils that are saturated prior to an extreme weather event will be more likely to result in a damaging flood than soils that are relatively dry. Fields that have recently been tilled and are devoid of vegetation are much more susceptible to soil erosion. Vegetation that is able to use much of the water and can act as a barrier to moving water (horizontally and vertically), can reduce flood severity and impacts. Water storage systems (rivers, lakes, reservoirs, etc.) that are able to capture and hold most of the incoming water, will be effective in reducing flood damage. Thus, water supply managers in snow-fed water supply regions of the world, typically draw down reservoirs as much as possible prior to the normal beginning of the snowmelt run-off season. In rain-fed agricultural systems, managers typically anticipate rainfall during the growing season sufficient to naturally or artificially irrigate crops. In both situations, however, there is often a balance needed between retaining enough water for agricultural production and environmental health and maintaining enough available storage volume to capture incoming water and prevent floods. Here, analysis of past weather and water data are critical for estimating average conditions and inherent variability. Crops like rice that can function effectively in saturated and even submerged conditions, are appropriate for locations that flood regularly and the system becomes dependent upon regular flooding. Many other crops (e.g. corn) would not be adaptable to such conditions and would not be appropriate alternatives to rice.

Geographical Information System (GIS) can be extremely useful tools in the analysis of flood-prone areas. A GIS in conjunction with digital elevation model (DEM) data can quickly determine slope and aspect of region and can be used to provide geospatial analyses of multiple spatial layers (elevation, slope and aspect-all at various scales or resolutions along with soil characteristics, precipitation, temperature, vegetation, and other factors). GIS are being used to develop new floodplain maps (at various frequency and severity levels) and delineate wetlands in many regions and countries, using these types of spatial layers as well as others, including aerial photos. Such information will certainly assist in the best design of agricultural systems, while accounting for reasonable risk.

Floodplain maps with appropriate information about probabilities (return periods) of certain amounts of precipitation and/or depth of flooding water should be developed and used in risk assessments and agricultural planning. Based on these analyses, economic studies can be conducted to determine if certain types of agriculture are appropriate and justified in certain flood regions. For instance, even though analyses indicate that a region is subject to significant flooding every five years, it may still be economically advantageous to farm the area, due to its high value of return. Structures and other items that would be damaged or destroyed by these frequent floods would need not be built in the flooded area, but in less vulnerable areas nearby. It is also important that future flood planning looks not only to structural solutions but also to land use planning, zoning and other solutions encouraging agricultural production in less vulnerable areas.

Prolonged flooding of soils causes several physical, chemical and biological changes, irreversible in nature. The application of Nitrogen, Phosphorous and Potassium in appropriate amounts could be especially important to build up soil nutrients in infertile lands under prolonged flooding with development of disease problems thus providing a perspective on the development of flood mitigation strategies in the future.

Protection of Crop against Wind

Crop damage by winds may be minimised or prevented by the use of windbreaks (shelterbelts). These are natural (e.g. trees, shrubs, or hedges) or artificial (e.g. walls, fences) barriers to wind flow to shelter animal or crops. Properly oriented and designed shelterbelts are very effective in stabilizing agriculture in regions where strong wind cause mechanical damage and impose severe moisture stress on growing crops. Windbreaks save the loose soil from erosion and increase the supply of moisture to the soil in spring.

Protection of Crops from Dust /Sand Storm

In most countries, field afforestation is the main measure to protect the soil from dust storm. Improving soil resistance to erosion can be achieved by careful selection of cultivation methods, applying mineral and organic fertilizers, sowing grass and spraying various substances which enhance soil structure. It is also important to reduce the areas where a dust can gather, especially in tracts in areas characterized by erosion. One major protection strategy is to establish well developed plant cover before the dust storm period. This can encourage a reduction in the wind speed in the layer next to the ground by forming an effective buffer.

When assessing the impacts of the dust storm on agricultural crops, it is necessary to take into account the degree of the development of the plants. The deposition of soil moved by airflow is observed more often than soil carried by wind erosion over long distances.

In dust storm susceptible area, it is evident that the measures to reduce the wind speed at the soil surface and to increase the hooking of soil particles are both crucial. Such measures include the establishments of tree belts and wind breaks. Leaving stubble in fields, non-mould board ploughing, application of chemical substances promote the hooking of soil particles. Soil-protective crop rotations using perennial grasses and seeding of annual crops are also important.

In regions with intensive wind erosion, especially on wind shock slopes or on light soils, strip cultivation may be used. On fallow lands ,bare fallow stripes of 50-100 m can be alternated with stripes of grain crops or perennial grasses; spring crops can be alternated with winter crops.

Protection of Crops from Cold Injury / frost

A key factor in protection of crops from cold injury is stable air temperatures and snow covers throughout the winter. Thaws, resulting in packing or disappearing snow cover, may worsen dormancy conditions and reduce or destroy the protective properties of snow cover. Long thaws can result in the renewal of vegetation, accompanied by the consumption of carbohydrates, increase in critical temperature and decrease in winter resistance.

The prevention of crop damage by frost can be controlled by breaking up the inversion that accompanies intense night time radiation. This may be achieved by heating the air by the use of oil burners which are strategically located throughout the agricultural farm. Other methods of frost protection include sprinkling the crops with water, brushing (putting a protective cover of craft paper over plant) and the use of shelterbelts (windbreaks).

Conclusion

Damage caused by extreme agrometeorological events to the agriculture sector may be significant, sometimes of the order of magnitude of the GNP growth. For many disaster-prone countries, agricultural losses due to exceptional weather events are a real constraint on their global economy. Indirect effects may continue to affect agriculture negatively long after the extreme event took place. The time needed to recover from some extreme agrometeorological events ranges from months to decades.

Environmental planning would be necessary to avoid or mitigate losses from disasters, by using such instruments as land-use planning and disaster management. Natural disaster reduction measures are in place in a significant number of the nations surveyed and ongoing research and development programmes to improve and expand these measures are also features of many national strategies to minimize adverse effects of extreme events on agriculture. Steps are being taken to significantly reduce the vulnerability of people and their communities to natural disaster; this can only be done through mitigation.

REFERENCES

Das, H. P. (1999). Management and mitigation of adverse effects of drought phenomenon. *In : Natural disasters – some issues and concerns.* Natural Disasters management Cell, Visva Bharati, Shantiniketan, Kolkata, India.

FAO, (1982). Salvage and rehabilitation in the forestry sector of the hurricane affected areas of the eastern region, Jamaica. FAO, Rome, pp.16.

Foth, H. D., (1978). Fundamentals of soil science. John Wiley and Sons, pp. 436.

Glantz, M. H., Katz, W. Magalhaes, A. R. and Ogallo, L., (1986). Drought follows the plow (draft). National Center for Atmospheric Research, Bouldev, Colorado.

Gommes, R. (1997). An overview (extreme agrometeorological events). CAgM Report No. 73, TD No. 836, WMO, Geneva.

McCann, J. (1986). The social impact of drought in Ethiopia. In : Glantz, M. H. (ed.), *Drought and hunger in Africa : denying famine a future,* Chapter 11, Cambridge University Press, Cambridge, England.

Midinra, (1988). Danos en el sector agropecuario provocados por el huracan "Juana". Evaluation preliminar. Managua, November 1988, pp. 39.

Oakley, W. J. (1993). A national disaster preparedness service. *In : Natural disaster,* Merriman, P. A. and Browitt, C. W. A. (eds.), Thomas Telford, London, pp. 270-281.

Osro, (1984). Madagascar, evaluation de la situation agricole a la suite des cyclones. OSRO. TCP/MAG/4404, FAO, Rome, pp.43.

Osro, (1989). Evaluation of the agriculture situation in eastern Caribbean countries affected by hurricane "Hugo". OSRO 03-89-E, FAO, Rome, pp.63.

Rahman, S. M. (1985). Agroclimatic methods for Bangladesh. Bangladesh Meteorol. Department, Dhaka, pp. 41.

Climate Change and Food Security, 2008
Edited by: M.Datta, N.P.Singh & Er.D.Daschaudhuri, pp.53-64
New India Publishing Agency, New Delhi (India)
E-mail : newindiapublishingagency@gmail.com
Web: www.bookfactoryindia.com

4

Agro Advisory Services A National Perspective

Y.S. RAMAKRISHNA, G.G.S.N. RAO
A.V.M.S. RAO AND V.U.M. RAO

INTRODUCTION

In spite of technological developments for the last few decades, Indian agriculture is still rain dependent. Extreme weather events such as drought in 2002, cold wave during 2002-03, heat wave in the year 2003, increased temperatures during rabi 2003-04 and 2004-05, deficit rainfall in many parts of the country during 2004 and excess rains in 2005 (100cm in Mumbai, floods in MP, Gujarat, Orissa, AP, Tamilnadu and Karnataka) could affect the food grain production considerably. The spatial and temporal changes in important weather parameters such as rainfall, temperature, wind, cloud cover, humidity, etc. would influence crop yields by affecting farmers' decision in selection of cultivar, use of inputs, crop management practices, etc. These decisions significantly influence the aggregate agricultural output, which , in succession has an impact on the economy of the country as a whole. Thus, the timely and accurate weather based agro - advisories is the need of the hour for sustainable agricultural production.

Further, the quality of utility of agro-advisories can be improved through value addition to the agromet information including forecast and demonstration of economic benefits accrued due to the following of the advisories. Thus making farmers to adapt weather-based advisories for (i) Increasing the quality and quantity of produce,

(ii) Efficient utilization of resources, (iii) Minimizing losses due to adverse weather, (iv) Effective utilization of favourable weather events, and (v) Reducing cost of cultivation to enhance profits.

ROLE OF AGROMET ADVISORY SERVICES IN AGRICULTURE

Right information at the proper time to the farmers will provide the necessary input for effective management of the risk associated in maintaining profitable agriculture on a sustainable basis. It will also help in conservation of the natural resources and protect the environment. Taking a cue from the developed countries about the benefits derived due to timely advises ,various developing nations have initiated weather-based agro-advisory bulletins for better agricultural production. In this endeavour, Farmers Weather Bulletins (FWB) prepared by the India Meteorological Department (IMD) was aired through All India Radio way back in 1945 in regional languages. With the expansion of agricultural research programmes at ICAR Institutions and State Agricultural Universities (SAUs), the Agro Advisory Network Programmes were strengthened. The IMD in cooperation with State Agriculture Departments has been providing the agro-advisory services at weekly intervals through its Regional Meteorological Centers located at State Headquarters. The agro-advisories were prepared based on the crop conditions, agricultural operations, prevalence of pest and diseases and weather. These bulletins were given wide publicity through mass media network.

The agro-advisory network was further improved in 1988 with the establishment of the National Centre for Medium Range Weather Forecast (NCMRWF) at New Delhi by the Ministry of Science & Technology to provide the medium range weather forecast for the next 3 to 10 days at a grid of size 75 km x 75 km (Rathore *et al*, 2006). The Agro-advisory Board operating at each Agromet Field Unit (AMFU) that are established at 127 agro-ecological regions, uses the forecast issued by NCMRWF, prepares the advisories based on the crop condition in regional languages and thus disseminates through effective mass media communication network. At present, 107 AMFUs are operating at various locations in the country (Ramkrishna *et al*, 2006).

Agro-meteorological Information Requirements for the user community are as follows:

- Rainfall is the main source of water, which causes instability in agricultural production. Therefore, information on the quantum of precipitation at various crop phenophases and at different probability levels, enables the farmer to manage the risk involved in crop management. The information on the probability of occurrence of dry and wet spells during particular week is beneficial in preparation of the agricultural crop planning for a region.
- Monitoring of soil moisture through water balance computation enables the planners to monitor the drought situation over a region. Similarly, the same information can be used by the farmers for efficient scheduling of the irrigation effectively.
- The occurrence of high and low temperatures during the crop season due to heat and cold waves significantly affect the crop growth and its yield by increasing the evapotranspiration, induction & sterility, destruction of cells and stunted growth. Therefore, the probability of occurrences of the above situations can help in preparation of crop calendars circumventing such events.
- Strong winds associated with weather disturbances can cause physical damages to the plants and also increases the water requirements considerably. The application of pesticides is also affected by the wind speed and direction. Therefore, information on the probability of occurrence of high windy days and the methods to reduce the impact of strong winds using shelterbelts on agricultural crops is very much essential.
- Crop pests and diseases are influenced by the prevailing weather conditions. Forewarning of onset and the spread of the pests and diseases is needed for timely management of these biotic stresses in a cost-effective manner and increase profitability. Hence, current efforts for development of forewarning models should be given priority.
- The network of various agencies through which information can be disseminated depending on the scale at which it is required. For example, general information about crop, soil and climate (Fig. 1) upto field level can be obtained from local weather station and the information for decision making can be sent through either fax or mobile. Similarly, the information from local to regional level can be met by TV, radio network whereas national

level, newsletter play a dual role of supplying general data and also advisory. However, TV and radio also play crucial roles in providing necessary data at national level too. At global and continental level, the Internet is the crucial information provider.

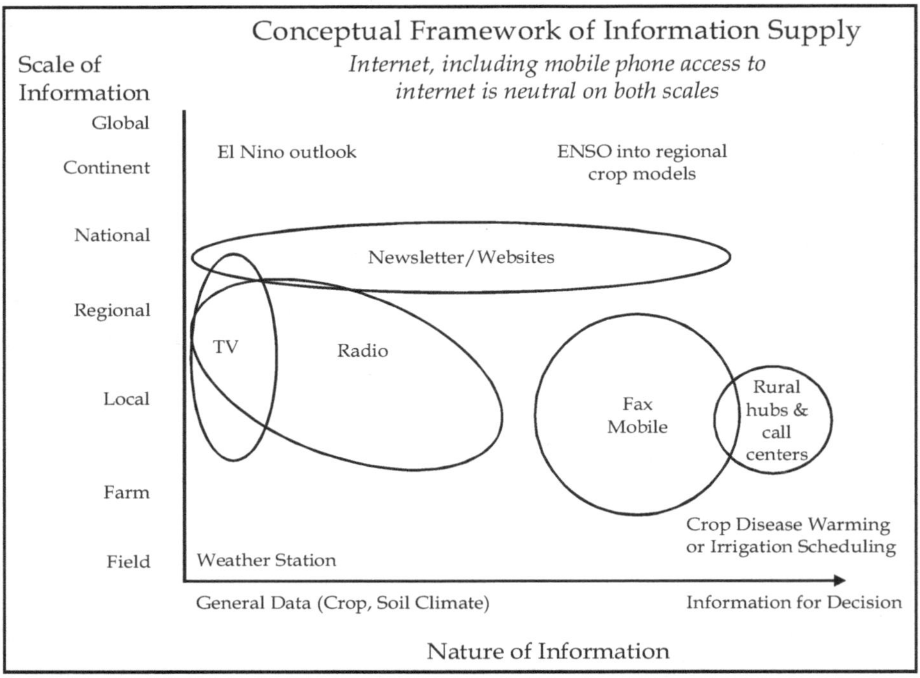

Fig. 1. The conceptual dissemination network mechanism

OPERATIONAL WEATHER SERVICES TO FARMERS BY IMD

a. India Meteorological Department

In order to advise the farming community on how best they can harness the advantages of the favourable weather parameters to increase the agriculture productivity and livestock by ploughing in additional resources. To enable this, operational weather services more effectively to be benefited for farmers with the following objectives are in operation at Regional Meteorological Centres located at State Headquarters:

- Perfection of the capability for weather forecasting
- Interpretation of weather forecast in terms of its effect on agricultural operations for each area

- Collection of up-to-date information on the crop status in each district / block level
- Quick and efficient dissemination of the advisories
- Mechanism for the evaluation of the effectiveness of advisories

The Agromet Advisory Bulletins that are prepared by the above agency consists of:

- Weather summary for the past three to four days
- Forecast valid for next two days
- Crop conditions in agricultural advisory including information on pests and diseases

b. National Centre for Medium Range Weather Forecast (NCMRWF)

The primary objective of NCMRWF is to prepare medium range (3 to 10 days in advance) weather forecast for 127 agro-climatic zones, covering 2 to 4 districts and an area as large as 40,000 to 50,000 sq km. It uses the general circulation model (NMC T-80) for preparation of five-day location-specific forecast. The model output consists of total precipitation, average wind speed, predominant wind direction, maximum temperature, minimum temperature, average cloud cover and relative humidity. These forecasts are sent to the AMFUs for preparation of the Agro Advisory Bulletins by the Agro Advisory Board .At each of the AMFU, the Agro Advisory Board consists of experts from various disciplines such as Agronomists, Entomologists meet twice a week and discuss about the current crop conditions and the weather likely to occur in the next four days and prepare suitable Agro Advisory Bulletins for the respective region. The bulletin usually consists of weather summary of the preceding week, the forecast of the next three days, the soil and crop conditions and pests and diseases infestation. The advisories also provide specific information on agricultural operations like time of sowing, harvest, irrigation schedule, fertilizer management and pest and disease control measures.

The Agro Advisory Bulletins reported by the AMFUs are disseminated through mass media and communication network for effective implementation of the programme. Based on the farmers' feedback and survey work conducted on the economic impact of agro-advisories, it was found that savings due to reduction in farm inputs has been noticed at many locations. The saving is much more especially in the vegetables and in cash crops.

NEW APPROACHES FOR IMPROVING THE AGRO-ADVISORIES

The common goal of minimizing the risks in agriculture and increasing the income of farming community cannot be achieved by single agency. Scientists, NGOs, extension agencies, media, agricultural officers and planners should cooperate and work synergistically for implementing Agromet advisory services so as to minimize the risk to agriculture and to achieve sustainable production. There are different approaches with which the weather-based agro advisories are prepared and made available to the farmer to take a right decision for higher and sustainable production.

Followings are some of the approaches that are envisaged for helping the farmers:

a. Consortium approach - Research Institutes, Line Dept., NGOs, Industrial Establishments, ICT, Extension Agencies
b. E-Sagu (at field and local level)
c. National Agro Advisory Network

a. Consortium Approach (local and regional level)

Under consortium approach, the local Agencies mainly working in the Agricultural research, extension, marketing and management, government, Industry and non-governmental organizations which have developed expertise in the above fields collaborate each other and bring out advisory on a collective basis and disseminate the advisories to the users i.e., farmers (Fig.2).

b. E-Sagu (e-cultivars at field and local level)

This particular approach operating by IIIT, Hyderabad uses recent advances that have come up in computer and Internet technologies. The two tire approach is mainly dependent on the coordinator of a village youth having minimum qualification for communicating with the expert who records the farmer's field conditions using digital camera and communicating the problem to the expert available at the central location on a weekly basis and obtain the appropriate expert advice and disseminating the same to the farmer (Fig.3.)

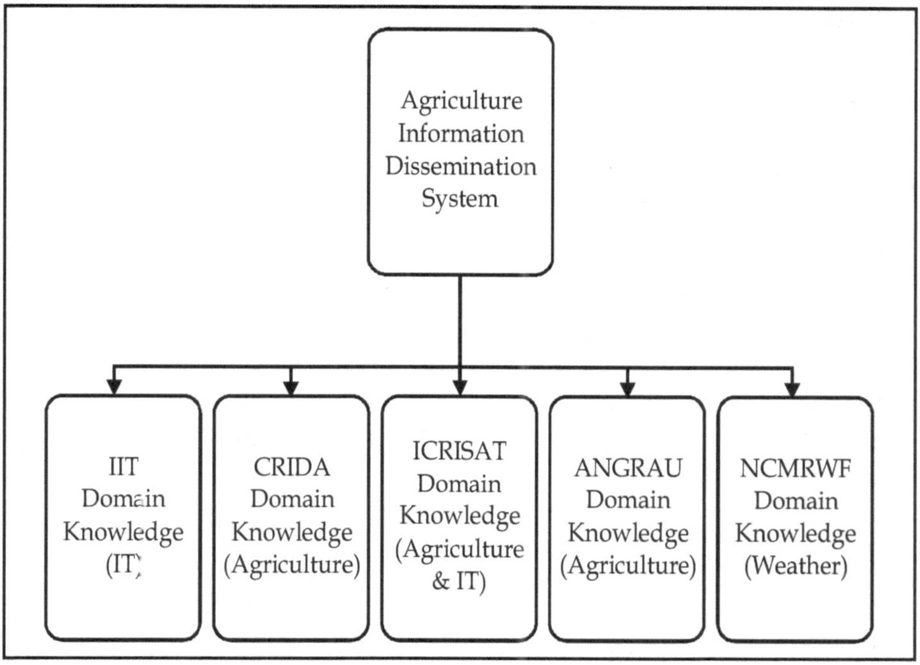

Fig. 2. Consortium approach – Pilot Program (AP State)

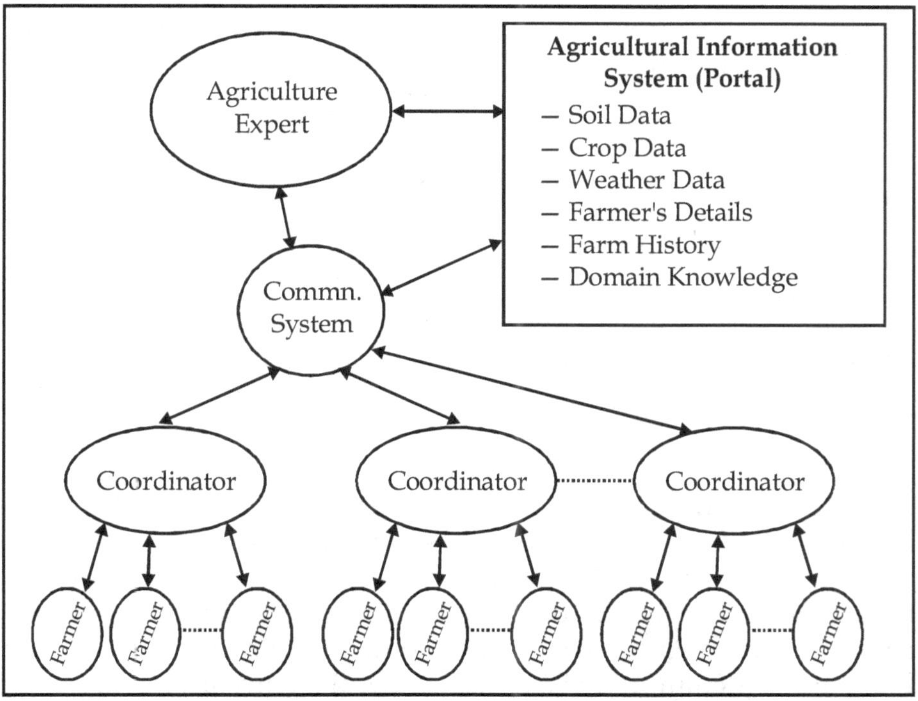

Fig. 3. Agricultural Information System (E-sagu)

c. National Agro Advisory Network

The revolutions in Information Technology (IT) over the last few years have paved the way in providing effective communication linkages between the farmers and the agricultural scientists. The proposed village knowledge centres under the Mission 2007 by the Government of India shall enhance the capacity of the farmers to get the necessary information for improving the farm yields in an interactive mode.

In this endeavor, Indian Council of Agricultural Research (ICAR) has already started a web-based dissemination of agro advisories through its Coordinated Project on Agrometeorology located at Central Research Institute for Dryland Agriculture (CRIDA) (Fig. 4).

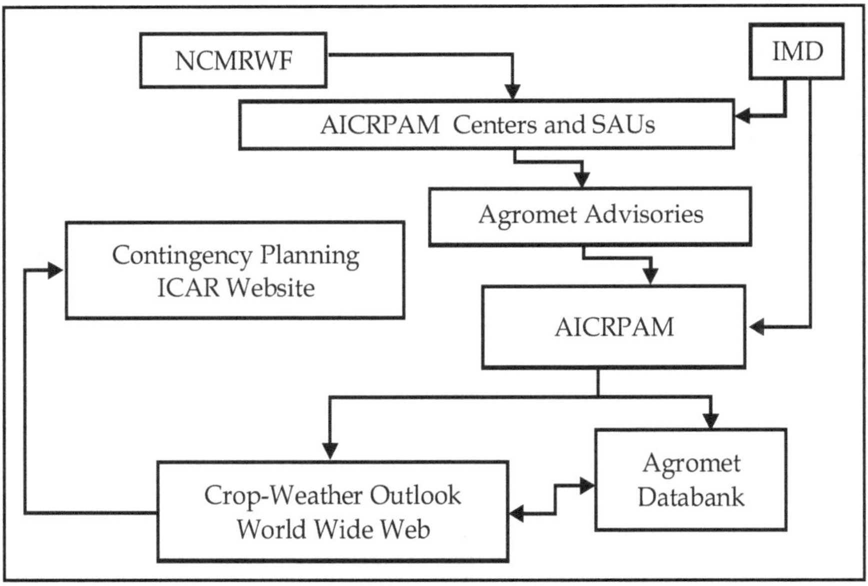

Fig. 4. Agro Advisory Services in ICAR

All India coordinated project on Agrometeorology (AICRPAM) is helping the Agro Advisory Services by providing necessary research backstop for improving the efficiency of agro advisory system through its technical programme operating at 25 centres spread across the country. All the 25 centres of the Coordinated Project located in all the State Agricultural Universities assess the current crop situation based on the NCMRWF forecast; the agro advisories are prepared and disseminated through mass media. They also make available the advisory through a website www.cropweatheroutlook.org operating at CRIDA.

Further improving the advisory system at local level and extrapolate to national level, AICRPAM has envisaged a web-based National Agro Advisory Services (Fig. 5) for effective and timely dissemination of advisories for the benefit of farming community (NAAN,2006).

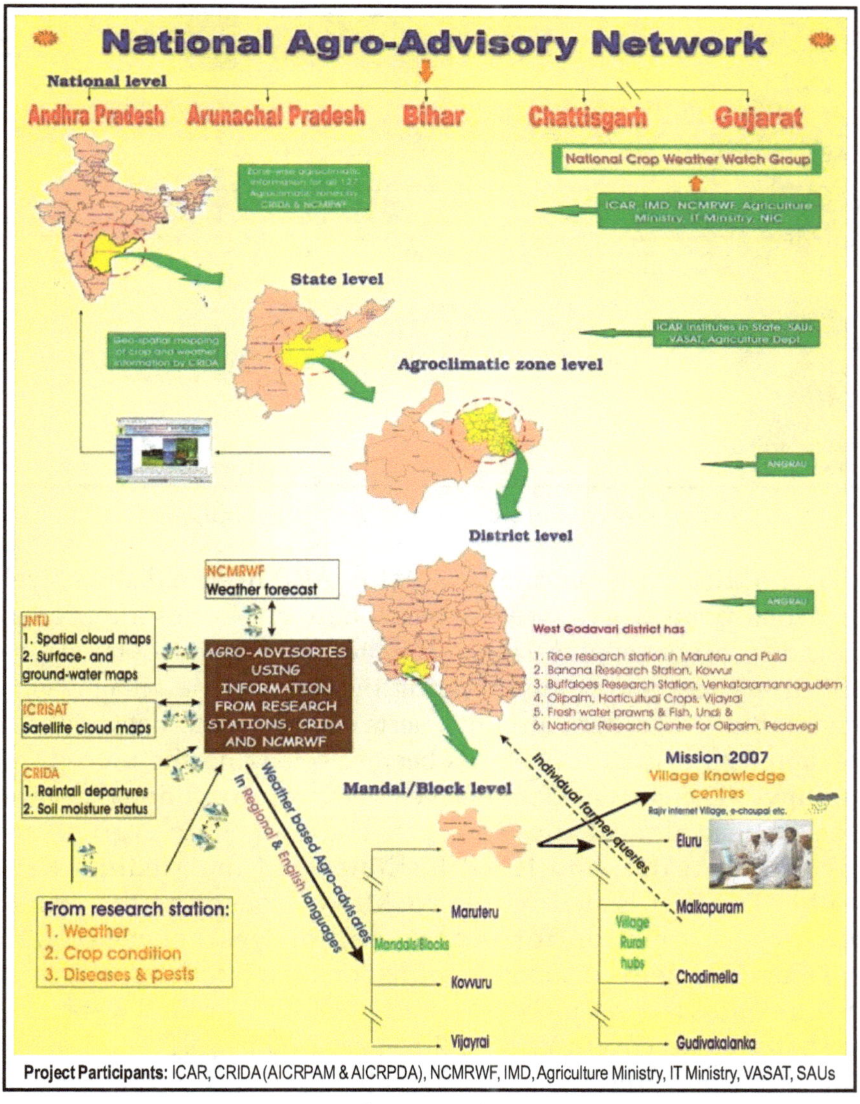

Fig. 5. National Agro -Advisory Network

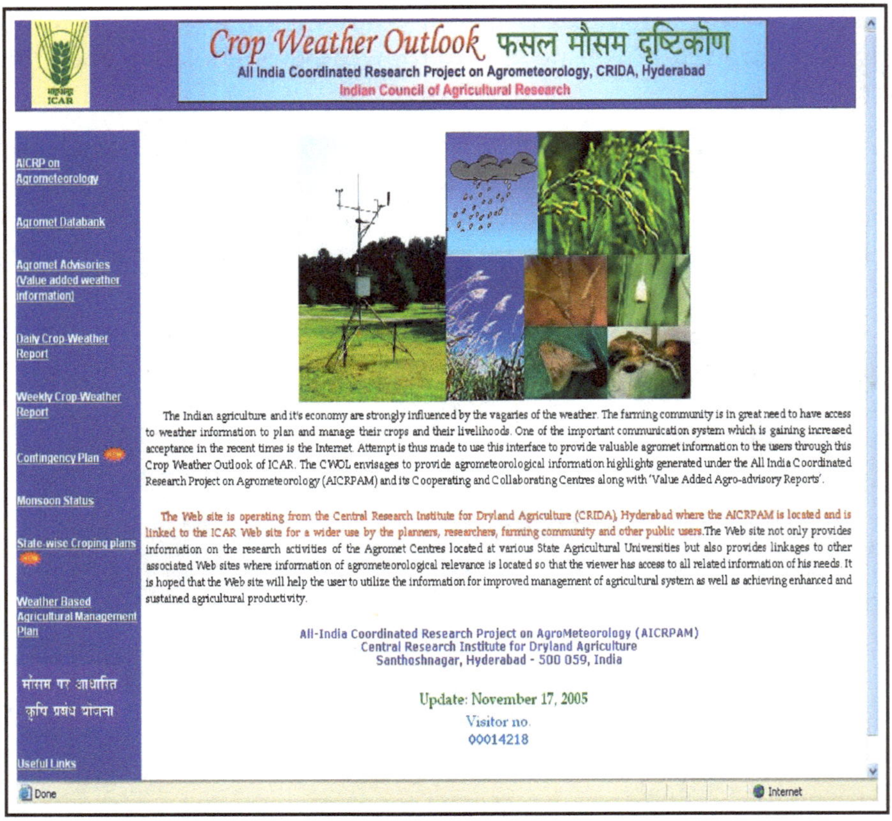

Fig. 6. A view of website of crop weather outlook

The Agricultural Information Centre (AIC) in each district level is the pivotal point where weather-based agro-advisories are prepared for the district after synthesizing the related information obtained from different sources and communicating the same to the villages via its Mandal/Block level agricultural experts using the Internet facility. As an example, a district in A.P. has been selected and the procedure is as follows. The current data on weather, crop conditions, pest and diseases recorded at the Research Units of State Agricultural Universities (SAUs) and ICAR Institutes and the Medium Range Weather Forecast issued by the NCMRWF constitute the primary source of information for preparation of weather-based agro-advisories. The derived spatial information on rainfall departures, soil moisture status maps shall be prepared by CRIDA, ground water status and spatial cloud cover maps shall be prepared by JNTU and satellite cloud maps prepared by ICRISAT shall be utilized in preparing the Mandal / Block wise agro-advisories.

The advisories thus generated at the District Centre will be made available through computer network to all the Mandals/blocks both in regional languages and in English. Mandal level hubs, in turn, flash the necessary information pertaining to each village through village rural hubs or Village Knowledge Centres.

A village knowledge centres is equipped with a computer kiosk linked with Internet. This is managed by a progressive farmer / trained personnel from the Department of Agriculture. Thus Department of Agriculture keeps on updating the farmers right from weather updates, sowing time, seed availability, information on hybrid seeds and their availability and management aspects, availability of fertilizer and pesticides and also on the issues like pest and disease attack and its management, time of application of nutrients and irrigation application, current market situation, etc., based on the Agro advisory services issued by the District AIC. It has unique facility where an individual farmer can interact online with the District AIC to obtain suitable advisories for better farm management operations and agricultural planning.

The agro advisory information generated at district level further synthesized to Agroclimatic zone level by the respective Agromet field units (AMFU) run under DST located at each of 127 agroclimatic zones spread across the country and the same is updated regularly in the website of AICRPAM. This synthesized information is pooled at state level by the SAUs and State Department of Agriculture for helping the state level planners and administrators. Like wise, the information of all the states will be further integrated at national level by the ICAR through its Coordinated Project, which will develop the crop contingency plans for different agro ecological regions.

Finally, this information thus made is available to the National Crop Weather Watch Group consisting of an Expert Body comprising members from ICAR, IMD, NCMRWF, Ministry of Agriculture, Information & Technology Ministry etc. monitoring the crop condition at national level and to take appropriate measures at the Government level.

STRATEGY FOR FURTHER STRENGTHENING THE AGRO ADVISORIES

- All the 127 centres of AMFUs situated in different NARP zones should be fully operational

- Improvement in the reliability of medium and long-range forecast and break monsoon conditions
- Advisories towards livestock management and fisheries need to be further strengthened
- Development of weather-based strategies and early warning systems
- Use of advanced tools like GIS for down scaling to the location
- Training of Agro meteorologists in weather forecasting
- Strengthening the network of Agromet Observation Units by involving the public and private agencies
- Impact of climatic variability on agricultural production needs to be addressed.

REFERENCES

Y.S.Ramakrishna, G.G.S.N.Rao, A.V.M.S.Rao and V.U.M.Rao (2006). Value added Agromet Services for managing Indian and Afghanistan deserts, Special Issue on "Converting Desert into Oasis" published by National academy of Agricultural Sciences, page no.12-18.

NAAN (National Agro Advisory Network) (2006). CRIDA Report submitted to the President of India on 26-04-06.

L.S. Rathore, Parvinder Maini and Sunil Kaushik (2006). Impact assessment of the agro-meteorological advisory service of the National Centre for Medium Range Weather Forecast (NCMRWF) published under Accounts of Operational Agrometeorology of International society for agro meteorology web site (www.agrometeorology.org/fileadmin/insam/repository/ncmrwf.pdf)

Climate Change and Food Security, 2008
Edited by: M.Datta, N.P.Singh & Er.D.Daschaudhuri, pp.65-78
New India Publishing Agency, New Delhi (India)
E-mail : newindiapublishingagency@gmail.com
Web: www.bookfactoryindia.com

5

National Facility of Agromet Databank - A Boon to Sustainable Agriculture

G.G.S.N. RAO, Y.S. RAMAKRISHNA
A.V.R. KESAVARAO A.V.M.S. RAO AND V.U.M. RAO

INTRODUCTION

The discipline of Agrometeorology, an applied science is less than fifty years old. Its services are operationally used by individual farmers in developed countries, and by agricultural or development services in more than half of the countries of the world. Its contribution to agricultural production by examining forecast is sought in virtually all countries. In many countries, the respective governments have expressed the desire to use meteorological information to a much larger extent in day-to-day farm planning and operations. The climate database is used for the development of practical criteria in planning and management of irrigated and rainfed crop production system (Martin Smith, 2000).

Though a huge set of Agromet data exists in India, its utility has not been effective due to the scattered location of these data sources and the impediments for free flow of data to the users to analyze and interpret for real-time decision making in agriculture. There is also a considerable time lag between the actual observation and data availability. In view of the importance and the immediate exigency to centralize the database system for helping the researchers and planners, a central agrometeorological databank facility funded by DST from 1998 to 2003 and later by ICAR with the following objectives was established at Central Research Institute for Dryland Agriculture (ICAR), Hyderabad.

- Collection of past weather data from the co-operating centers of the AICRPAM and AICRPDA and observatories maintained by the various ICAR Institutes and State Agricultural Universities.
- Collection of data on crop morphology, phenology and yield characters along with soil data.
- Scrutiny and computerisation of data using RDBMS.
- Computation of derived agroclimatic parameters.
- Development of software for retrieval of data in the various formats *viz.,* weekly, monthly, seasonal and yearly.
- Providing Agromet data to NCMRWF, DST, ICAR and other users through direct access, E-mail and also through ARIS network of the ICAR.

DATA COLLECTION AND STORAGE

The Input data for the databank is collected from various sources shown in Fig.1. The daily meteorological data of all the twenty five co-operating centres of the AICRP on Agrometeorology and twenty two co-operating centres of the AICRP on Dryland Agriculture, besides data from the agrometeorological observatories maintained by the various ICAR Institutes and the State Agricultural Universities since their inception has been collected. From the total data sets, the basic parameters useful for agricultural research purpose, viz., maximum temperature, minimum temperature, relative humidity (morning), relative humidity (afternoon), Wind speed, Evaporation and Sunshine hours. Air temperature, relative humidity and wind speed are recorded everyday at 07:00 h and 14:00 h IST were collected and updated regularly.

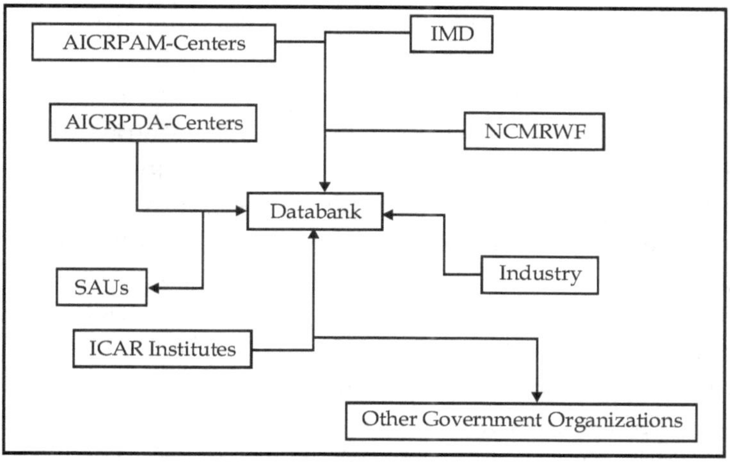

Fig. 1. Sources of data collection for databank

Temperatures are measured using dry bulb, wet bulb, maximum and minimum thermometers, housed in a Single Stevenson Screen kept at height of 1.2 m above ground. Relative humidity is computed from dry bulb and wet bulb temperatures using Hygrometric Tables. Rainfall is measured by FRP non-recording rain gauge and evaporation by open pan evaporimeter. Wind speed is measured with an Anemometer installed at height of 3 metres above ground. Campbell Stoke's Sunshine Recorder installed at height of 3 m above ground is used to record sunshine hours. Wind speed is expressed in km hr^{-1}, mean of past 24 h ending at 07-00 h IST. Rainfall and evaporation are recorded in mm at 08-30 h Indian Standard Time (IST). Sunshine is recorded in hours per day.

Weather as well as climate data also were collected from the state agricultural departments and bureau of economics and statistics. Further, the data from IMD is purchased for improving the databases for better research analysis. Data from Agromet Field Units (AMFUs) working under NCMRWF were also collected. The data recorded from Automatic weather Stations established in various locations in the country by private agencies whose mandate is to collect the met data for crop insurance purpose from interior areas where IMD or other organization do not have any observatory ,has also been added to the databank.

CROP DATA

Crop data for the last five years from the various centres of AICRPAM has been collected as hard copy and processed for computerization. The data includes experimental data from the various field experiments on crop morphology, phenology and yield characters for crops like rice, wheat, mustard, soybean, chickpea, groundnut, pearl millet, sorghum, sunflower, safflower, amaranths and tomato.

Since, the field experimental data differs from location to location, the data are kept in the form of individual tables, providing information about the season, crop, variety, no. of treatments, date of sowing, pod yield, grain yield, etc. This data is useful for the researchers / scientists to study the characteristics of the particular crop / variety at a specific agroclimatic zone and for developing crop-weather relationships leading to better crop weather models.

Light trap pest data with respect to rice from various locations for the pests viz., yellow stem borer, gall midge, leaf folder, brown plant hopper and green leaf hopper for the period 1990-2000 has been collected and computerized. The data on crop morphology and phenology, pest and disease are very complex in nature due to the non-uniformity of generation of data and non-homogeneity of experimental treatments and replications. Hence, these data are available in MS Excel and Word formats.

The crop statistical data on rice were collected from more than 1000 blocks spread across the seven eastern States, viz., Orissa, West Bengal, Chhattisgarh, Jharkhand, Assam, East Uttar Pradesh and Madhya Pradesh, where rice is grown predominantly under rainfed conditions.

DATA BASE DESIGN AND DEVELOPMENT

The raw data including the eight basic parameters viz., Maximum temperature, Minimum temperature, Rainfall, Relative humidity (morning), Relative humidity (afternoon), Wind speed, Evaporation and Sunshine hours collected from the AICRP Centres, Research Institutes and Universities were in the form of both soft copy and hard copy. In general, the data was not in uniform format i.e., in terms of Data Structure. Hence, as a first step all the collected data were made into one uniform format. The data files in the various formats were first imported into Microsoft Excel and then formatted with respect to the order of parameters and numeric format.

Further, a database in Oracle has been designed and Entity Relationship (E-R) Diagram has been prepared and deployed. Database consists of data and data dictionary viz. table structures, column constraints and PL/SQL programs. The E-R Diagram consists of five entities namely Station, Daily, Weekly, Monthly and Normal Data in which, Station and Normal Data entities are stronger entities while the remaining are weak entities.

Quality Check

Quality of Agromet data is essential for derived parameter computations and meaningful interpretations. The raw data as collected from the various locations suffers from errors mainly due to typography. Application program in Visual Basic was developed for Data Quality Check. The program uses Weekly / Monthly Normal Database as reference to check the quality of data collected (Table.1). There are two main functionalities of this program i.e., it identifies the data not satisfying predefined rules and comparing the raw data with the normal data of the same station or nearby station (if normal data is not available for the specified station). The identified erroneous data were marked for further verification. These marked data were checked manually and corrected based on the feedback from the original location.

Table 1. Quality controls used to maintain the quality of data

Parameter	Quality control value
Max.T	± 3 °C
Min.T	± 3 °C
RH1 and RH2	± 10%
Rainfall	Half to two-times the normal
Wind speed	± 3 km/h
Sunshine	± 3 hours
Evaporation	± 3 mm

This improves the quality of the data by containing the redundancy and abnormal values from the database (Fig.2).

ORACLE DATABASE MAINTENANCE

All the data was stored and maintained in the Oracle Database. The core part of the Oracle Database contains data dictionary and the lateral part consists of PL/SQL programs i.e. stored procedures for data insertion / updation and derived parameter computations. Since the data entrance into the Database is vast, data entry through a front-end application is not desirable. Oracle provides SQL *Loader utility for huge data insertion into the database (Fig.3). The Database requires not only data insertion but also updation. Unfortunately, SQL *Loader does not support vast data updation. To overcome this particular problem, PL/SQL programs were developed for data insertion, quality check and updation of the database.

DERIVED PARAMETERS

Visual Basic program for computing Seven derived parameters namely potential evapotranspiration, actual evapotranspiration, estimated soil moisture, water surplus, water deficit, Index of moisture adequacy and soil moisture index were developed and stored in the Oracle database.

The derived parameters are available on weekly and monthly basis (conversion program in VB was developed) and for all the stations in the database. Widely accepted methods were used to compute the PET (modified Penman) and water balance (Thornthwaite and Mather). Both these methods are fairly complex and the users are greatly benefited by the ready availability of these data sets for interpretation. These parameters help the users to characterize the thermal and moisture regimes for understanding the agroclimate of a region. This further helps in determining the Length of Growing Period and planning of irrigation schedule.

DATA RETRIEVAL

The data bank's final architecture is presented in (Fig.4). An Internet Website is designed and developed for easy retrieval of the weather data including derived parameters. It is seen from the Fig. 4 that the raw data in different formats obtained from different centres are subjected to quality test using application programmes and filling the data gaps. The data are later stored in an Oracle Database through SQL loader. The derived parameters can be computed and the required data can be retrieved through Agromet Databank dynamic website by date browsing techniques.

AGROMET DATABANK - DYNAMIC WEBSITE

A dynamic website was designed and developed in JAVA (Java server pages) with Apache web server for the benefit of users to browse data (Fig.5). Registered users can access full data, while registration is free and has to be done only once for a user. After registration and validation, the user can logon and access the agromet data by selecting station, weather parameter, period, and retrieval format viz., daily, weekly, monthly or seasonal data. Users can view data in either tabular or graphical (line graph or bar graph) formats.

Users can access the seven derived parameters on weekly / monthly formats (Fig.6). Data on weekly rainfall probabilities for 308 stations distributed all over India is available in both tabular and graphical forms. Data on various crop and yield characters from the field experiments conducted at the cooperating centres of AICRPAM is also made available in the form of HTML files through the Agromet Databank Dynamic Website.

This website is currently available on Local Area Network. Once hosted on the Internet; this website will be available to all scientists through ARIS network of ICAR and Internet. Users can also request the agromet data for any station, period and parameter, for providing through E-mail.

Some of sample pages for retrieval of data in the Databank website are shown below. This website which is non functional at present will be made operational on Internet after finalizing the 'Data Policy Issues' with the approval of ICAR.

Some applications developed using data from database

The Mandal-wise rainfall data of Anantapur and Mahaboobnagar was analysed (Final report of RRPS-32, 2003).

RAINFALL ANALYSIS

Anantapur

The annual rainfall in the district ranges between 300 to 750 mm (Fig. 7-III). The coefficient of variation of annual rainfall varied between 20 to 50 percent. High rainfall region has been found associated with low variability. While, highest C.V. of above 40% is found at Atmakur mandal, which happens to receive lowest rainfall in the district (350mm). In most of the areas, the rainfall variability ranges between 30 to 40 percent. The annual rainfall is seen increasing in eastward direction with high rainfall (more than 700mm) at Mudigubba mandal. The seasonal rainfall (June-September) in the district varied between 200 mm to 450 mm. The monsoon season rainfall variability in the district ranged between 30 to 70 %. The trend of monsoon season rainfall is similar to that of annual rainfall. Post monsoon season's total rainfall ranged between 50 to 250 mm in the district whereas the rainfall variability ranged between 40 to 90 percent. The high rainfall (>200 mm) was observed at Kadiri, Tanakal mandals, whereas low rainfall (<100mm) areas are comprised of Atmakur mandal. From the monthly rainfall maps of the district for the period June to November, it is observed that September is the highest rainfall month followed by October. It has been observed that the rainfall increased from June to September and thereafter, it started decreasing.

Mahaboobnagar

In Mahaboobnagar district ,the annual rainfall is greater in the North parts of more than 700 mm at Achampeta, Kodangal, Wanaparthy. While the southwestern parts receive less rainfall of 600-650 mm (Fig.7-IV) the mean annual rainfall in the district varies between 600 to 800 mm with a coefficient of variation of annual rainfall of 20 to 40 percent. The seasonal rainfall in the district varied between 425 to 600 mm. The monsoon season rainfall variability in the district ranged between 20 to 50 percent. The post monsoon season's total rainfall ranged between 100 to 150 mm in the district whereas the rainfall variability ranged between 60 to 90 percent. The data of the mean monthly rainfall distribution shows that August is the highest rainfall-contributing month followed by September and July. The rainfall increased from June to August and thereafter, it started decreasing in the district.

AGRICULTURE DROUGHT

Anantapur

The annual meteorological drought probabilities estimated based on rainfall departures are shown in Fig.7 (I & II). It is observed that annual drought probability of Anantapur district varies between 10 to 40%. The probabilities are low in the southern and western portions of the district, while in rest of the portion of the district, it was slightly on the higher side. Monthly drought probabilities indicate that chances of getting rains during September, October are assured, hence the probability for getting droughts during these months is also less. The probability of getting droughts between 30 to 80% is more in the months of July and again in November. The probabilities of getting severe drought conditions are very less among all the mandals. The highest probability is only 11 per cent recorded at Atmakur mandal.

The combined probabilities of mild and moderate is above 50% at all the centres indicating deficit condition once in every two years.

Probability of occurrence of agricultural drought in respective of groundnut crop (Fig.7-I) indicates that the central portion and a part of western part of the district are more prone to drought conditions that affects groundnut productivity while in the southeastern parts, it is slightly better. Highest probabilities of agricultural droughts of above 80% are seen in Kudair, Atmakur and C.K.palle.

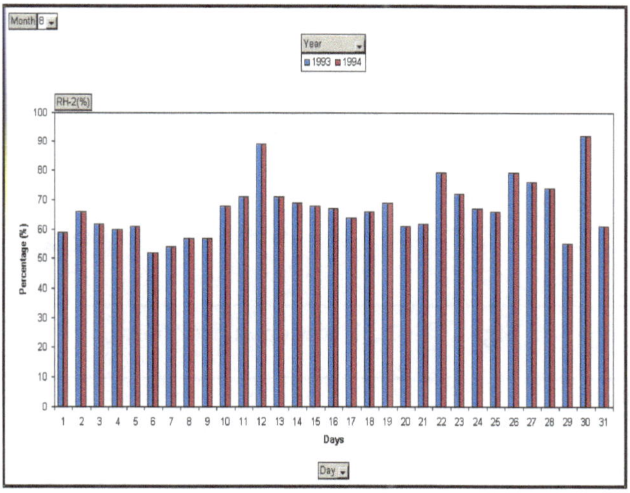

Error data in Sunshine hours

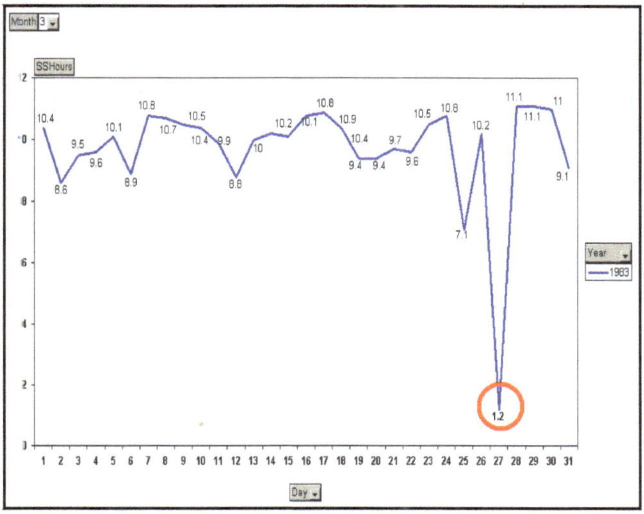

Fig. 2. Quality check for redundancy in data and error data

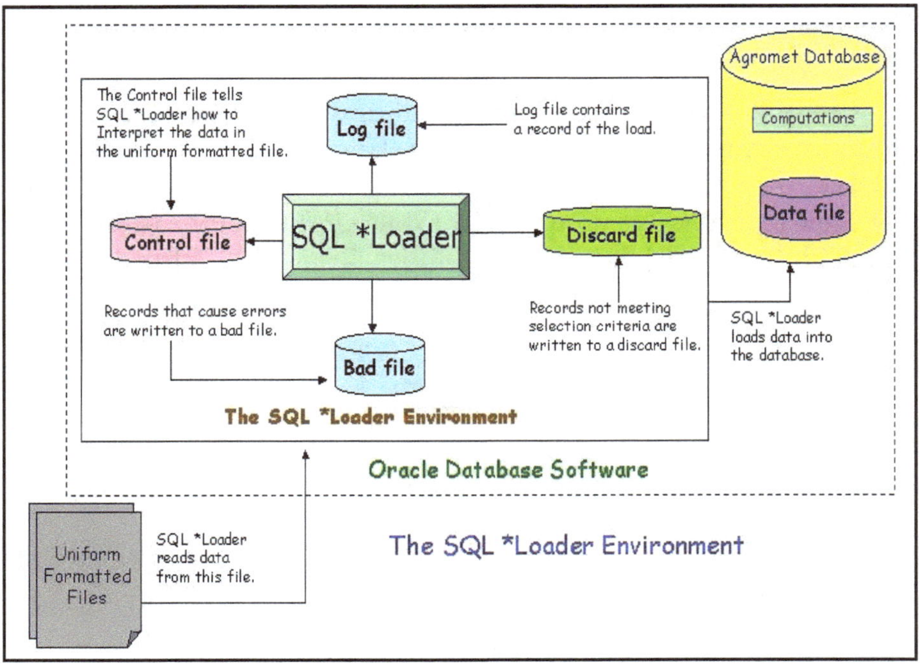

Fig. 3. The SQL loader environment

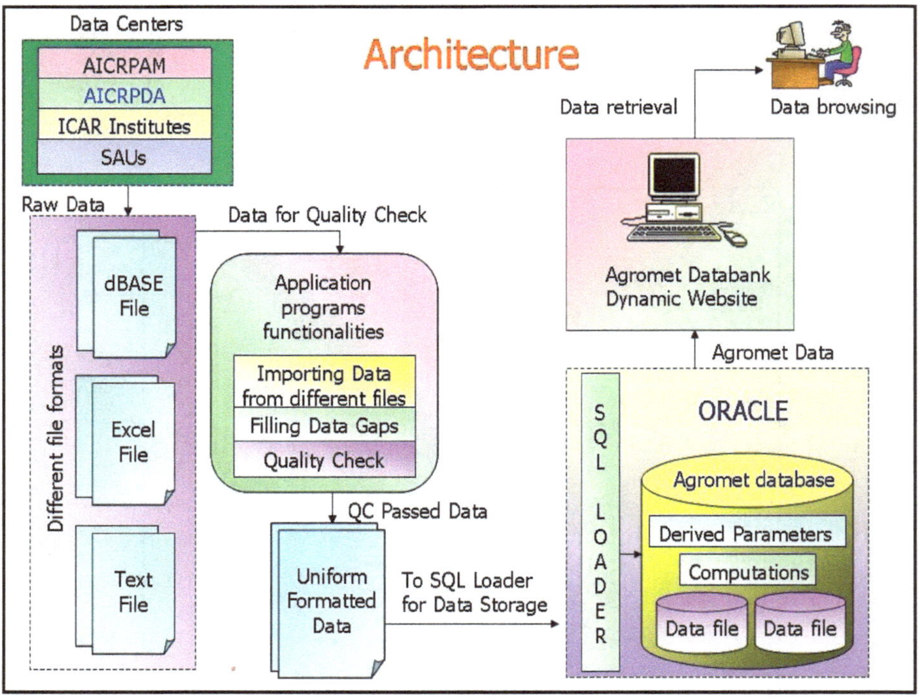

Fig. 4. Final architecture of the databank

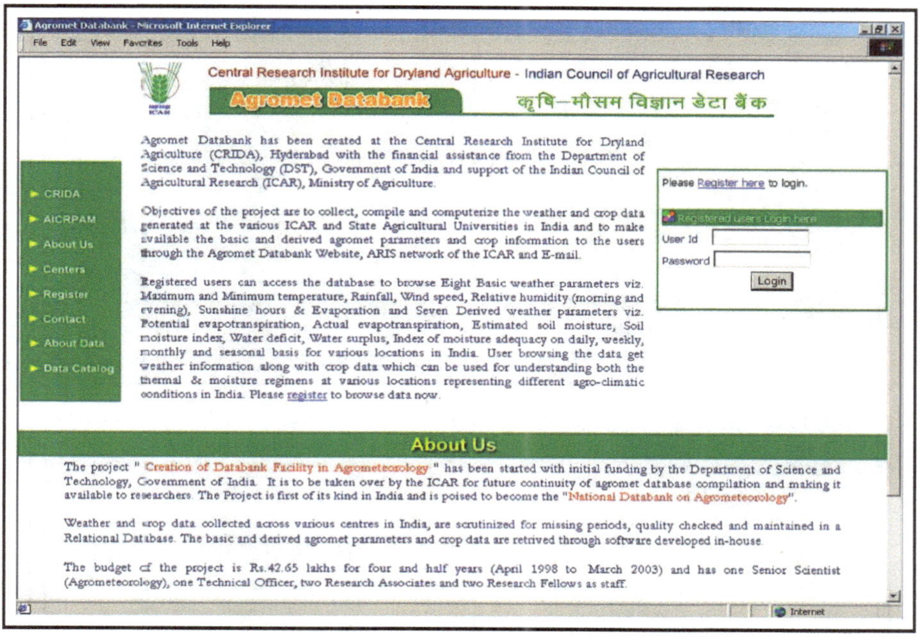

Fig. 5. Opening page of Agromet databank web site

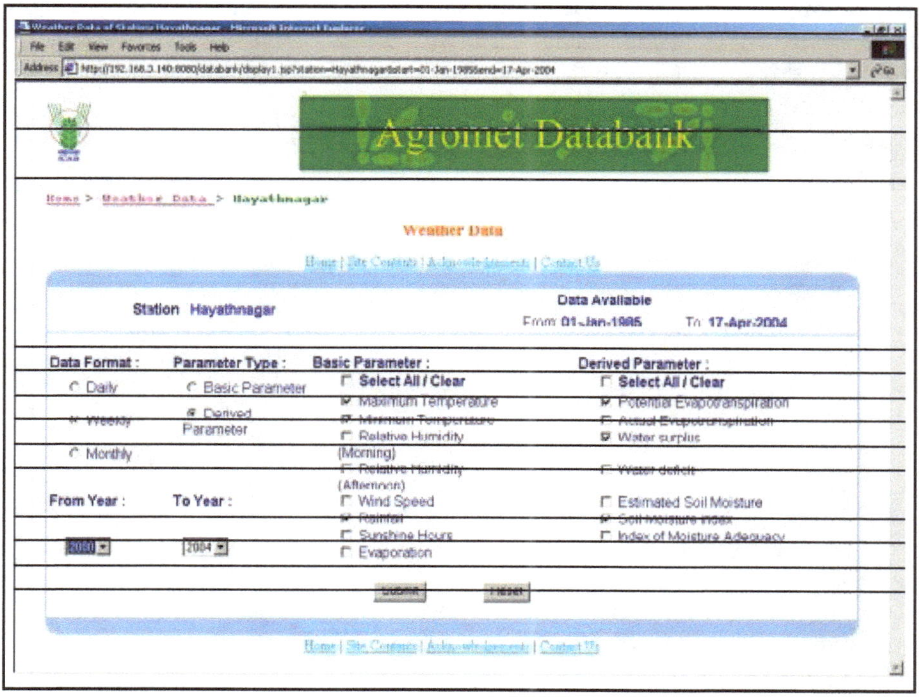

Fig. 6. Weather data page of Agromet databank web site

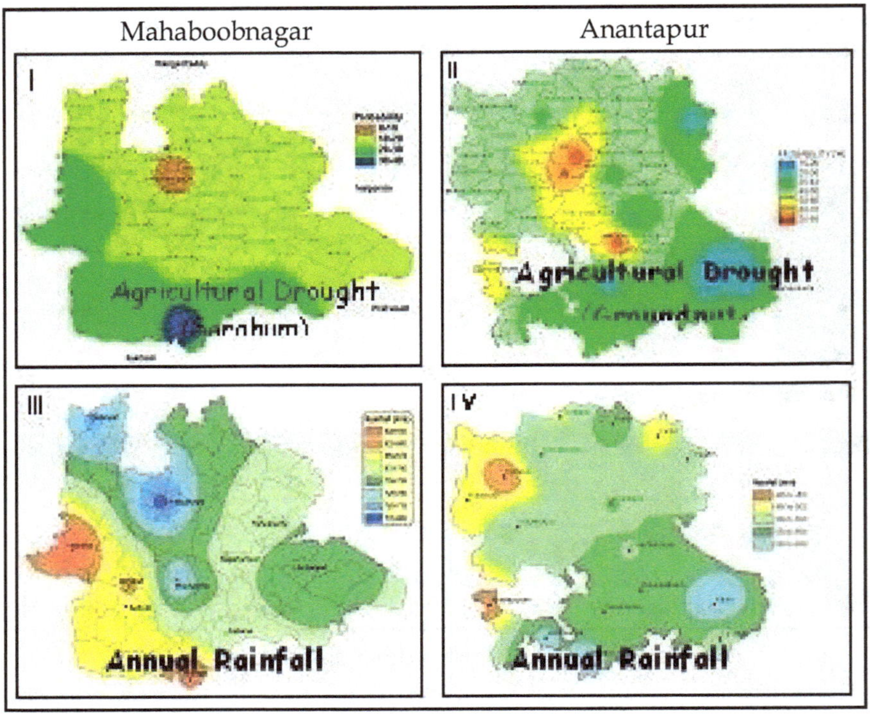

Fig. 7. Rainfall, Agricultural drought analysis of Anantapur & Mahaboobnagar

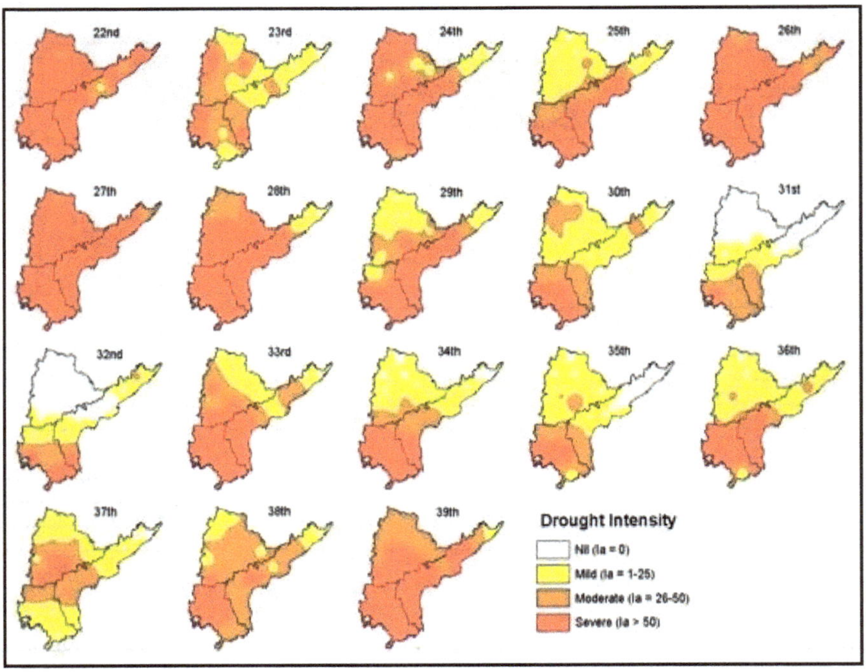

Fig. 8. Drought affected areas in different weeks during kharif season in AP

Mahaboobnagar

It is observed that annual drought probability of Mahaboobnagar district varies between 10 to 40%. The probability of experiencing droughts is more in Southern mandals compared with Northern mandals in which slightly higher rainfall is recorded. Monthly drought probabilities indicate that chances of getting rains during July, August and September are assured, so, probability for getting droughts during these months are less. As the contribution from post monsoon season ,rainfall is less, the analysis has been restricted to monsoon season months (June-September) only. It is seen that chances of getting severe droughts are more at these mandals compared to mandals at Anantapur. Perhaps, the paucity of long-term data may be responsible for such results. Probability of occurrence of agricultural drought in respective of Sorghum crop is shown in Fig.7-II. The probabilities vary from 0 to 40%. It is observed that southern portion of the districts may be subjected to higher probabilities of agricultural drought compared with northern portion.

MONITORING DROUGHT CONDITIONS IN THE STATE OF ANDHRA PRADESH

Weekly rainfall deviation maps (Fig.8) for the period from 4th June to 1st October indicated that rainfall was favorably distributed in many of the districts during the season up to end of August. Scanty to deficit conditions existed for a 3 week period during September and the situation improved at the end of September. The week wise soil moisture index maps (SMI) for the above period indicate that the rainfall received up to first week of July was not sufficient to initiate any agricultural operations under rainfed conditions. Also the districts in Rayalaseema viz. Anantapur, Kurnool, parts of Cuddapah experienced severe water stress throughout the season indicating its impact on agricultural crops. The districts situated in southern portion of coastal region were also subjected to water stress.

FUTURE REQUIREMENTS

Application to animal husbandry, fisheries and forestry sectors needs to be strengthened considerably. Basic knowledge about practical application of meteorology in agriculture exists but needs to be adapted for local use. Awareness of the economic benefits and practical possibilities of its use need to be increased among information producers, disseminators and users.

REFERENCES

Martin Smith (2000). The Applications of climatic data for planning and management of sustainable rainfed and irrigated crop production. *As. For. Met.* 103: 99-108.

Final report of NATP-RRPS-32 (2003). Final report of " Near real-time monitoring of agroclimatic conditions – A case study for Andhra Pradesh (RRPS-32), CRIDA , Hyderabad.

Climate Change and Food Security, 2008
Edited by: M.Datta, N.P.Singh & Er.D.Daschaudhuri, pp.79-101
New India Publishing Agency, New Delhi (India)
E-mail : newindiapublishingagency@gmail.com
Web: www.bookfactoryindia.com

6

Agroclimatic Variability A thought on Sustainable Agriculture

N.P. SINGH, M. DATTA AND D. DASCHAUDHURI

INTRODUCTION

India is endowed with a vast and rich diversity of natural resources particularly soil, water, climate and agro-biodiversity. To realize the optimum potential of the agricultural production system on a sustainable basis, there is a need to bestow a paramount importance in the arena of efficient management of natural resources. India with a geographical area of 329 M ha presently supports 17% of the world's population on merely 2.5% of world's land area and 4% of world's freshwater resources. India's economy is predominantly agrarian in nature with about 72.2% of its 1028 million population (census 2001) living in villages and about 54.4% people (cultivators) directly depending on agriculture. Indian population would surpass China by 2035 reaching 1460 millions. People living below poverty line are 260 million (18.75%). Presently, there are about 1167 million farm holdings, average farm size being 1.4 ha. About 100 million people make a living with forests constituting an important component of their livelihood. Literacy is 65.2%. So far, about 14.2 M ha of forest area is covered under Joint Forest Management (JFM). Per capita available land is 0.14 ha and that of forest is 0.09 ha (Anonymous, 2002). The contribution of agriculture and allied activities to GDP is 26.8% of which 1/3rd is contributed by livestock enterprises. The livestock

(Bovines and Ovines) population as per FAO estimates stands at 452 million by 2002 (Pradhan 2003) with 413 million poultry. These work out to be 333.3 million adult cattle units (ACUs).

The present food grain production is about 211 MT and by 2020 the demand would move up to 286 MT (Bhalla and Hazell, 1997) keeping in view the requirement of 200 kg foodgrain per capita per annum. There is a great diversity in food habits calling for more fruits, vegetables and livestock based products. As of new, there is a fodder shortage to the extent of 31% dry roughage, 23% green fodder and 47% of concentrates (Pradhan, 2003).

AGROCLIMATIC ZONE

India's agriculture is broadly demarcated under 15 resource development regions (Planning Commission's creation of agro climatic Zones with the objective of regionalization of agricultural economy) or 131 Agro-climatic zones (NARP Concept of zoning mainly based on ecological land classification considering soil, climate, topography, vegetation, crop as major influencing factors) or 21 agro-ecological regions (NBSS& LUP considering length of growing period).

The regional approach in agricultural development planning had not received the due attention in the past and therefore, the location-specific needs of various regions remained neglected. For balanced regional growth, agroclimatic zonal planning has been considered as an effective approach. Therefore, the Planning Commission felt it desirable to formulate development strategies based on identified constraints for each of the relatively homogeneous agro-climatic zones. For this purpose, Planning Commission launched the Agroclimatic Regional Planning (ACRP) Project in mid- 1988, mainly with a view to help in formulating the VIII Five Year Plan for the agriculture and allied sectors, on the basis of agroclimatic regions. The potential for growth and diversification would be fully exploited taking a holistic view of the climate, soil type, topography, water resources and relating them to production requirement and employment.

Fifteen agro climatic zones were delineated on the basis of homogeneity of various agroclimatic factors (Table 1). Each zone was subsequently sub-divided into more homogeneous sub-zones for the purpose of realistic planning. In this innovative approach, an overall development profile of each region is formulated through on optimal mix of land management, crop production, animal husbandry,

Table 1. Salient agro-climatic feature of 15 zones

Zone No.	No. of sub zone	Rainfall (mm)	States represented	Climate	Major Crops
I	3	165-2000	J&K, HP, UP.	Cold arid to humid	Wheat, Maize, Rice, Potato
II	5	1840-3528	Assam, WB, NE states.	Per humid to humid	Rice, Maize, Jute, Rapeseed.
III	4	1302-1607	West Bengal.	Moist sub-humid to dry sub-humid	Rice, Maize, Jute, Rapeseed
IV	6	1211-1470	UP,Bihar.	Moist sub-humid to dry sub-humid	Rice, Wheat, Maize
V	3	721-979	UP.	Dry sub-humid to semi-arid	Wheat, Rice, Maize, Tur
VI	3	360-890	Punjab, Haryan, Delhi, Rajasthan.	Extreme arid to dry sub-humid	Wheat, Maize, Rice, Sugarcane
VII	5	1271-1436	Maharashtra,MP, Orissa,West Bengal.	Moist sub-humid to dry sub-humid	Rice, Wheat, Maize, Ragi
VIII	14	490-1570	MP, Rajasthan, UP.	Semi arid (drier half) to dry sub-humid	Wheat, Gram, Jowar, Rice, Bajra.
IX	4	602-1040	Maharashtra, Rajasthan, MP.	Semi-arid	Jowar, Bajra, Cotton, Wheat
X	6	677-1001	AP, Karnataka, Tamil Nadu.	Semi-arid	Jowar, Rice, Ragi, Groundnut
XI	6	780-1287	Orissa, AP, Tamil Nadu, Pondicherry.	Semi-arid to dry sub-humid	Rice, Groundnut, Ragi, Jowar, Bajra
XII	4	2226-3640	Tamil Nadu, Kerala, Goa, Karnataka. Maharashtra	Dry sub-humid per humid	Rice, Ragi, Groundnut, Tapioca
XIII	7	340-1793	Gujarat	Arid to dry sub-humid	Rice, Groundnut, Cotton, Bajra, Wheat
XIV	-	395	Rajasthan	Arid to extremely arid	Bajra, Gram, Wheat, Rapeseed
XV	2	1500-3086	A&N and Lakshadeep Islands	Humid	Coconut

aquaculture, horticulture, forestry and agro- processing activities. A Zonal Planning Team (ZPT) is set up for each zone and these ZPTs have submitted development profiles and made recommendations on the above aspects at the Zonal and Sub-zonal levels.

A critical analysis of these reports has clearly and rightly brought out that "Land and Water Resources Management" is the key issue for agricultural development in all the zones. It is also quite evident that inspite of endowment of good soil, abundant sunshine and ample water resources in alluvial tracts, average crop productivity is very poor as compared with other countries (Borlaug and Dowswell, 1993). As 90% of the food requirements come from land -based farming, integrated management of this resource coupled with improved crop production techniques to increase crop production and productivity without disturbing ecological balance needs priority. This underlines the need in management of this precious natural resource to feed the burgeoning population.

On the basis of climatic variability, the country can be broadly grouped into 11 zones (Velayatham *et al.,* 1999).

Table 2. Agroclimate zones of India

S. No.	Region	Annual avg. rainfall (mm)	Moisture index (as per Thornthwaite classification)	Growing period (days)	Total land area	
					M ha	%
1	Glaciers and others	-	-	-	5.2	1.5
2	Hyper arid	< 100	< - 83.3	0-60	22.9	7.0
3	Typic arid	100-500	-66.7 to -83.2	60-90	22.7	7.0
4	Semiarid (dry)	500-750	-50 to -66.6	90-120	51.2	15.6
5	Semiarid (moist)	750-850	-49.9 to – 33.4	120-150	72.2	22.0
6	Subhumid (dry)	850-1000	-33.3 to 0	150-180	54.1	16.5
7	Subhumid (humid)	1000-1500	0 to 20	180-270	39.8	12.1
8	Dry/ moist subhumid transition	1000-1500	0 to 20	210-270	21.0	6.4
9	Humid	-	21 to 99.9	210-330	16.6	5.1
10	Perhumid	> 2500	>100	> 300	20.5	6.3
11	Transition humid/ perhumid	-do-	> 100	-do-	1.8	0.5
	Total	-	-	-	328	100

As per the estimate of NBSS & LUP, India has 21 agro-ecological regions with variable length of growing periods.

As stated in Fig.1, the regions are as follows:

1. Western Himalayas cold arid regions with shallow skeletal soils and length of growing period (GP) < 90 days.

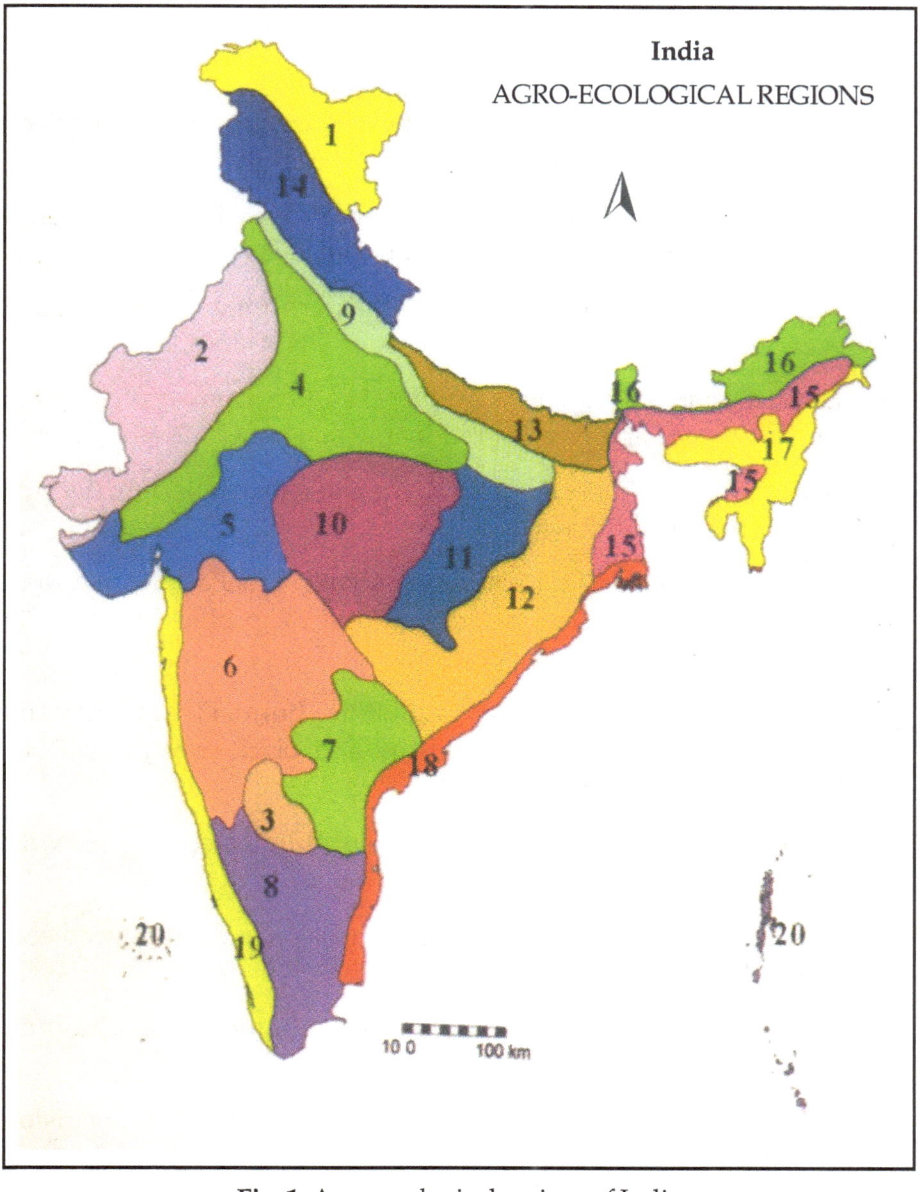

Fig. 1. Agro-ecological regions of India

2. Western plain Kachchh and part of Kathiawar peninsula hot arid eco regions with desert and saline soils having GP < 90 days.
3. Deccan plateau, hot arid regions with red and black soils having GP <90 days.
4. Northern plain and central highlands including Aravallis, hot semi-arid eco-region with alluvium derived soils of GP 90-150 days.
5. Central (Malwa) highlands, Gujarat plains and Kathiawar peninsula hot semi -arid ecoregions with medium and deep black soils of GP 90-150 days.
6. Deccan plateau, hot semi-arid ecoregions with shallow and medium (with inclusion of deep) black soils of GP 90-150 days.
7. Deccan (Telangana) plateau and Eastern ghats,hot semi-arid ecoregions with red and black soils of GP 90-150 days.
8. Eastern ghats, TN uplands and Deccan (Karnataka) plateau,hot semi arid eco regions with red loamy soils of GP 90-150 days.
9. Northern plain, hot sub humid dry regions with alluvium derived soils of GP 90-150 days.
10. Central Highlands (Malwa,Bundelkhand and Eastern Satpura), hot sub eco-regions with medium and deep black soils of GP 150-180 days.
11. Deccan Plateau and central highlands (Bundelkhand) hot sub humid eco-regions with mixed red and black soils of GP 150–180 days.
12. Eastern plateau (Chattisgarh), hot sub humid eco-region with red and yellow soils of GP 150-180 days.
13. Eastern plateau (Chattisgarh), Eastern Ghats, hot sub-humid eco-region with red and lateritic soils of GP 150-180 (to 210) days.
14. Eastern plain, hot sub-humid (moist) regions with alluvium derived soils of GP 180-210 days.
15. Western Himalayas, warm sub humid to humid with inclusion per humid eco–region, with brown forest and podzolic soils of GP 180-210 days.

16. Bengal and Assam plain, hot sub-humid (moist) to humid (inclusion of per humid eco-region) with alluvium derived soils of GP 210 + days.
17. Eastern Himalayas,warm per humid eco-region with brown and red hill soils of GP 210 +days.
18. North Eastern Hills (Purvachal),warm per humid regions with red and lateritic soils of GP 210 +days.
19. Eastern Coastal plain,hot sub humid to semi arid regions with coastal alluvium derived soils of GP 90-210 + days.
20. Weatern Ghats and Coastal plain, hot humid and per humid eco-regions with red / lateritic and alluvium derived soils of GP 90 -210 + days.
21. Island of Andaman - Nicobar and Lakshadweep,hot humid and per humid island eco-region with red loamy and sandy soils of GP 210 + days.

IMPACT OF CLIMATE CHANGE

India is a large developing country with nearly 700 million rural population directly depending on climate sensitive sectors (agriculture, forests and fisheries) and the natural resources (water, biodiversity, mangroves, coastal zones and grasslands) for their subsistence and livelihood. Further, the adaptive capacity of dry land farmers, forest dwellers, fishermen folk and nomadic shepherds (Shukla et al.,2003) is very low. Thus, the probable impact of the climatic change in India is summarized as below.

Water resources : The hydrological cycle is likely to be altered and severity of droughts and intensity of floods in various parts of India is likely to increase.Further,a general reduction in the quality of available runoff is predicted.

Agriculture : Simulations using dynamic crop models indicate a decrease in yield of crop as temperature increases in different parts of India. However, this is offset by an increase in CO_2 at moderate rise in temperature and at higher warming, negative impact on crop productivity is projected due to reduced crop duration.

Forest : It is projected that 77 and 68 per cent of the forested grids in India are likely to experience a shift in forest type under A2 and B2 emission scenario, respectively. Indications show a shift towards wetter forest type in the north eastern region and drier forest type in the north western region in the absence of human influence.

Coastal Zone : Simulation models show an increase in the frequencies of tropical cyclones in the Bay of Bengal .Intense events are projected during the post monsoon period. Sea level rise is projected to displace people, increase floods in low lying coastal areas, loss in crop yields, inundation and salinization etc.

Desertification : Land degradation is on the increase due to climate change.Globally, an area of 1900 M ha with 500 M ha in Africa and Asia -Pacific and 300 M ha in Latin America may be subjected to degradation through emergence of desert like situation.

Health : Health hazards in India may be on the rise and particularly malaria may turn out to be epidemic in many parts of the country due to water logging, stagnation and poor sanitation.

AGRICULTURAL SCENARIO

Two hundred million tonnes of foodgrain produced by India's farmers is intended to feed the country's billion plus people-one-sixth of the world's population. With a population growth rate roughly estimated at two per cent per year, this translates into more than 18 million new mouths to be fed each year or 26 babies born per thousand people every year. If this rate of population growth continues unabated, India will overtake China in population, but not in Agricultural and Industrial progress. This is a matter of great concern.

Notwithstanding the vastly greater population, India's population on an average, is better fed now compared to the 1980s. The availability of food grain (cereals plus pulses) has increased (11 %), from 430 to 475 g per person per day (157 to 174 kg per year). However, the availability of pulses declined by 54 %, from 66.4 g per person per day in the 1980s to 35.9 g in the 1990s.

Declining pulse availability in a vegetarian -dominated place must be a clear indication of nutritional insecurity. The compound growth rates in respect of production and productivity of principal crops also present a gloomy picture (Table 3).

Undoubtedly, India is proud of its "Green Revolution', since it enabled a nearly four-fold increase in food production in the last 50 years. But this achievement has been followed by browning and loud protests degrading environmental effects. Balanced fertilization is expected to help diminish environmental concerns while sustaining a strong food production capacity.

The target growth rate of GDP, population, and agricultural crops during the 9th, 10th and 11th plan periods (Table 4) show that

Table 3. Compound growth rates (% per year) of production and productivity of principal crops

Crop	Production		Productivity	
	1981-82 to 1989-90	1990-91 to 2000-01	1981-82 to 1989-90	1990-91 to 2000-01
Rice	3.55	1.74	3.47	0.92
Wheat	3.57	3.27	3.10	2.21
Total cereals	3.03	1.86	2.90	1.36
Total pulses	1.52	-0.04	1.61	0.55
Total food grains	2.85	1.66	2.74	1.28

Table 4. Target growth rate of GDP, population, and agricultural crops during 9th, 10th and 11th plan periods

Item	Per annum growth rate(%)		
	9th Plan, 1996-97 to 2001-02	10th Plan, 2002-03 to 2006-07	11th Plan, 2007-08 to 2011-12
GDP	6.5	7.70	8.1
Population	1.57	1.57	1.45
Agricultural Crop	3.82	4.54	4.27
Foodgrains	3.05	3.57	2.73
Rice	2.75	3.08	2.73
Wheat	3.75	4.31	3.64
Coarse cereals	2.20	2.40	2.43
Pulses	3.50	4.93	5.66
Oilseeds	5.25	8.85	5.16
Sugarcane	4.00	6.16	6.07
Fruits and Vegetables	7.00	8.01	7.89
Tea	5.00	6.16	6.07
Coffee	5.00	6.16	6.07
Other agricultural products	2.64	3.08	3.04

achievement of these successive production targets will be possible only with increased efficiency blended with positive government policy.

Projected population, foodgrains production and consumption, and fertilizer use by the end of 11th plan pose a great challenge for the policy-makers, scientists, industry and farmers (Table 5).

Table 5. Projected population, foodgrains production and consumption, and input use by the end of the 11th Plan (2011-12).

Item	2011-12
Population, M	1,196.4
Per capita foodgrain consumption, kg	223.4
Foodgrains, Mt	337.3
Rice, Mt	128.2
Wheat, Mt	130.4
Coarse cereals, Mt	48.9
Pulses, Mt	29.8
Oilseeds Mt	58.6
Sugarcane, Mt	679.6
Net area sown, Mha	142.0
Gross area sown, Mha	219.2
Gross irrigated area, Mha	105.5

Although gains in food grain production have kept pace with the population growth, there is a little room for complacency because of the following reasons.

The productivity performance picture is not at all encouraging when compared with the China and world average yield levels (Table 6).

- Any depression in foodgrain production by an abnormal monsoon is likely to produce disastrous effects.
- The need for food will continue to rise in the absence of any reduction in population growth.
- Projected foodgrain productivity targets can not be met based on current growth rates.

- The staple rice-wheat and coarse cereal cropping systems are showing signs of fatigue in large areas of the country.

Table 6. Yield levels (kg ha^{-1}) of some selected crops in India, China and World in 2001

Crop	Productivity (kg ha^{-1})		
	India	China	World
Rice	2,964	6,361	3,912
Wheat	2,742	3,832	2,725
Maize	1,807	4,933	4,427
Cereals	2,318	4,902	3,110
Pulses	586	1,587	798
Foodgrains	2,038	4,789	2,904
Potato	18,643	15,238	15,968

Production targets can be met by increasing yields to levels 2 to 2.5 times higher than current yields. Land is limited and shrinking whereas the human and animal population are increasing. The land to man ratio has fallen rapidly in the past half century from 0.34 in 1950 to 0.14 and is projected to be 0.10 in 2025. It is also disheartening that during the 1990s, India witnessed depressed rates of growth in yield and production compared to world averages.

NATIONAL LAND USE POLICY

Land use statistics in India is presented in Table 7 for the period from 1950-51 to 2000-01.

Following are the observations as emanated from the change in land use statistics.

- Area under forests increased from 14.2 to 22.7% of the reporting area.
- Gross area available for grazing decreased from 66 to 38 M.ha.
- There is an increase in current fallows, a point of concern.
- Net sown area increased from 41.80 to 46.11%.
- Net irrigated area also increased from 20.85 to 54.68 M.ha.
- Cropping intensity also improved from 111 to 133 %.

Table 7. Land use pattern in India over time scale

S. No.	Classification/year	1950-51 (M ha)	1980-81 (M ha)	2000-01 (M ha)
1.	Geographical area	328.73	328.73	328.73
2.	Reporting area	284.32	304.15	306.26
	Forests	40.48	67.47	69.41
	Area under non agricultural uses	9.36	19.66	22.57
	Barren and uncultivable land	38.16	19.96	19.26
	Permanent pastures and other grazing areas	6.68	11.97	10.90
	Land under miscellaneous trees	19.83	3.60	3.37
	Cultivable wastelands	22.94	16.74	13.68
	Fallow land other than current fallows	17.44	9.92	10.19
	Current fallows	10.68	14.83	14.80
	Net sown area	118.75	140.00	141.10
	Total cropped area	131.89	172.63	187.94
	Area sown more than once	13.14	32.63	46.84
	Cropping intensity	111.1	123.3	133.20
3.	Net irrigated area	20.85	38.72	54.68
4.	Gross irrigated area	22.56	49.78	75.14

For sustainable livelihood security, there is a need for proper land use. The degraded areas need utmost attention. To meet these , a National Land Use Policy was evolved by the Government of India with the following objectives.

- To meet the consumption needs of the growing population by increasing productivity of the integrated land resource in the country.
- To prevent any further deterioration of the land resource by appropriate preventive measures.
- To restore the productivity of the degraded land by an appropriate package of practices.
- To provide necessary technological and extension support to all concerned and to the farming community in particular, for obtaining the maximum production through increased productivity.
- To allocate land for different uses based upon land capability, land productivity and national production goals.
- To install efficient and effective administrative structure for prescribing and regulating land use by all concerned, including the Government Departments and to revitalize the Land Use Boards in this regard.

- To involve the community for adoption of appropriate land use for increased productivity by ensuring that the land use policy provides adequately for the consumption and energy needs, generally improves their income levels and provides them a better quality of life.
- To create a greater awareness of the advantages of national land use policy at all levels through appropriate educational, extension and training programmes.
- To restructure the livestock production programme in such a manner that the livestock population is gradually limited to economically productive stock and to prevent degradation of grassland by promoting measures for increasing stall feeding of livestock in rural areas.
- To provide for optimum use of land which is under agriculture by promoting the concept of mixed-farming systems in which the production programme will include fodder and tree crops and also on the marginal and sub-marginal land of farmers.
- To motivate farmers by organizing input supplies and marketing support for encouraging them to cultivate the appropriate crops or fodder or trees in conformity with land use policy.
- To take up on priority basis the completion of land and soil surveys and to complete the inventory of land resources on the basis of the prescribed land use classification so that that the resource allocation is based on a reliable data base.
- To examine the legal support available for enforcement of land use policy in the form of existing State and Central laws and to consider the need for comprehensive legislation to provide some teeth to any machinery entrusted with implementation of the national land use policy.
- To coordinate the formulation and implementation of water resource management policies, forest management policies and urban planning within the overall resource allocation needs dictated by a comprehensive national land use policy.
- To prepare a plan for action at all levels covering a time-frame relevant to all the objectives specified above and to continuously monitor action taken there on in an effective fashion.

LAND DEGRADATION

There are several estimates available. In all these, estimates on soil erosion by water and wind are at the centre stage. The others include water logging (7.6 M ha), salt infestation (7.2 M ha), ravines (4.0 M ha) and shifting cultivation (4.9 M ha) besides the fatigue in soil health. The CSWCR&TI estimated that if the present processes continue unabated, by 2020 the degradation scenario will be alarming (Table 8). This works out to 3.0 ha of degradation per minute in the country.

Table 8. Land degradation scenario in India in 2020

Process	Mha	
	1997	**2020**
Soil erosion	166.0	189.0
Waterlogging	12.7	22.3
Salt infestation	11.0	15.1
Total	**189.7**	**226.4**

As given in Table 9, soil erosion can be grouped into at least 4 categories (NBSS&LUP, 1994; Kar and Joshi,1995). The loss in productivity of major soil groups under different degrees of water erosion has been given in Table 10.

The overall loss in production as per FAO (1993) estimates has shown that about 2.5 to 3.0 Mt of foodgrain production are lost annually through water erosion alone which works out to Rs. 15,000 million $annum^{-1}$ considering the modest value of food grain @ Rs. 6000 per tonne. In fact, if soil erosion is not controlled, more and more loss in productivity would occur as most of the top fertile soil would be lost.

Table 9. Categories of annual soil erosion ($t\ ha^{-1}$)

Category	Water erosion	Wind erosion
Slight	5-10	Upto 29
Moderate	10-20	30-50
Strong	21-40	60-119
Very strong	> 40	>120

Table 10. Loss of productivity (%) of major soil groups under different degrees of erosion

Degree of erosion	Annual erosion (t ha^{-1})	Loss in agricultural productivity (%)		
		Alluvial Soils	Black Soils	Red Soils
None to very slight	< 5	None	< 5	< 10
Slight	5-10	<5	5-10	10-25
Moderate	11-20	5-10	11-25	26-50
Strong	21-40	11-25	26-50	> 50
Severe	>40	26-50	> 50	(?)

Table 11. Highly degraded areas in different states (NBSS &LUP)

Degradation process	Severe to very severe degradation
Water erosion	Himachal Pradesh, Uttar Pradesh, Rajasthan, Madhya Pradesh, Maharahstra, Andhra Pradesh, Karnataka, Kerala, Tamil Nadu, Orissa, Arunachal Pradesh.
Wind erosion	Rajasthan
Waterlogging	Uttar Pradesh, Assam, Rajasthan, Gujarat, Andhra Pradesh
Salt affected	Uttar Pradesh, Andhra Pradesh, Tamil Nadu, Gujarat, Rajasthan, West Bengal.

Similar would be the problems with other components in soil degradation. For instance annual loss of crop production with ravines is estimated to be 3.0 Mt. Added to this, there is an annual loss in cultivable land by ravines to the tune of 8000 ha year^{-1} (Anonymous, 1990).

So is with shifting cultivation practised (2.4 M ha annually) in 10 states of which NE part has 84% area. About 40-50 tonnes of soil loss is reported from such areas. The problem is becoming serious as the jhuming cycle is reduced from 20-25 year to 2-5 years (Toky and Ramakrishnana, 1981).

The overall loss due to land degradation in the country as per (1998) Human Resource Development Report for 1992 year is about US $ 10-13.8 billion (about 4.5 to 6.0% of GDP).

Unfortunately, most of these degraded areas are the habitat of the disadvantaged communities except in waterlogged/salt affected areas where there would be other communities as well. They need correction for not only improving the sustainability in production systems, but also to meet the growing demands for food /food products. The spread of problem soils under various land resource category is presented in Table 12.

Table 12. Problem soils and their extent of occurrence

Land resource category	Percent of TGA*	Remarks
Cold land	5.6	Mean temperature of less than 5 °C steeply sloping land
Hot arid land	9.7	Hot desert and semi-desert areas
Steeply sloping land	10.0	More than 30% slope
Land with shallow soils	5.7	Less than 50 cm soil depth
Poorly drained land	2.5	Waterlogged / flood prone
Land with coarse textured soils	1.2	Less than 18% clay; more than 65% sand / gravel / stones
Land with heavy cracking clays	20.4	Soils with more than 30% clay with cracks (1.0 cm wide up to 50 cm depth)
Land with severe fertility limitations	2.4	Exhibit nutrient deficiencies
Land with saline/ sodic soils	4.1	High salt and exchangeable sodium
Land with acid sulphate soils	0.1	pH of less than 3.5; brackish and ill-drained
Peat land	0.1	High organic matter in first 80 cm depth
Constraint free land	38.2	No physical constraints
Arable and permanently cropped land	56.8	
Total agriculture land	60.8	

* Total geographical area

APPRAISAL OF NATURAL RESOURCES

Soil Resources

Soils are the most vital and precious natural resource for the existence of mankind. Of late, the pressure on this vital resource has increased to such an extent that the relationship between the living beings and the soil has become critical. This has resulted in various kinds of land degradation, environmental pollution, and decline in crop productivity and sustainability. Shrinking resources of prime lands, deforestation and accelerated erosion, deterioration of soil physical environment, increasing waterlogging and salinity in canal irrigated areas, declining water table in the well-irrigated areas, declining water table in the well-irrigated areas, poor land areas, lower efficiency of inputs such as water, fertilizer and agrochemicals, rapid industrialization coupled with pollution and environmental degradation and hazards have aggravated the problem. Soils of India and their distribution (Velayutham and Bhattacharyya, 2000) are presented in Table 13.

Table 13. Soil of India and their distribution

Soil Order	Area (Mha)	Per cent
Entisols	80.1	24.37
Inceptisols	95.8	29.13
Vertisols	26.3	8.02
Aridisols	14.6	4.47
Mollisols	8.0	2.43
Ultisols	0.8	0.24
Alfisols	79.7	4.25
Oxisols	0.3	0.08
Non classified (uncultivable)	23.1	7.01
Total	328.7	100

The land surface is represented by red and lateritic soils (Alfisols, Oxisols, Ultisols and Inceptisols etc.; 172.2 Mha), black soils (Vertisol and their associatios; 73.5 Mha), alluvial soils (Entisols and Inceptisols, 58.4 Mha), desert soils (Mostly Aridisols and Entisols, 30 Mha) and the rest other soil types (Table 13).

Water Resources

Water is the most critical component of life support system. India shares about 16% of the global population but it has only 4% of the total water resource. The national water policy gives priority to drinking water (domestic) followed by agriculture, industry and power. The basic needs as liter per capita per day (l pcd) of human beings, bovines and small ruminants are 40, 30 and 1.5, respectively. Scarcity of water in rainfed areas is causing serious hardship. Average annual surface water flow available in the country is estimated at around 1869 billion cubic meters (bcm). But only about 690 bcm of this can be utilized in addition to the annual replenishable ground water resources of 432 bcm. Thus, the total utilizable water resource in the country is assessed at 1122 bcm. At the national level, the availability of water is declining. Annual internal renewable water resources in India fell by 652 cubic meters pre capita just in two years - from 1896 cubic meters in 1998 to 1244 cubic metres in 2000. It is estimated that our total annual renewable water resources are of the order of some 350 million hectare metres (mhm) of which around 160 mhm find their way back to the sea as river flow (Anonymous 2001).

The single most important task before the country in the field of India's water resource management is to pay special attention to rainwater conservation, especially which falls on our vast rainfed land but most of which flows away from it. If out of this, adequate precipitation is stored in the soil profile it self, as sub-surface water and wells as ground water, it would save the soil from erosion which occurs due to surface runoff, mitigate the impact of droughts and floods, and minimize siltation of reservoirs in addition to optimizing the productivity of rainfed lands.

Most of the rain falls in 4 months (June to September) in high intensity storms (wet spells) on a few rainy days interspersed with long dry spells. Rainy season is followed by dry winter, and dry and hot summers with very high evapo-transpiration. The hydrological challenge is how to relocate rainwater in time so that 4 months' rainwater can be utilized for 12 months for multiple uses such as human and livestock drinking, domestic consumption and life-saving irrigation, on which depends food-security in rainfed areas. Under the low rainfall situation in tropical/sub-tropical conditions and in the present scenario of depleting water resources, urgent attention is needed for sustainable development and utilization of water resources.

Forest Resources

Indian forestry has undergone a phase of very depressing scenario during the last two centuries. The major landmarks in the history of forest depletion include the impact of large scale teak harvesting to enrich the British crown, the two great wars, the expansion of Indian Railways, and the Green Revolution besides the continuous pressure of rapid urbanization and non-forestry uses of forest areas. The commercial forestry to augment the national economy under the banner of scientific management of forests during the last century based on the principal of sustained yield had no element of sustainability, and further added to the pace of forest depletion. The latest date on diversion of forest land as well as the encroachment (MoEF, 2004), shows that a total of 9,54,839 ha forest land has been diverted for non-forestry uses and only a very meager percentage of this has been covered up as compensatory afforestation and a total forest area of 13,43,346 ha is still under encroachment by various agencies. Recent reports (NFAP 2003) show that :

- India having just 1% of the global forest area is laden with the 16% human and 18% cattle population of the world.
- Per capita forest area and biomass availability in India is very low to the tune of just 0.08 ha and 6.0 t as against 0.5 and 82 t average figures respectively, for developing world.
- The biomass and growing stock of wood in natural forests of India is 93 t ha^{-1} and 47 m^3 ha^{-1} as against average figures of developing world to the tune of 163 t ha^{-1} and 113 m^3 ha^{-1} respectively.
- Around 86 Mt of fuel wood is removed from the forest in excess of the productivity of existing growing stock.
- Over 41% of forest cover has already attained a degraded stage and the dense forests are continuously losing on their crown density and productivity.
- Above data reflect a gigantic over-load on the Indian forests giving way to devastative momentum of forest degradation. Major challenges faced by the forestry sector of the country (FAO 2004) are as follows.
- Devising suitable interventions involving government and non-government institutions to expand the forest and tree cover in the country to 25% by the year 2007 and 33% by 2012.

INTER-SECTORAL CO-ORDINATION AND POOLING OF RESOURCES

Public-private partnerships

The micro-plans prepared by JFM committee are as follows.

- Upgrading of skills and training of forestry personnel to get them in tune with the participatory mode of forest planning and management.
- Growing shortage of financial resources for implementation of forestry development schemes and conservation programmes.
- Fragmentation of wildlife habitats and the consequent rise in man-animal conflict.

The NFAP (2003) clearly mentions that "sustainability of forest ecosystem is an essential component of the environmental conservation efforts and any degradation of forest will have adverse impact on various systems such as water sources, agriculture, biodiversity, environment, climate and human health, besides the subsistence living of tribals and other communities living in and around forest areas".

The established role of forest as an important base for the healthy status of soil, water, environment and the rural economy necessitates expeditious revival of healthy status of forests for sustainable development, but the extension of tree plantation on fresh areas of land seems quite impossible task because of the scarcity of land, so the only way out to solve this problem appears to be the integration of tree planting into conventional agriculture through balanced and synergistic agro-forestry systems (Quli ,2005).

SUSTAINABLE DEVELOPMENT

It is the development that meets the needs of the present without compromising the ability of future generations to meet their need (Burtland Commission, WCED 1987).Sustainability has many definitions. But all of them have at least four commonalities (Box ,1993). These are: (i) Equity for today's stakeholders of the natural resources, (ii) Equity for future generations, (iii) Long-term approach, by living on the interest without depleting the principal, and (vi) Environmental enhancement. UNEP (1989) proclaimed that sustainability implies programme toward equity and the existence of a supportive international economic development leading to sustained economic growth and development in all the countries. Thus the maintenance, rational use and enhancement of the natural resource base came into sharp focus. Even the World Commission on Environment and Development (WCED) in 1987 voiced concern on the vastly deteriorated resource base through processes like soil erosion, deforestation, air pollution, ozone depletion and water pollution. WCED further expressed its anxiety on the widening gap between the rich and the poor. Sustainability is a grass -root movement. People are an important part in this endeavour. It primarily revolves around sustainable land use which is linked to cultural demands (Box ,1993). Cultures change. Land use changes. They both have to be compatible.

Sustainable agriculture implies the management of resources for agriculture to satisfy changing human needs while maintaining or enhancing the quality of the environment and conserving the natural resources. Agriculture remains the engine of economic development in most developing countries. And a more sustainable agriculture is more likely to provide the long-term benefits required to achieve sustainable development and poverty alleviation (Dasmunshi,1997). The foundation for sustainable agriculture is maintenance of biological production potential, particularly maintenance of land and water quality, and genetic diversity.

Sustainable Land Management (SLM) combines technologies, policies and activities aimed at integrating socio-economic principles with environmental concerns so as to simultaneously : i) Maintain or enhance production , ii) Reduce the level of production risk, iii) Protect the production potential of natural resource and prevent degradation of soil and water quality, iv) Be economically viable, and v) Be socially acceptable (Smyth and Dasmunshi, 1993).

ASCE (1998) defined Sustainable Water Resource Systems as those designed and managed to fully contribute to the objectives of the society, now and in future, while maintaining their ecological, environmental and hydrological integrity. They must be planned, designed and managed in such a way that the life-support system at all biological levels remains functional and that the water and related land resources are not irreversibly degraded over time (Loucks *et al.*, 2000).

In summary, sustainable agricultural systems (SAS) imply that agriculture will remain the principal land use over long periods relative to human life span, as it is economically competitive (Hamblin ,1991). This is more so in developing countries. The SAS also imply that resource base necessary for supporting them will not decline in extent, productivity or health (quality).

CONCLUSION

Let us consider the national scenario for food and its nutritional security. In view of the optimistic population growth rate of 1.98 %, the human demand for food grain will be 245 Mt by 2010. The demand for milk, fruits, vegetables, meat and fish is estimated at 103.7, 56.3, 112.7, 3.7 and 5.7 Mt ,respectively by the end of 2010.This is not going to be an easy task to increase the production of all kinds of food under variable agro climatic regions in India. Location specific crop planning under the umbrella of Integrated Plant Nutrition Management (IPNM) and Integrated Pest Management (IPM) may have to be introduced to combat the threat of climate change in the coming years. New crop varieties evolved through apt biotechnological tools, may find a promising role considering the sustainability in agriculture. Keeping in view, the rise in world organic market from $ 4.0 billion in 1992 to about $ 21 billion, there may be a need to raise the present area (75,000 ha) under organic agriculture in India. Water management to reduce the impact of drought as well as flood like situations may be given top priority so as to give relief to the farmers in the best possible way. Diversification of the cropping pattern may also be in the agenda to avert the crop losses in various parts of the country. Animal husbandry along with fisheries has great prospects for development through introduction of elite breeds and suitable management practices in variable climatic regions of the country. Finally, a close watch may be in necessity to monitor the changing pattern in climate and immediately act to formulate the remedial measures to sustain the level of productivity to feed the millions.

REFERENCES

Anonymous (1990). Indian Agriculture in Brief. 22nd edition. MoA, GOI, New Delhi, pp. 460.

Anonymous (2001). Report of the Working Group on Watershed Development, Rainfed Farming and Natural Resource Management for X Plan, Planning Commission, GOI, New Delhi.

Anonymous (2002). Empowering People for Sustainable Development, Ministry of Environment and Forests, GOI, New Delhi, pp. 55.

ASCE (1988). Sustainability Criteria for Water Resource Systems. Division of Water Resources. Plg and Mgt., ASCE, Reston, Va.

Bhalla, G.S. and Hazell, P. (1997). Foodgrain demand in India by 2020. *Economic and Political Weekly*. December Issue, 33 (A), 150-154.

Borlaug, N.E. and Dowswell, C.R. (1993). *Fertilizer News*, 38(7),11.

Box,Thadis W (1993). Sustainable landuse and Cultural Change, XVII. The William P. Thompson Memorial Lecture Series. School of Forestry, North Arizona University,pp. 20.

Dasmunshi, J. (1997). Agriculture, Rural Development, 4,15.

FAO (1993). Guidelines for Landuse Planning, FAO,Rome, pp. 1-10.

FAO (2004). Country Report India. Report of Asia – Pacific Forestry Commission 20th Session, Nadi,Fuji 19-23 April CD-ROM by Patrick B.Durst.

Hamblin, A. (1991). *Australian Journal of Soil Research*, 29, 709.

Kar, A. and Joshi, D.C. (1995). Sand movement and control of aeolian hazard. In : Land Degradation and Desertification in Asia and the Pacific Region. (A.K. Sen and A. Kar, eds) Scientific Publishers. Jodhpur, pp. 19-40.

Loucks, D.P., Stakhiv, E.Z. and Martin,L.R. (2000). Journal of Water Resources Planning and Management. March/ April, 43-47.

MoEF (2004). Forests and Wildlife Statistics, India. Ministry of Environment and Forests, GOI,New Delhi.

NBSS & LUP (1994). Annual Report, Nagpur, pp. 48-57.

NFAP (2003). National Forestry Action Plan, Ministry of Environment and Forestry, GOI,New Delhi.

Pradhan, K. (2003). Feed Resources and Technological Intervention for Enhanced Livestock Productivity. In: Feed Technology (G.V.N. Reddy, N.Krishna, V.L.K.Prasad eds.) ANGRAU & NATP, pp.1-7.

Shukla,P.R., Sharma, S.K., Ravindranath, N.H., Bhattacharyya, S. and Garg, A.(2003). Climate Change and India : Vulnerability Assessment and Adaptation, University Press, Hyderabad.

Smyth, A.J. and Dasmunshi, J. (1993). FESLM. An International Framework for Evaluating Sustainable Land Management. *World Soils Report*, 73, FAO, Rome, pp.74.

Toky, O.P. and Ramakrishna, P.S. (1981). Agro-ecosystems, 7,11.

UNEP (1989). G.C. Decision 15/2 on Sustainable Development.

Velayutham, M., Mandal, D.K., Mandal,C. and Sehgal, J. (1999) Agroecological subregions of India for Planning and Development, NBSS & LUP, Nagpur, Publ. 35, pp.372.

Velayutham, M. and Bhattacharyya, T. (2000). Soil Resource Management. In Natural Resource Management for Agricultural Production in India (J.S.P. Yadav and G.B. Singh eds), ICMNR,Feb.14-18,2000, New Delhi.

Climate Change and Food Security, 2008
Edited by: M.Datta, N.P.Singh & Er.D.Daschaudhuri, pp.103-117
New India Publishing Agency, New Delhi (India)
E-mail : newindiapublishingagency@gmail.com
Web: www.bookfactoryindia.com

7

Effect of Climatic Variability on Crop Productivity in North East India

M. DATTA, D. DASCHAUDHURI AND N.P. SINGH

INTRODUCTION

Any crop has its own optimal requirements or selective response to the environmental factors such as rainfall/ soil moisture supply, day/night air temperature, soil temperature, duration of sunshine and its intensity. The growth and development of a plant which ultimately governs the agricultural productivity of a region, is closely related with climatic factors. Weather phenomenon has a profound influence on crop growth stage. A moderate wind or a light rain, when the crop growth is near to harvest, may do more damage than even heavy rain and high wind in the vegetative phase. So land use planning by matching crop growth with weather parameters forms a basis for successful agriculture (Samra, 2003) in a particular region. In order to achieve sustainable agricultural productivity in the 21st century, the climate and weather systems are being recognized as critical agenda and input.

The agroclimatic zones in North East India are of six types namely Alpine, Temperate sub alpine zone, Sub tropical hill zone, Mild tropical hill zone, Mild tropical plain zone and Subtropical plain zone.

DISTRIBUTION OF CLIMATIC ZONES

The climatic zones of NEH region are presented in Table 1.The maximum area in NEH region is in alpine zone mainly present in Arunachal Pradesh and Sikkim followed by temperate subalpine, mild tropical plain, subtropical hill, mild tropical hill and subtropical plain. Alpine zones are present in places having an altitude more than 3500 m and the places having an altitude upto 200 m msl are mild tropical plain. It may be pertinent to mention that Assam has mild tropical climate in hill and plain ecosystem.

Table 1. Areawise Distribution of Agroclimatic Zones in NEH Region

Zones	Altitude above mean sea level	Approx. geographical area (sq. km)	Place
Alpine	More that 3500 m	47,068	Arunachal Pradesh, Sikkim
Temperate sub alpine	1500-3500 m	33,564	Meghalaya, Manipur, Sikkim, Mizoram, Nagaland
Sub-tropical hill	1000-1500 m	29,021	Arunachal Pradesh, Meghalaya, Sikkim, Mizoram, Nagaland.
Sub-tropical plain	400-1000 m	812	Manipur, Nagaland, Meghalaya.
Mild tropical hill	200-800 m	26,349	Arunachal Pradesh, Meghalaya Manipur, Sikkim, Tripura, Nagaland.
Mild tropical plain	0-200 m	29,333	Arunachal Pradesh, Meghalaya, Mizoram, Tripura, Nagaland.

CLIMATE AND AGRICULTURAL DEVELOPMENT

Agricultural development of a particular area is dependant on climatic variables and crop planning is to be monitored in accordance with weather conditions prevailing in different places of NE India. Agricultural development potentials of NE states undergoing variation with change in agro- climatic features are presented in Tables 2-7.

In Arunachal Pradesh, there are 5 agroclimatic zones and rice is the main crop grown in this state.Both tropical and temperate fruits are also grown in Arunachal Pradesh. In Assam having 8 climatic zones, three successive crops of rice are grown in plenty. Fish farming in large marshy bodies is also promising in Assam. In 3 agro-ecozones of Manipur, rice, fruits, vegetables, spices and forest trees are grown. Meghalaya has got 5 different agro-ecozones and rice followed by maize,oilseeds and horticulture / plantation crops are grown.

Table 2. Agricultural development potential in Arunachal Pradesh

Sub regions	Agro-climatic features	Agricultural development potentials
I	Humid hyperthermic with rainfall varying between 1000mm and 2000mm in the Western part and between 2000mm and 3000mm in the Eastern part : Max. 36^{0}C and Min. temp. 8^{0}C	Rice , maize, sesamum, tropical fruits, tea, black pepper etc.
II	Per humid hyperthermic, rainfall varying between 2000mm and 4000mm. Max. Temp. 32^{0}C and Min 8^{0}C temp.	Rice, maize, ginger, oil seeds, tea, arhar, large cardamom, tropical fruits, piggery, poultry etc.
III	Thermic perhumid mid-hills, and valleys, rainfall varying between 1000mm and 3000mm. Max. temp. 19.4^{0}C and Min. temp. 3^{0}c	Rice, maize, ginger, oil seeds, tea, arhar, large cardamom, tropical fruits, piggery, poultry etc.
IV	Thermic humid mid hills and valleys, rainfall varying between 2000mm and 5000mm, Max. Temp. 28^{0}C and Min Temp. 5^{0}C	Two crops of rice, maize, potato, tea etc.
V	Alpine and high hills, rainfall between 1000mm and 2000mm. Min. Temp. - 0^{0}C to - 7^{0}C	Summer maize, French bean, soybean, temperate fruits, potato and sheep rearing.

Rice is grown in terraces of Mizoram with horticultural crops in sloppy areas. Nagaland has got 4 climatic zones where rice is cultivated in valleys and horticultural/plantation crops are grown in hills.In Sikkim, there are 4 climatic zones where agriculture is well established in bench terraces on 30- 50 % slopes. Horticultural /plantation crops are ideally suited to the existing environmental conditions of Sikkim.

Large cardamom and temperate orchids are plenty in Sikkim. In Tripura, there are 3 eco- subregions, namely humid dissected mounds and valleys, subhumid hills, Subhumid denudation hills where rice, oilseeds, pulses, fruits, vegetables, coconut, arecanut, cashewnut are grown.

Table 3. Agricultural development potential in Assam

Sub Zones	Agro-climatic features	Agricultural development potentials
I	Humid and per-humid with rainfall varying between 2800-3000mm, Max-Min. Temp of 32-9^{0}C	High potentials for two rice crops, pulses and mustard in younger alluvial flood plains; Horticulture in piedmonts and foot hills, Livestock, Poultry, Sericulture and Fisheries in water bodies.
II	Humid and per-humid with rainfall varying between 2100-2800 mm, Max-Min temp of 36-7^{0}C	High potential for two rice crops, orchards, dairy, tea, orange and fisheries in water bodies.
III	Humid and sub-humid with rainfall varying between 1100-1800 mm, Max-Min Tem of 37-8^{0}C	Potential for three successive crops of rice, oilseeds, wheat, Jute, pulses, vegetables, Fish farming in marshy and waste land and poultry.
IV	Humid rainfall varying around 2100mm, Max-Min Temp of 32-11^{0}C.	Potential for three crops of rice, oilseeds, pulses, livestock and poultry and improvement of natural fisheries.
V	Humid with rainfall varying between 2400-3000mm, Max Min Temp of 32-12^{0}C.	Potential for three crops of rice, oilseeds, pulses, livestock and poultry, fodder in foot hills and piedmonts, sericulture, citrus and pineapple.
VI	Humid with rainfall varying between 1600-2500mm, Max-Min Temp of 35-7^{0}C	Two crops of rice, horticulture and vegetables, livestock and poultry, and commercial fisheries.
VII	Humid with rainfall varying between 3200-3500 mm, Max-Min Temp of 37-9^{0}C.	Three crops of rice with irrigation, pineapple, banana and fruits, spices and commercial fisheries.
VIII	Sub-Humid and Humid with rainfall varying around 1200-1700 mm Max-Min Temp of 28-10^{0}C	Intensive farming system by scientific management of soil and enterprises like fishery, horticulture, poultry, piggery, and field crops. Improvement in yield of rice in valleys, vegetable development in hills.

Table 4. Agricultural development potential in Manipur

Agro-eco-zone	Agricultural potentiality
Warm humid	Forest trees and bamboos in upper reaches of hills, wild banana, Jhum cultivation, in medium hill range, orange, pineapple, lemon in hill slopes, rice cultivation in valley areas.
Hot humid	Forest trees, sesamum, potato, maize, ginger, tapioca, orange, lemon, rice, maize etc.
Warm per humid	Jhum cultivation, fruit trees and rice in narrow valleys.

Table 5. Agricultural development potential in Meghalaya

Areas	Agro-climate	Agri-development Potential
North-Eastern Jaintia Hillsto NW parts of West Garo Hills	Humid/Warm, 1270 to 2050 mm rainfall.	Maize, Pulses, Cotton, Tea, Cashewnut, Pineapple etc.
Plateaus of East and West Hills	Humid, Hyperthermic, Moderate Cold in winter and warm in Summer, rainfall of 2800-4000mm.	Oilseeds, cotton and plantation crops.
Parts of East and West Khasi Hills & Jaintia Hills	Humid/Moderately Warm in summer, severe winter, rainfall of 2800-6000mm	Rice, Pulses, Maize, Potato, Citrus, Fruits, Vegetables
South of Khasi and Jaintia Hills	Humid and Warm, rainfall of 4000-10,000mm	Maize, Oilseeds, Citrus and arecanut.
West Garo Hills in southern slopes.	Humid and Hot. Rainfall of 2800-4000mm	Rice, Oilseeds, Rubber & Horticultural crops.

Table 6. Agricultural development potential in Mizoram

Ecoregion	Agricultural Development
Warm perhumid, thermic purbanachal ecosystem	Maize, millets, vegetables in hill slopes, rice in terraces, orange and coffee
Warm humid thermic regime	Jhum cultivation, maize, millets, vegetables, rice.
Warm humid hyperthermic temperate	Forest in upper cover, horticulture in hill slopes and rice and narrow valleys.

Table 7. Agricultural development potential in Nagaland

Ecoregion	Agricultural Development
Hot Per humid hyperthermic	Jhum cultivation, maize, millets, vegetables and rice.
Hot moist subhumid hyperthermic regime	Horticultural/plantation crops, forestry.
Warm humid thermic regime	Horticulture/tea,forestry.
Warm per humid thermic regime	Forest, potato, maize, vegetables.

Table 8. Agricultural development potential in Sikkim

Sub Zones	Agro-climatic features	Agricultural development potentials
I	Alpine type in area having altitude above 4000 m amsl with snow covered mountain peak	No cultivable land
II	Subalpine type in area having altitude above 2700-4000 m amsl with snow clad winter and high rainfall	Temperate fruits
III	Temperate type in area with altitude varying from 1500-2700 m amsl	Large cardamom, ginger, tea, medicinal plants, orchids etc.
IV	Subtropical type with elevation upto 1500 m amsl	Maize, rice,fingermillet,ginger, potato, mustard and vegetables, guava, banana etc.

Table 9. Agricultural development potential in Tripura

Sub Zones	Agro-climatic features	Agricultural development potentials
I	Humid dissected mounds	Rice, oilseeds, pulses, fruits, vegetables, coconut, arecanut, cashewnut
II	Subhumid hills	- Do -
III	Subhumid denuded hills	- Do -

ATMOSPHERIC FACTORS

The principal factors of the aerial environment thus strongly influencing the crop growth and yield are : air temperature, moisture content of the air, duration and quality of light, radiation from sun, wind velocity, cloudiness and precipitation including rain, dew, snow and hail. The main edaphic factors in the order of increasing importance are soil aeration, soil temperature and soil moisture.

Many of the physical and chemical processes vital for crop growth are greatly influenced by temperature. Mean day and night temperature as deduced from maximum and minimum temperature assume great significance to crop growth (Venkataraman, 1968). Various types of plant disorders like leaf chlorosis, leaf scorch, necrosis, stem lesions, heat spots in fruits, frost injury, albinism and malformation are associated (Tsnuoda,1959) with high and low temperature. Duration of day light period can have a profound influence on plant flowering behaviour and plants are thus classed as ' Long day', 'Short day' and ' Day neutral'. Light intensity, on the other hand, is also one of the most critical variables influencing crop production with its influence on gross photosynthesis and leaf area ratio. Thus humidity exerts indirect effects on crop growth through its influence on the availability of net energy, precipitation phenomenon, outbreak of pest and disease epidemics etc. The direct effects of humidity are on evapotranspiration rates, stem elongation; flower expression of plants.Wind can exert both direct as well as indirect effect on plant growth. Winds do the maximum damage in association with rain at the time of flowering. Rainfall largely manifests itself through its effect on edaphic factors like soil moisture, soil temperature and soil aeration. The agricultural productivity under dryland condition is closely related with rainfall. Dew, fog, hail, snow and frost have also their respective effects on crop growth.

WATER BUDGET

Water budget of NE states (Thansanga and Saxena, 2000) is presented in Table 10. The average annual rainfall undergoes a variation from 1972 to 2930 mm in NE states and the annual evaporation losses are 807 -905 mm thus accounting of 31 to 44 % of the total rainfall received in NE states. The potential evapotranspiration, the measure of which is utilized for estimating the irrigation need of the crops, was correlated with pan evaporation in Tripura and the following equations as developed (Bhattacharya *et al*.., 1996) are mentioned below.

PET 1 (Penman) = 2.269 + 0.555 pan evaporation (R^2=0.764)

PET 2 (Blaney -Criddle) = 0.991-0.863 pan evaporation (R^2=0.465)

PET 3 (Thornthwaite formula) = 0.776 +0.861 pan Evaporation (R^2= 0.509)

The Penman equation showed the highest correlation value and such equation can be utilized to find out PET through the value of pan evaporation. Water deficit (PET-AET) and surplus { (P-PET)-Change in Water Storage} may be estimated and thus utilized (Bhattacharya *et al.,* 1997) to have a proper monitoring of irrigation need to crops. Nagaland and Manipur have shown the highest evaporation losses of the received rainfall. Surface water runoff is observed to be appreciably high varying from 51 to 59 % thus indicating the presence of undulated terrain and low retentive capacity of moisture in soils. On the other hand, the recharge to ground water is extremely low with 5 to 10 % of rainfall received. Water harvesting structure can be a reality to conserve the surface water runoff and water remaining in water harvesting structure can be utilized for irrigation to crops during non rainy periods and fisheries.

Table 10. Water Budget of North-Eastern States

Item	AP	Assam	Manipur	Meghalaya	Mizoram	Nagaland	Tripura
AV rainfall (mm)	2,930	2,336	1,972	2,253	2,565	1,986	2,516
ET losses (mm)	905 (31%)	873 (37%)	864 (43%)	807 (36%)	976 (39%)	872 (44%)	857 (34%)
Recharge to ground water (mm)	205 (7%)	234 (10%)	118 (6%)	113 (5%)	128 (5%)	99 (5%)	151 (6%)
Surface Water runoff (mm)	1820 (62%)	1229 (53%)	1008 (51%)	1313 (59%)	1431 (56%)	1015 (51%)	1508 (60%)

WEATHER AND CROPS

Rice

Rice is an important cereal crop grown widely in NE states under variable weather conditions and is raised both as direct seeded and transplanted crop. Pre-monsoon rainfall is highly essential to grow direct sown rice. There is a wide variability in pre-monsoon rainfall and 7 to 20 % of total rainfall is in pre-monsoon period with AP (20%), Assam (8%), Manipur (7%), Meghalaya (7%), Mizoram (12%), Nagaland (11%), Sikkim (10%) and Tripura (12%). The water need of rice is quite high (750-2500 mm) and 37 % yield variability may be noted depending on pre-monsoon/monsoon rainfall. Wet spells if continuing during the period from flowering to maturity of rice remain detrimental to productivity. Rice requires a mean temperature of 20⁰C and not less than 15⁰C during the entire growing period. The sunshine requirement in rice is 1000 hours with 220- 240 hours in the last 30 days. Strong winds are observed to induce sterility. Growth of rice crop is affected by the relative humidity which is known to monitor occurrence of disease and pest attack. Flowering of rice is inhibited below 40 % RH and is optimum around RH of 70 -80 %(Venkataraman and Narasimhamurthy, 1973). Low relative humidity of 60 % may induce a decrease in moisture content, loss of chlorophyll and faster senesecence of leaves (Hirai *et al*, 1985).On the other hand, the higher the RH, the higher would be the photosynthesis rate suggesting that the stomatal aperture may be increased with high RH.

As reported by Samra (2003), boro rice in Assam could experience a long spell of cold stress in 2002-2003.Average night temperature remained below 20⁰C for 183 days and the lowest daily temperature reported was 12.5⁰C. Seedling mortality (Table 11) was found to vary from 20.7 to 50.8 % with a high proportion of albinism (16.8 to 35.5 %). The average crop recovery after planting at the vegetative stage was around 71 %.

Leaf yellowing at the vegetative stage (Table 12) was quite high (5-8) and 50 to 60 % grain sterility was observed at the reproductive stage. The unproductive tillers due to cold stress was also strikingly high (14 -20 %).

Table 11. Cold injury in rice at seedling and vegetative stage during *boro* season in Assam

On farm location	Seedling stage		Vegetative stage
	Albinism (%)	Seedling mortality (%)	Crop recovery after planting (%)
Shillongani	35.5	50.8	66.9
Tokowbari	30.2	33.3	70.2
Bhakatgaon	31.0	25.4	66.4
Raha	16.8	29.7	75.8
Tezpur	18.2	20.7	82.2
Kaliabor	20.4	42.2	76.5
Kawoimari	37.9	45.6	60.6
Morigaon	18.9	40.0	70.8
Juria	26.8	46.2	65.5
Mean	26.2	37.1	70.5

Source: Assam Agricultural University

Table 12. Cold injury in rice at vegetative and reproductive stage during *boro* season in Assam

On-farm location	Vegetative stage	Reproductive stage	
	Leaf yellowing (1-9 score)	Grain sterility (%)	Unproductive tillers (%)
Shillongani	8	51.3	18.8
Tokowbari	5	48.9	14.2
Bhakatgaon	5	48.9	18.7
Raha	5	40.2	14.9
Tezpur	4	68.4	12.8
Kaliabor	8	52.1	16.4
Kawoimari	6	61.2	19.0
Morigaon	6	52.6	20.2
Juria	7	60.6	16.9
Mean	6	53.8	16.9

Source: Assam Agricultural University.

In contrast with the susceptible varieties, some promising boro rice varieties when subjected to cold stress in Assam showed (Table 13) low albinism, low seedling mortality and high crop recovery at the vegetative stage.

Leaf yellowing (Table 14) underwent a variation from 3-8 thus indicating that the variety 'Gautam' was susceptible to cold stress. A high grain sterility (45.6 %) and unproductive tillers (44.4 %) was noted in Gautam. All other varieties due to low grain sterility and unproductive tillers were having cold tolerance behaviour.

Pulses /Oilseeds

Pulses and oilseeds are highly susceptible to weather vagaries in NE states and productivity shows a decline in trend due to wide range of weather variations. Heavy rains may lead to rotting of seed if sown during kharif (July to Sept.). The kharif season is congenial for luxuriant vegetative growth with low pod formation. As such pre and post kharif season is the best period for higher productivity of pulses.

Table 13. Variability in tolerance of some promising *boro* rice varieties at seedling and vegetative stage to cold wave in Assam

Boro rice Varieties	Seedling stage		Vegetative stage
	Albinism (%)	Seedling mortality	Crop recovery after planting (%)
Dhanalaxmi	10.5	16.0	88.0
Saroj	9.4	10.1	87.3
Prabhat	10.5	11.4	80.3
Gautam	19.0	25.4	37.7
Joymati	9.0	16.8	86.0
Bishnuprashad	15.6	19.0	85.2
Mahsuri	11.1	11.6	90.0
Satabdi	10.7	12.2	85.0
Jagilee boro	18.9	21.9	80.1
PSB RC 2	12.3	20.1	80.6
Mean	12.7	6.5	80.0

Source: Assam Agricultural University.

Table 14. Variability in tolerance of some promising *boro* rice varieties at vegetative and reproductive stage to cold wave in Assam

Boro rice Varieties	Vegetative stage	Reproductive stage	
	Leaf yellowing (1-9 score)	Grain Sterility (%)	Unproductive tillers (%)
Dhanalaxmi	4	35.6	20.0
Saroj	4	29.0	26.4
Prabhat	3	32.6	14.2
Gautam	8	45.6	44.4
Joymati	5	29.9	12.0
Bishnuprashad	5	17.5	15.3
Mahsuri	5	18.6	15.0
Satabdi	4	14.2	15.6
Jagilee boro	3	20.2	38.1
PSB RC 2	5	14.1	12.2
Mean	5	25.7	21.3

Source: Assam Agricultural University

At crop maturity during the time of heavy rainfall and high humidity in kharif, groundnut shows germination thus declining the productivity. Fungal growth during seed storage in groundnut is a common phenomenon. Waterlogging or saturated moisture condition being prevalent during kharif season is highly detrimental to sesamum. North-west monsoon is scanty in NE region and thus soil moisture stress as generally noted in rabi season restricts the production of pulses/oilseeds. In the alpine, sub alpine and some parts of temperate hilly regions of NE India, the occurrence of frost becomes injurious to rabi pulse *viz.* peas, lentil and gram. Mustard, an important oilseed crop in NE region is also susceptible to frost in rabi season. Low germination and cold injury of groundnut are generally noted during winter in hills.

Vegetables/ Fruits

Weather requirements for production of vegetables and fruits in a profitable way are quite variable and as such, care is to be taken to achieve the productivity at the desired level. Cabbage can withstand extreme cold and frost but 12.8 to 15.6^0 C soil temperature is required for seed germination. So poly house structure is necessary for raising cabbage seedlings in winter. The temperature requirement of cauliflower is generally 5-25^0 C and air temperature not in this range may affect vegetative, curding and reproductive phase. In tomato, the air temperature required is 15-27^{0}C and low seed germination, poor plant growth and flower drop both at high and low temperature are noted. Excessive rain in kharif in NE states is injurious to tomato causing flower drop. For capsicum, ideal temperature for flowering is 26-28^{0}C (day) and 16-18^{0}C (night). Low temperature in hills hamper growth. The crop is unable to withstand heavy rains during flowering.

Among the tropical fruits grown in NE states, cashew, citrus, banana, litchi, pomegranate and coconut are important. Cashew is unable to tolerate frost and extreme cold. Temperature should not fall below 20^{0}C. High temperature (39 to 42^{0}C) may cause fruit drop. Heavy rains and cloudy weather are affecting yield to a large extent Citrus is susceptible to cold injury, hot winds and excessive heat, low humidity (colour development).

In banana, temperature required is 10-40^{0}C. The plant is susceptible to high wind, low temperature (delay in ripening). In litchi, prolonged rain or draught is not desirable at the time of flowering. Fruit splitting of litchi is caused by alternate spells of rain and dry heat. In pomegranate, there is a requirement of hot and dry climate during fruit development. Fruit quality may be affected by humidity.

So, high quality pomegranate may not be grown in NE states. In coconut, optimum temperature is 20-32^{0}C and is thus ideal for subtropical plain areas of NE states. High temperature and heavy rain may interfere with pollination. On the other hand, severe draught at flowering period may cause inflorescence to abort. The effect of frost on the mortality of fruit plants as recorded in Meghalaya is reported in Table 15. Mango and pineapple are highly susceptible to frost as indicated by high mortality (80 - 100%). Litchi showed the least mortality (10%) followed by peach (16%) and guava (20%).

Table 15. Effect of forst on one-year-old fruit/ forest plants at Barapani, Meghalaya

Fruit species	No. of saplings planted	No. of saplings survived	Mortality (%)
Mango (Amrapali)	15	Nil	100
Sapota	20	04	80
Chuckrasia tabularis	36	24	33.9
Michelia champaca	50	20	60
Jack fruit	60	25	58.9
Guava	50	40	20
Peach	50	42	16
Pear	24	16	33
Rose apple	20	07	65
Assam lemon	100	36	64
Litchi	10	09	10
Pineapple (Kew)	3500	600	83
Citrus	30	16	47

Source: ICAR Research Complex for NEH Region Barapani, Meghalaya.

Temperate Fruits are grown in Arunachal Pradesh, Sikkim, some parts of Meghalaya, Manipur and Nagaland. In apple, 1000-1500 hours of chilling is needed but temperature below (-) 2^{0}C may kill open blossoms. Low temperature, rain and cloudy weather are affecting cross pollination in apple. Strawberry is mainly temperate fruit but may also be grown in subtropical climate. Short day/long day/day neutral varieties are available. Peach is moderately winter hardy and sensitive to low temperature. Swelling buds become injured at (-) 6-5^{0}C. Plum is a fruit of subtropical to temperate climate in high hills, thus requiring 1000-1200 hours below 7 ^{0}C during winter. Forest free spring, good drainage and adequate nutrients during summer are required. Sweet orange which needs dry semiarid to subtropical climate may not be suitable in NE states. In Kiwis, 700-800 chilling hours below 7°C are needed to break the rest period. The impact of cold wave on the growth of one year old fruit plants is presented in Table 15. Most of the fruit plants besides litchi, peach, pear, guava are highly susceptible to frost attack during winter in Meghalaya.

Floriculture

Ornamental crops are classified as temperate and tropical ones with variability from shade (5,000-10,000 lux) to sunny plants (50,000-80,000 lux). In crysanthamum, temperature required is 12 to 18 °C. It shows bronze colouration of leaves, pink colouration in white flowers in low temperature. Bleached inflorescence may be visible at high temperature. In rose, high temperature(> 25° C) causes yellow/brown colouration and low temperature kills flower buds and dieback. In anthurium, gerbera and carnation, the optimum temperature is 16-19°C, 14-27°C and 10-24°C , respectively. Humidity variations are also wide in ornamentals. Cacti/Succulants need low humidity, anthurium/orchids require high humidity. Most foliage plants need more than 50% RH.

CONCLUSION

Weather relations of some of the important crops grown in NE India are reported herein. The study of such weather relations is, it is hoped, useful to select the crop or to take some appropriate measures to avert the crop losses generally incurred due to weather vagaries. Factors of weather in each growth stage of the crops need to be determined to elucidate the predominant influence on the yield. Such determination when compared along with the detailed climatological features would help in i) Avoiding the crop being raised in unsuitable seasons and areas and ii) Evaluating the risks to crop production due to inter- seasonal and intra- seasonal weather vagaries. So climate and weather systems are recognized as a critical agenda and input for achieving sustainable agricultural productivity in the 21st century.

REFERENCES

Bhattacharya, B.K., Datta, M. and Sharma, U.C. (1996). Computation of potential evaporation (PET) in Tripura. *Journal of Indian Society of Soil Science* 44 (4) :763-766.

Bhattacharya, B.K., Mitra, S. and Datta, M. (1997). Identification of water deficit and water surplus periods and prediction of profile water content from surface water status in upland of Tripura. *Journal of Indian Society of Soil Science* 45 (1) :698-701.

Hirai, G., Tanaka,O., Takahaski, M. and Koshi, Y.(1985). Studies on the effects of relative humidity on growth and physiology of rice.V. Influence of humidity on senescence of rice leaves. *Japan Journal of Crop Science* 54 :146-151.

Samra , J. S. (2003). Impact of climate and weather on Indian Agriculture, *Journal of Indian Society of Soil Science*, 51 (4): 418-430.

Thansanga, R and Saxena,R.S.(2000). Water resource availability and management in north eastern region of India for agricultural development.Extended summaries (Vol-2). *International Conference on Managing Natural Resource for Sustainable Agriculture Production in the 21st Century*. New Delhi,457-459.

Tsnuoda, S. (1959). A developmental analysis of yielding ability in varieties of crops II. The assimilation system of plants as affected by form, duration and management of single leaves. *Japan Journal of Breeding* 9:237-244.

Venkataraman,S (1968). Climatic consideration in cropping pattern, *Proceedings of Symposium on Cropping Patterns of India*, Indian Council of Agricultural Research, New Delhi,pp.251-260.

Venkataraman, S. and Naraisimhamurthy, K. (1973). On the flowering behaviour of rice crop at crop weather stations in India. *Indian Journal of Meteorological Geophysics* 24 : 293-298.

Climate Change and Food Security, 2008
Edited by: M.Datta, N.P.Singh & Er.D.Daschaudhuri, pp.119-127
New India Publishing Agency, New Delhi (India)
E-mail : newindiapublishingagency@gmail.com
Web: www.bookfactoryindia.com

8

Agrometeorological Scenario of North East India

K.K. SATAPATHY, U.S. SAIKIA AND P. CHOUDHURI

INTRODUCTION

The agriculture in north eastern hill region of India is mainly rainfed in nature and shifting cultivation, known as jhuming, is the dominant farming system. The shifting cultivation is detrimental to both soil and vegetation as it is connected to heavy soil and nutrient losses when practised in steep slopes. The burning and slashing operations are responsible for reduced floral and faunal diversity in the region. It is a well known fact that climate is the main driving factor that governs the sustainability of other natural resources such as land, water and vegetation. So far the NE region gets abundant rainfall for survival of the crops at least during the kharif season. The rainfall starts early March and continues up to the late October. The jhum cultivators follow the rainfall trend and get their field ready for sowing of seeds by early March itself so that vegetative stage of crops gets plenty of rainwater for growth and development. Due to increased population pressure, the hills of NE region are over exploited and the jhum cycle has been reduced from 10-12 years to 4-5 years now-a-days. Hence the soils do not get sufficient time to revitalize it and are subjected to more and more exploitation and get degraded permanently.

Besides the population pressure, it is the climate which is going to take its toll on the natural resources of the region. Temperature plays an important role in seed germination, crop growth and development, root functioning etc. Faster decomposition of organic matter is also linked to rise in temperature. Gadgil (1996) reported sharp increase of global temperature since 1970's. The trend of rainfall also shows a grim scenario on agricultural productivity. The increase in temperature is definitely going to reduce the yield of many field crops, particularly maize, over the years by shortening the vegetative and reproductive stages. Though the trend of amount of total annual rainfall is increasing over the years, there is a decrease in amount of rainfall in the pre-monsoon season which is very much vital for commencement of jhum cultivation. When sufficient rainfall is not received during the month of March, the jhum crop period is bound to reduce drastically leading to poor production and productivity of crops. The post monsoon season has observed an increase in amount of rainfall over the years and thereby deteriorating the quality of matured grains, delaying the ripening period etc. Although not much difference could be observed in overall rainfall amount during the monsoon period, the intensity of rainfall has been changed drastically over time. High intensity rainfall events have become more frequent over the moderate ones. The rainfall associated with increase in temperature will cause quick degradation, wash out of the important organic matter of the soil and thus reduce the soil productivity very fast as no artificial fertilization is done in jhum lands. The jhum cultivators will have to explore more and more new forest lands by burning and slashing them.

AREAS OF CONCERN

In recent times, the agriculture in India and north east India in particular, is dominated by the following:

- Overproduction in short-term, yet food insecurity for a large population
- Stagnation/decline in yields
- Absence of crop diversification
- Soil organic matter decline, reduced input use efficiencies and narrow genetic base
- Increasing cost and reduced profitability
- Deteriorating quality and quantity of water resources

The much talked about global climate change is affecting agriculture in different ways as :

- Increase in CO_2 concentration in the atmosphere
- Increase in temperature as a whole
- Change in precipitation pattern (amount and distribution)
- Variability and extreme events such as floods and droughts becoming more and more frequent

The above factors are directly or indirectly affecting food production including crops, livestock, fisheries and their stability as well as sustainability. Lal *et al.* (2001) reported from predictive research that climate change will affect Indian agriculture in the following ways:

- Rainfall during the KHARIF season will be most affected
- More then 15% increase in kharif season rainfall by 2080 over present rainfall
- RABI season rainfall will be decreasing simultaneously.
- Annual temperature is predicted to be increased in this century

CLIMATE OF NORTH EASTERN REGION

The regional orographic structure of North Eastern region is so varied and different that the weather elements can produce different climatic characteristics. The vast hills interspersed with fertile valleys represent agro climates of unique diversity ranging from extreme temperate to typical tropical, falling within the altitude range of 0 to 5000 m above mean sea level. The tropic of cancer passes just south of Aizawl of Mizoram state and as a result the greater part of northeast lies in the sub tropical belt. The weather and climate of NE India is itself a type, which is not comparable with any other parts of the sub continent. The border ranges, the hills and the topography of the regions exert both mechanical and thermo- dynamic influences on the general and seasonal circulation of wind distribution of pressure, temperature and precipitations.

There are definite pockets representing sub-alpine, temperate, subtropical and tropical areas. The Himalayan mountain range acts as a strong barrier for the circulation pattern, as a heat source in summer and acts as a sink in winter. The Himalayan range on the Northern boundary and the Patkai range with its innumerable outliers in different states act as regulator of the air mass movement; The Meghalaya plateau, Khasi-Jaintia and Garo hills, to some extent the Karbi Anglong hills living on the South of Brahmaputra valley act as a divide for in blowing the air mass. (Satapathy and Das Chowdhury, 2001).

CLIMATE OF NORTH EASTERN REGION - UNDERGOING ANY CHANGE?

The climatic scenario of last 20 years at Barapani, Meghalaya has been studied and it has been found that climate follows some sort of change during this period with respect to rainfall and temperature (Saikia *et al.* 2005[c]) (Figure 1).

The following observations have been made though it could not be concluded whether these changes are due to some local variability factors or part of overall global climate change.

- Mean annual maximum temperature is increasing significantly
- Slight decrease of annual mean minimum temperature
- Difference between maximum and minimum temperature is increasing, denoting a warmer climate with greater variability.
- The annual mean rainfall is also increasing slightly though not significant

Impact of climate change on agricultural production at north eastern region

From Table 1, it is observed that per capita availability of foodgrain at NE India has been reduced from 0.442 kg (1984-85) to 0.418 kg (200-01) which emphasizes a situation of more and more food shortage in this region for an ever growing population.

Table 1. Food grain production, population and per capita availability of food grain at NE India

Foodgrain production (000 tonnes)			Population (000 thousands)	
1984-85	3,989.2		1981	24,752
2000-01	5,875.0		2001	38,495
% change	47.27		% change	55.52
	Per capita availability (kg/person/day)			
	1984-85		0.442	
	2000-01		0.418	

PREDICTION OF ANNUAL FOODGRAIN PRODUCTION IN NORTH EAST INDIA BASED ON CLIMATIC PARAMETERS

Based on climatological parameters, a study has been conducted to predict the foodgrain production of north east India and following equations have been derived meteorological sub-division-wise (Saikia *et al.* 2005[a]; Ramakrishna *et al.*, 2003).

For North Assam Meteorological Sub-division

$PI = 103.631 + 0.012 * RI_{june} - 0.006*RI_{july} - 0.014*RI_{august} - 0.003 * RI_{september} - 0.022 * SOI$ ($R^2=0.61^{**}$, **Significant at 1% probability level)

For South Assam Meteorological Sub-division

$PI = 99.756 + 0.006 * RI_{june} + 0.001 * RI_{july} - 0.006 * RI_{august} - 0.002 * RI_{september} - 0.018 * SOI$ ($R^2=0.32^{*}$, *Significant at 5% probability level)

where,

PI = Production Index, RI = Rainfall Index, SOI = Southern Oscillation Index

The equations were found worthy in the following ways (Table 2):

- The model predictions for foodgrain production are within ±10% of the actual values
- It will make the foodgrain production estimates available 7-8 months ahead (Just after September) of official final estimates which are available during April next year
- It can be used as an essential tool to mitigate any shortfall of foodgrain for the region well in advance by the planners of the region

Table 2. Simulated Values of Annual Food grain Production for NE India

Year	North Assam Sub-division			South Assam Sub-division		
	Model Estimates	Actual Production	Deviation (%)	Model Estimates	Actual Production	Deviation (%)
1997	3948.83	3577.6	9.40	1636.26	1681.4	-2.76
1998	3741.88	3848.7	-2.85	1644.66	1683.6	-2.37
1999	3677.83	3934.4	-6.98	1586.01	1718.3	-8.34
2000	3769.64	4167.0	-10.54	1642.64	1708.0	-3.98

As the climate of this region is showing increasing trend of variability, it is very much essential to bring more and more area under sustained irrigation of any form for maintaining the agricultural production and productivity. Hence we should have a device to determine how much potentiality the atmosphere has to withdraw water from soil as well as plant. Potential Evapotranspiration (PET) determination is very much essential for irrigation scheduling, crop planning etc. The Papadakis method has been found most accurate to determine the PET for hilly terrains of NE region (Saikia and Satapathy 2005[b]) (Table 3).

CROP CALENDAR FOR MAJOR CROPS OF MEGHALAYA

A Crop calendar for different major crops of Meghalaya has been prepared based on weekly rainfall at 50% probability level and mean weekly temperature that is critical for growth and development of a particular crop (Table 4).

Table 3. PET estimation using different methods at Barapani, Meghalaya

Month/ Season	PE ± s.d.	PET (mm)		
		Papadakis	Blaney-Criddle	Thornthwaite
January	71.5 ± 6.4	71.1 ± 8.0	52.1 ± 1.7	25.2 ± 4.1
February	89.4 ± 13.8	80.5 ± 11.7	58.5 ± 2.9	34.8 ± 7.7
March	123.3 ± 20.3	100.8 ± 13.2	84.2 ± 2.6	66.1 ± 8.4
April	139.1 ± 32.8	103.2 ± 12.7	101.0 ± 3.9	92.8 ± 13.0
May	129.9 ± 24.3	93.2 ± 9.2	120.2 ± 2.9	108.5 ± 9.7
June	87.5 ± 9.8	85.4 ± 9.4	131.6 ± 2.6	120.9 ± 8.3
July	97.1 ± 14.9	78.0 ± 8.4	144.6 ± 2.0	127.2 ± 6.2
August	76.1 ± 10.8	85.6 ± 11.1	138.4 ± 1.5	120.7 ± 4.7
September	73.0 ± 10.1	84.7 ± 12.7	108.3 ± 1.7	100.3 ± 5.2
October	86.4 ± 7.6	86.9 ± 8.6	86.0 ± 1.7	80.9 ± 5.5
November	67.0 ± 5.5	82.8 ± 14.6	65.8 ± 2.5	51.1 ± 7.6
December	60.7 ± 8.6	77.2 ± 9.1	54.1 ± 1.7	32.2 ± 4.3
Winter	73.9 ± 15.5	76.3 ± 10.3	54.9 ± 3.4	30.7 ± 6.9
Summer	128.8 ± 26.8	98.8 ± 12.5	101.8 ± 15.2	89.1 ± 20.5
Monsoon	83.4 ± 14.9	83.4 ± 10.8	130.7 ± 14.0	117.3 ± 11.9
Post-Monsoon	76.7 ± 11.8	84.8 ± 12.0	75.9 ± 10.5	66.0 ± 16.4

Table 4. Crop Calendar for major crops of Meghalaya

Crop	Mar	Apl	May	June	July	Aug	Sept	Oct	Nov	Dec	Jan	Feb
Kharif Rice		1	2	3	4	5	6	7	8	9		
Karif Maize	2	3	4	5	6							1
Groundnut			1	2	3	4	5	6	7			
Rape & Mustard	7						1	2	3	4	5	6
Soybean			1	2	3	4	5	6	7			
Spring Potato	3	4	5	6							1	2
Winter Potato					1	2	3	4	5	6	7	8
Turmeric	1	2	3	4	5	6	7	8	9	10		
Ginger	1	2	3	4	5	6	7	8	9	10		
Lentil	7						1	2	3	4	5	6
Jute	2	3	4	5	6	7	8	9				1
Cotton	2	3	4	5	6	7	8	9	10			1
Summer Vegetables	5	6	7	8	9	10			1	2	3	4
Winter Vegetables						1	2	3	4	5	6	7

Figures 1-10 indicate the decreasing order of preference

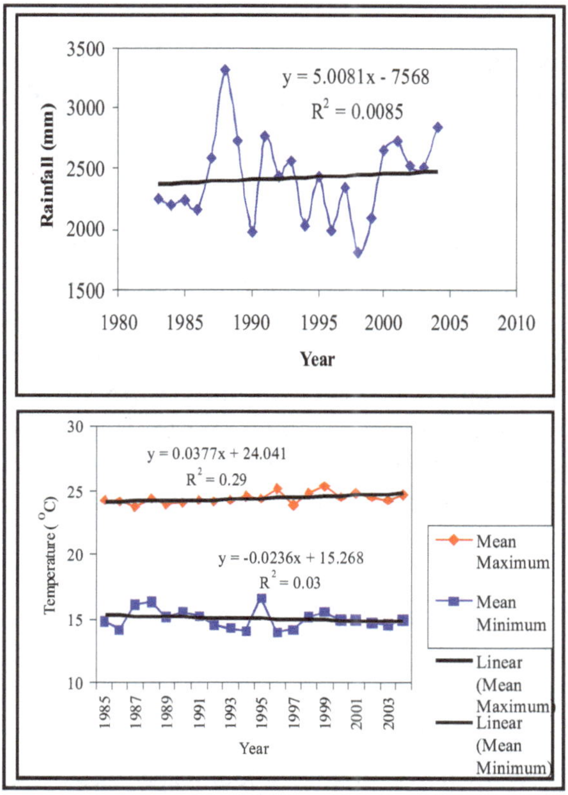

Fig. 1. Trend of annual rainfall and temperature at Barapani, Meghalaya

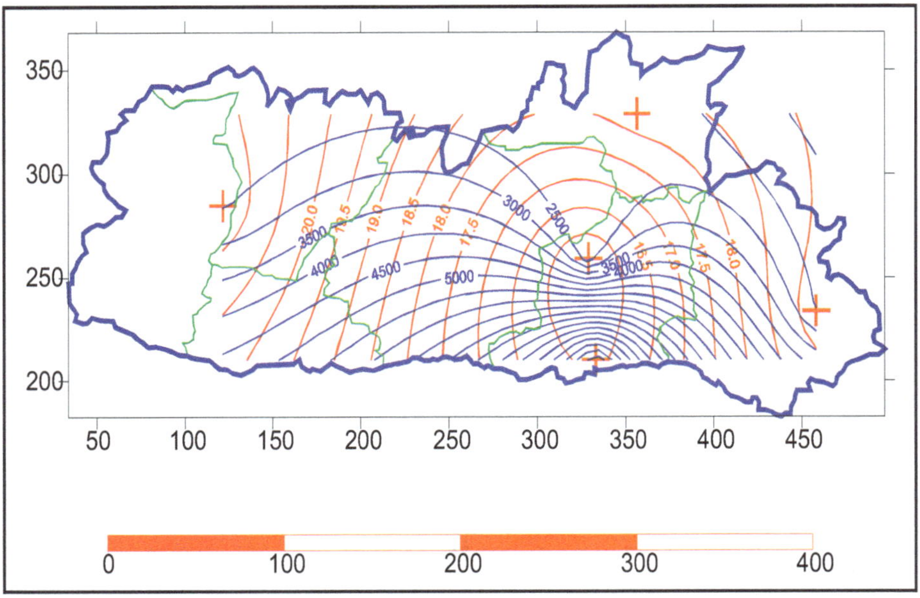

Fig. 2. Mean Annual Isohytes and Isotherms for Meghalaya

Time bound agricultural operations are possible if a crop calendar is followed properly. Over all cropping intensity can also be increased accordingly.

CONCLUSION

From the above analysis, it may be concluded that climate of north eastern region is undergoing some sort of change and steps have to be taken up to cope up with the situation in a more scientific and rational way keeping in mind of the limited resource base of the region. The water resource of this region is not unlimited and judicious application of water following improved management tools is the most important challenge lying in front of the decision making as well as scientific bodies to sustain the agricultural productivity and ascertaining food security of the region.

REFERENCES

Gadgil S, (1996). Climate change and agriculture-an Indian perspective. pp. 1-18. *In*: "*Climatic Variability and Agriculture*". Abrol, Y P *et al.* (eds), Narosa Publishing House, New Delhi.

Lal, R. *et al*, (2001). www.ficci.com/media-room/speeches-presentations/ 2002/ oct/ oct-climate-aggarwal.ppt.

Ramakrishna Y S, Singh H P and Nageswara Rao G, (2003). Weather based indices for forcasting foodgrain production in India. *Journal of Agrometeorology*, **5** (1), 1.

Satapathy, K. K. and Daschaudhuri, D, (2001). *Weather and Climate*. pp. 25-41. *In*: "Steps towards modernization of agriculture in NEH region". Verma N D and Bhatt B P (eds), ICAR Research Complex for NEH Region, Umiam, Meghalaya.

Uday S. Saikia, K. K. Satapathy, Anil Kumar Singh and B. Goswami, (2005[a]). A Weather Based Method to Forecast Foodgrain Production of North East India. *Journal of Assam Science Society*. Vol 46 (December). pp. 4-8.

Uday S. Saikia, K. K. Satapathy, B. Goswami and T. D. Lama, (2005[b]). Estimation of PET by Empirical Models for North Eastern Hill Region of Meghalaya. *Journal of Agrometeorology*,7 (2). pp. 268-273.

Uday S. Saikia, K. K. Satapathy, B. Goswami, R K. Singh and B. K. Rao, (2005[c]). Trend of Rainfall and Temperature Change at Umiam, Meghalaya (Submitted paper).

Uday S. Saikia and K. K. Satapathy, (2005). *Climatic Atlas of Meghalaya*. Unpublished report. ICAR Research Complex for NEH Region, Umiam, Meghalaya.

Climate Change and Food Security, 2008
Edited by: M.Datta, N.P.Singh & Er.D.Daschaudhuri, pp.129-147
New India Publishing Agency, New Delhi (India)
E-mail : newindiapublishingagency@gmail.com
Web: www.bookfactoryindia.com

9

Agro-climatic Characterization A Necessity for Scientific Crop Production And Crop Planning in Dryland Agriculture-A Case Study of Varanasi Region

TARKESHWAR SINGH, M.K. SINGH AND S.R. SINGH

INTRODUCTION

Of an estimated 142 mha of net cultivated areas in India, about 67 % is rainfed and only 45 mha is irrigated. The NCA (1976) predicted that even when the full irrigation potential of the country is exploited by 2013 AD, over 50% of arable land would continue to remain rainfed in future. It is the rainfed belt where cultivation of 91 % coarse cereals (Sorghum, Pearl millet and Finger millet), 91% pulses (Chickpea and Pigeon pea), 80% oilseeds (Groundnut, Rapeseed, Mustard and Soybean) and 65% cotton have predominated. Even the crop that formed the backbone of green revolution such as rice and wheat, still have 50% and 19% area under rainfed condition, thus emphasing the importance of rainfed agriculture in India. Inadequate resource (soil and rain water) management coupled with traditional farming practices are responsible for subsistence farming. Monocropping, traditional crops and their varieties, low and imbalance nutrients and inadequate crop establishment, in past, had been the major cause for poor development of dryland agriculture in the eastern region.

One of the worst uncertainties in dryland agriculture is the climate. Besides development of various management technologies for specific location, successful crop production in dryland areas requires a thorough understanding of different crop growing environments. Risk to dependable crop production varies with the place and natural resources in agronomic terms, it is possible to provide land use recipes for agroecological sub regions. The variability of seasonal rainfall ,in particular the uncertainty of the amount and distribution of rainfall in any given season ,forms the greatest source of risk to crop yields in dryland agriculture.

Rain is the only source of water for dryland crops. The southwest monsoon contributes 74% of total rain water. The fate of rainfed agriculture oscillates with the quantity, onset, progress and spatial distribution of the south west monsoon. At times , rain is limited in volume so that it is either inadequate for preliminary land preparation and sowing or it cannot offset the evapotranspiration induced water deficits. So, the southwest monsoon becomes a matter of serious concern due to its uncertaintity. It can start late thus squeezing the growing period . On the other hand, it can withdraw early leading to water deficiency during the reproductive phase and critically influencing productivity.

Rainfall-its intensity, amount and distribution, temperature variations, relative humidity, sunshine hours, solar radiation, potential evapotranspiration and the water holding capacity in soils are all important parameters which need concerted attention. These elements could combine in a wide variety of ways and thus may be thought as the 'Climatic resource'. Such interactions would help to find the best match between the climatic resource and the crop which can be grown using the resources.

Climatologically, the region in Varanasi is characterised as Dry-Sub Humid, covering large areas belonging to different soil groups. The average annual rainfall ranges between 1050-1120 mm (997.9 mm dependable) and evapotranspiration between 1400-1535 mm. The CV of rainfall is 19.1%. About 87 per cent of total annual rainfall is received during the monsoon season. The potential evapotranspiration during monsoon period is 574 mm. The length of growing period varies from 27-32 weeks of which 14 weeks are humid. The area is characterised as a potential region for enhancing and stabilising crop production through sequential/intercropping systems based on climatic and edaphic features. Under these conditions, upland rice, maize, pearl millet, sorghum, pulses, oilseeds during *kharif* and barley, wheat, chickpea, lentil, mustard and linseed in post rainy season are commonly raised.

RAINFALL CHARACTERISTICS

Pertinent data in respect of rainfall characteristics are presented below.

Table 1. Rainfall Characteristics of Varanasi region.

Sowing rain	-	June 10 to July 04 (spread over a period of approximately 4 weeks)
Cessative rains	-	September 23 to October 4 (spread over a period of 2 weeks)
Mean annual rainfall (mm)	-	1080 (909 mm dependable)
CV (%) of annual rainfall	-	19.1
Mean annual PET (mm)	-	1525
Median date of onset (SW monsoon)	-	20/6
Mean date of onset (SW monsoon)	-	22/6
Variability of commencement	-	CV (%) - 37.5, SD (days) - 9
Earliest date of onset of SW monsoon	-	31/5 (1933)
Latest date of onset of SW monsoon	-	15/7 (1926)
Withdrawal		
Median date of withdrawal of SW monsoon	-	27/9
Mean date of withdrawal of SW monsoon	-	27/9
Variability of withdrawal - CV (%)	-	35, SD (days) - 12
Earliest withdrawal date on record	-	17/8 (1925)
Latest withdrawal date on record	-	30/10 (1929)

Table 2. Probability of onset of SW monsoon on weekly basis (Varanasi)

Class interval	No. of years	% Probability
Before 1/6	1	1
1/6 - 7/6	5	7
8/6 - 14/6	8	11
15/6 - 21/6	20	27
22/6 - 28/6	24	32
29/6 - 5/7	11	15
After 5/7	6	8

Table 3. Probability of withdrawal of SW monsoon on 10 days interval (Varanasi)

Class interval	No. of years	% Probability
After 4/9	2	3
5/9 - 14/9	6	8
15/9 - 24/9	19	25
25/9 - 4/10	25	33
5/10 -14/10	15	21
15/10 - 24/10	6	8
After 24/10	2	3

RAINFALL PATTERN

Daily

From Table 4, it is seen that on an average, out of 59 rainy days in a year with measurable rainfall, probability of rainfall exceeding 100 mm will be only on a single day and exceeding 50 mm on 4 days. However, the maximum amount of rainfall received on a single day ranged between 165-681 mm during the months of June to November (1971-2000) although such occasions are expected to be of rare.

Table 4. Distribution of rainy days at Dryland farm, B.H.U., Varanasi (1970-2000)

Month	Average number of rainy days with the corresponding rainfall					
	1 mm	3mm	10mm	20mm	50mm	100mm
January	2.5	1.7	0.6	0.3	-	-
February	2.3	1.2	0.6	0.2	-	-
March	1.3	0.9	0.3	0.1	-	-
April	1.0	0.7	0.2	0.1	-	-
May	1.5	0.9	0.4	0.1	-	-
June	5.2	4.1	2.3	1.2	0.2	-
July	5.3	13.1	7.9	4.9	1.3	0.2
August	14.5	12.2	7.8	5.0	1.4	0.3
September	10.8	8.9	5.5	3.4	1.3	0.5
October	2.5	1.8	0.8	0.3	0.1	-
November	1.1	0.6	0.1	0.1	-	-
December	1.2	0.8	0.3	0.2	0.1	-
Annual	59.2	46.9	26.8	15.9	4.4	1.0

Weekly

The association between rainfall amount and date of onset, rainfall pattern and date of onset were studied by analysing the weekly rainfall data (over the available years). On the basis of normal rainfall values, it is easier to assume longer growing season in association with early start and shorter growing season in association with late start of rains. However, it might happen in a different way during each year as data set out in Table 5 is based on average value over years. In this context, identification of major rainfall patterns associated with different dates of commencement of rainy season will be useful. It is evident from the Table 5 that if monsoon commences between June 4-10 (week 23), there is a possibility of not getting adequate rains during weeks 24 and 25. Therefore, it would be better to take up sowing only if the moisture build up is adequate to withstand the dry spell likely to occur around 24 and 25th weeks. If the rains commence between June 18 to 24, possibility of dry spell will be the least. Further late commencement of SW monsoon rains beyond July 2 is likely to result in its early withdrawal i.e. before 38th week (September 17-23) causing moisture stress at later stage of crop growth. In this situation, one may not like to. plan for long duration crop(s). In this system, a suitable alternative like short duration crop and early maturing crop may be considered.

Table 5. Mean weekly rainfall (mm) associated with different dates of onset of South-West Monsoon in Varanasi (1941-2000)

Standard Week Meteorological	Dates of onset of South- West Monsoon				
	June	June	June	June	July
	4-10	11-17	18-24	25-1	2-8
	Weekly Rainfall (mm)				
23	29.9	-	-	-	-
24	17.3	52.5	-	-	-
25	7.0	28.4	67.8	-	-
26	48.7	48.8	66.6	37.1	-
27	100.2	45.5	37.3	42.6	49.2
28	59.2	91.1	24.2	66.3	27.8
29	70.8	53.8	115.2	93.3	73.2
30	71.2	50.8	97.4	31.5	35.7
31	59.1	42.5	84.7	31.5	35.7
32	114.3	87.7	78.2	62.6	130.1
33	51.0	71.5	42.7	50.0	20.7
34	33.7	68.2	105.5	49.4	57.4
35	39.7	65.6	10.6	91.29	2.6
36	109.1	55.7	99.5	30.6	16.23
37	83.6	120.6	14.8	57.1	86.67
38	34.2	50.1	26.0	48.1	0.0
39	49.7	22.6	39.1	23.3	4.9
40	23.1	2.0	15.2	40.7	11.5

(The association between rainfall amount and date of onset were studied by analysing the weekly rainfall data).

Ten Days

Characteristics of ten day rainfall distribution (over a period of 1951-93) are given in Table 6. August 11-20 are most reliable expecting rains between 12-191 mm followed by July 21 to 31 (between 0-275 mm) and the period between August 1-10 and 25-31 with expected rains upto about 300 mm.

Seasonal

Seasonal-rainfall pattern during Southwest Monsoon at Varanasi during the three decades 1971-80, 81-90 and 91-2000 are given in Table 7. It is clear from the table that with delay in start of rainy season in the decade 1981-90 and 1991-2000, the end of monsoon season advanced by 10 and 13 days. This pattern resulted in shortening of monsoon season by 13 days and reduced quantity of rainfall by about 47mm. This is an indicative of recent shift in seasonal pattern of rainfall.

Table 6. Characteristics of ten day rainfall distribution in varanasi (1951-1993)

Duration of ten days	Mean rainfall	Coefficient of variation	Extreme of rainfall (mm)		Rainfall (mm) that can be expected with probability (in per cent of)					
		(per cent)	High est	Low est	30	40	50	60	70	80
May 21-31	3.3	284.8	50.0	0.0	4.1	2.6	1.6	0.5	0.0	0.0
June 01-10	15.1	190.7	107.9	0.0	16.6	10.4	6.4	2.8	0.3	0.0
11-20	18.2	130.2	88.6	0.0	22.3	15.9	10.8	6.3	3.1	0.6
21-30	48.2	103.9	176.8	0.0	58.1	43.1	30.8	21.2	13.0	6.2
July 1-10	88.3	86.5	297.4	0.0	103.6	82.5	63.1	47.3	33.2	20.9
11-20	126.0	84.1	525.0	0.3	152.4	119.7	93.7	71.9	51.0	34.5
21-31	106.0	71.2	275.8	0.0	128.3	103.4	80.4	63.1	45.9	31.7
Aug. 1-10	87.9	75.5	292.7	0.0	104.9	84.1	69.2	52.9	39.6	27.7
11-20	84.7	57.8	191.5	12.0	105.0	86.3	71.1	60.2	46.0	34.0
21-31	98.1	75.2	303.3	0.0	116.7	93.0	71.2	53.4	37.6	23.7
Sept. 1.10	106.2	119.0	700.0	0.0	131.3	102.8	76.6	57.5	39.7	25.4
11-20	97.4	124.9	542.9	0.0	116.9	87.2	62.9	43.9	27.7	14.2
21-30	57.7	101.7	231.1	0.0	67.4	50.1	38.5	26.9	18.2	11.0
Oct. 1-10	25.8	131.8	141.8	0.0	29.0	19.8	12.9	8.0	4.2	0.7
11-20	7.4	214.8	75.2	0.0	9.0	5.7	3.3	1.5	0.2	0.0
21-31	1.6	237.5	17.6	0.0	1.4	1.6	1.0	0.6	0.0	0.0

Table 7. Rainfall pattern during SW monsoon season (1971-2000)

Period	Start of SW monsoon		End of SW monsoon		Duration of monsoon season		Seasonal rainfall		Number of rainy days	
	Mean (Julian days)	CV (%)	Mean (Julian days)	CV (%)	Mean (Julian days)	CV (%)	Mean (mm)	CV (%)	Mean (No)	CV (%)
1971-80	176	4.9	278	3.5	102	6.3	947.7	26.5	46	17.9
1981-90	178	6.2	271	3.7	92	14.7	909.1	23.8	48	27.5
1991-2000	180	7.0	269	4.2	89	15.3	890.3	24.2	49	20.3

Relationship between Onset of Monsoon and Seasonal Rainfall

The seasonal rainfall bears non-significant relationship with onset of rainy season and it was observed that if the onset of monsoon takes place during 170-175 Julian days, then good amount of rainfall (948 mm) is received and the delay in onset had decreased the seasonal rainfall.

Relationship between onset of monsoon and duration of the rainy season

The relationship between onset of monsoon and duration of the rainy season showed that the years with early and normal onset of

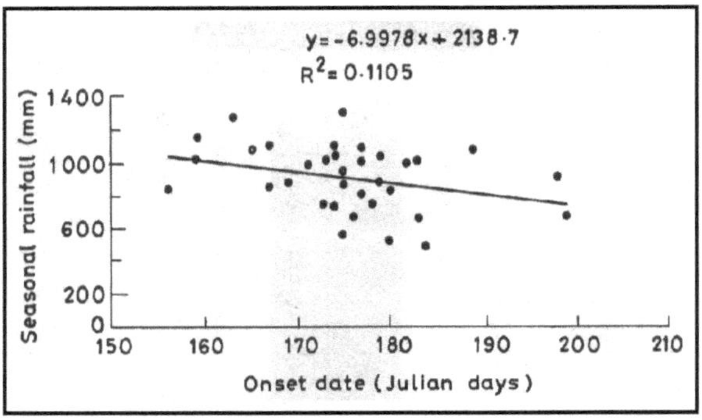

Fig. 1. Relationship between onset of monsoon and seasonal rainfall

monsoon had longer duration of rainy season and the delay in onset has resulted in curtailed duration of the season. (Fig 2)

There is a high variability in the seasonal distribution of rainfall, the coefficient of variability increasing with decreasing annual rainfall, prolonged and large amount of single day rains normally damage pulses and oilseeds during its crop growth period. The rainfall pattern

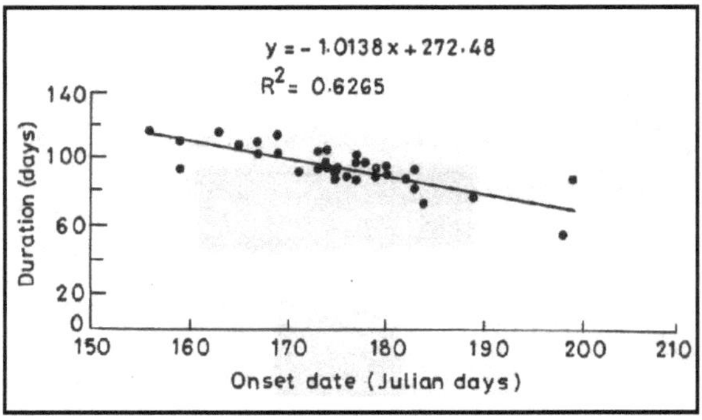

Fig. 2. Relationship between onset of monsoon and duration of rainy season.

indicates that out of 52 weeks in a year, weekly rainfall is highly dependent only in 7 weeks and dependable in 7 weeks (Table 8).

The distribution of annual rainfall is presented in Table 9 and it is revealed that 88.43 % of rainfall is generally received in monsoon but only 4.19 % of rainfall is available in winter season.

Table 8. Dependable and highly dependable weeks of the year in respect of rainfall variability

Highly dependable weeks (SMW)	Mean rainfall (mm)	SD (mm)	CV (%)	Dependable weeks (SMW)	Mean rainfall (mm)	SD (mm)	CV (%)
27	46.2	46.0	93	25	17.0	23.1	141
29	98.9	81.6	84	26	45.2	50.8	114
30	48.1	43.9	88	28	80.1	102.9	130
32	99.8	78.6	81	31	65.0	74.1	117
33	87.6	56.6	68	34	60.4	70.3	104
35	61.1	59.5	99	38	39.2	55.8	150
36	60.0	44.9	83	40	25.1	29.4	118

Table 9. Distribution of annual rainfall in the region

Period	Amount (mm)	Per cent	Period	Amount	Percent (mm)
Monsoon	958.6	88.43	Winter	45.4	4.19
Post- monsoon	50.8	4.68	Summer	29.2	2.69

CLIMATIC WATER BALANCE

Assessment of water availability for any place is essential in order to know the agricultural potential of the place. Such an assessment indicates periods of water deficit, water surplus, moisture utilization and recharge pattern for contingent planning to minimize risk on account of weather aberrations. For mandated area i.e. Varanasi, Mirzapur and Sonbhadra districts, five elements of climatic water balances e.g. precipitation, potential evapotranspiration, actual evaporation, water deficit and water surplus have been presented on weekly basis (from Standard Week, no. 1 to 52) for Varanasi (Table 10). Pictorial view of the water balance for Varanasi region is given in figure 3.

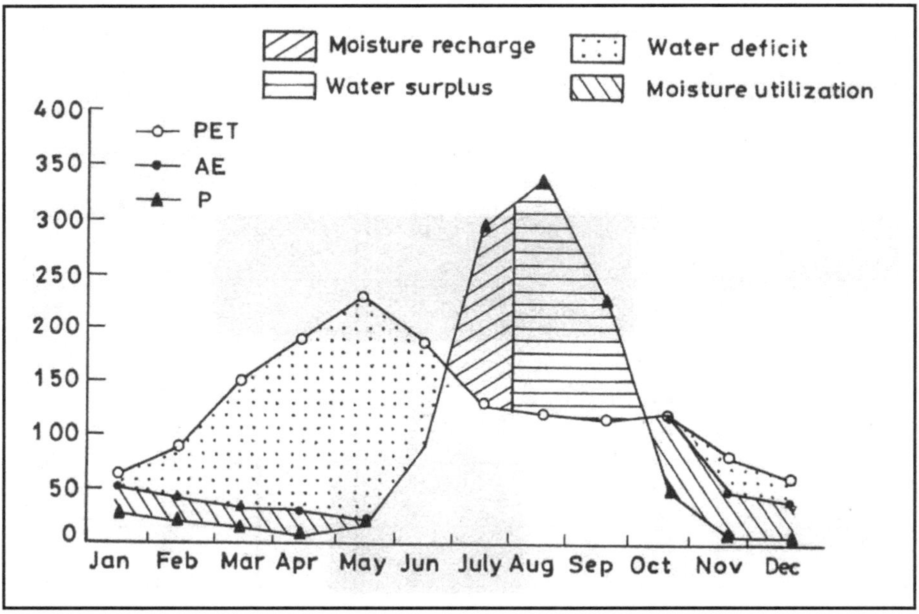

Fig. 3. Climatic water balance of Varanasi

Table 10. Weekly water balance 1971-2004

Week No.	Month	Rainfall (mm)	Potential evaporation	Actual Evapo-transpiration	Deficit (mm)	Surplus (mm)
1	January	4.4	13.1	13.1	0.0	0.0
2		2.3	14.1	14.1	0.0	0.0
3		7.8	14.1	14.1	0.0	0.0
4		5.3	14.4	14.4	0.0	0.0
5		6.1	18.4	15.7	2.7	0.0

Contd...

Table 10. Contd...

6	February	5.3	26.0	5.3	20.7	0.0
7		2.9	26.5	2.9	23.6	0.0
8		7.8	26.3	7.8	18.5	0.0
9		2.1	26.7	21	24.1	0.0
10	March	1.8	29.1	1.8	27.3	0.0
11		1.6	32.3	1.6	30.7	0.0
12		2.1	34.7	2.1	32.6	0.0
13		5.8	37.5	5.8	31.7	0.0
14	April	1.0	40.5	1.0	39.5	0.0
15		1.0	43.9	1.0	42.8	0.0
16		0.4	44.7	0.4	44.3	0.0
17		1.0	47.1	1.0	46.1	0.0
18		2.2	48.2	2.2	46.0	0.0
19	May	4.9	48.6	4.9	43.7	0.0
20		2.5	49.4	2.5	46.9	0.0
21		1.0	54.4	1.0	50.4	0.0
22		5.8	50.0	5.8	44.2	0.0
23	June	8.2	50.6	8.2	42.4	0.0
24		17.6	47.2	17.6	29.6	0.0
25		16.3	44.0	16.3	27.0	0.0
26		44.7	39.3	39.3	0.0	0.0
27	July	49.3	31.5	31.5	0.0	0.0
28		79.4	31.8	31.8	0.0	0.0
29		96.7	31.6	31.6	0.0	0.0
30		49.8	29.7	29.7	0.0	0.0
31		63.1	29.7	29.7	0.0	0.0
32	August	101.1	30.0	30.0	0.0	0.0
33		83.5	30.8	30.8	0.0	39.7
34		67.5	30.8	30.8	0.0	29.2
35		60.4	29.1	29.1	0.0	28.9

Contd...

Table 10. Contd...

36	September	54.4	27.1	27.1	0.0	41.4
37		82.9	27.9	27.9	0.0	62.8
38		37.2	28.6	28.6	0.0	31.4
39		36.0	28.0	28.0	0.0	0.0
40	October	24.8	27.9	27.9	0.0	0.0
41		9.6	25.4	25.4	0.0	0.0
42		10.4	25.3	25.3	0.0	0.0
43		1.2	22.4	22.4	0.0	0.0
44		5.1	20.9	20.9	0.0	0.0
45	November	0.3	19.9	19.9	0.0	0.0
46		2.6	17.6	17.9	0.0	0.0
47		1.9	16.6	16.6	0.0	0.0
48		0.4	16.2	16.2	0.0	0.0
49	December	2.0	13.7	13.7	0.0	0.0
50		0.2	13.5	13.5	0.0	0.0
51		1.2	13.3	13.3	0.0	0.0
52		1.2	13.0	13.0	0.0	0.0

ONSET OF SOWING RAIN AND CESSATION OF RAINY SEASON

It is important to know the dates of onset of sowing rains and cessation of rainy season for crop planning and refinement owing to weather aberrations. A simple graph (Fig. 4) incorporating potential evapotranspiration (PET) and rainfall can be of great help. For example at Varanasi the sowing rains commence, at the earliest around June 10 and the latest by the first week in July. The start of sowing rains has been considered as the week which had more than 30 mm of rain in one or two consecutive days, provided, that the probability of at least 15 mm of rain in the subsequent week is greater than 0.7. The cessation of rainy season occurs between September 23 to October 4. This is an indicative that recessive rains are more reliable than sowing rains requiring for a contingent planning for *kharif* sowing.

Annual Rainfall (mm)	:	1081.4
Annual PET (mm)	:	1525.2
Annual AET (mm)	:	843.0
Annual Water Surplus (mm)	:	238.2
Beginning of soil moisture recharge	:	28th June
End of soil moisture recharge and beginning of water surplus	:	1st August
End of water surplus and beginning of soil moisture use and water deficit	:	10th October

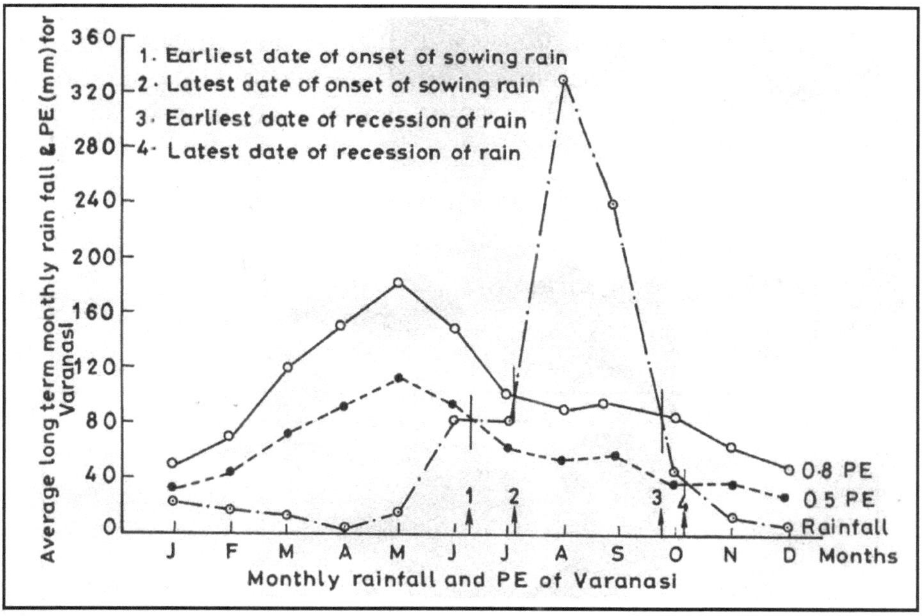

Fig. 4. Average long term monthly rainfall & PET (mm) of Varanasi region.

MOISTURE AVAILABILITY AND CROP PLANNING

The levels and duration of moisture availability in rainfed situations depend on the rainfall and potential evapotranspiration parameters. The time of onset and cessation and the amount distribution pattern of rainfall along with the pattern of PET determine the moisture availability (Fig. 5).

In the beginning of the rainy season, the PET is quite high, even with low crop canopy due to high temperatures and longer day length.

Most of the crop establishment operations are done at this time. The period when rainfall is less than PET but greater than ½ PET can be designated as 'Moist I' (mid May to mid June in Fig. 5). With the advancement of the rainy season, rainfall exceeds PET and inspite of an increased crop canopy, the PET becomes low due, to lowering of the temperature and shorter daylength. In this duration, there is surplus water after satisfying the crop water requirement and the soil moisture recharge. This period can be termed as 'Humid' period (mid June to mid Sept in Fig. 5. This period is considered best for the crop growth. At the end of the rainy season, both the rainfall and PET decreases, and again when rainfall is lower than PET but greater than 1/2 PET, this period is considered 'Moist II' one. The cropping practices, especially the duration of the varieties and their sowing time of main *kharif* crop paddy need to be adjusted in a way that the crop maturity coincides in this period and that the harvesting of rice and field preparation of next *Rabi* crop in sequential cropping which is going to be sown on residual soil moisture, are also done in this period.

Beginning of soil moisture recharge	:	28th June
End of soil moisture recharge and beginning of water surplus	:	1st August
End of water surplus and beginning of soil moisture use and water deficit	:	10th October

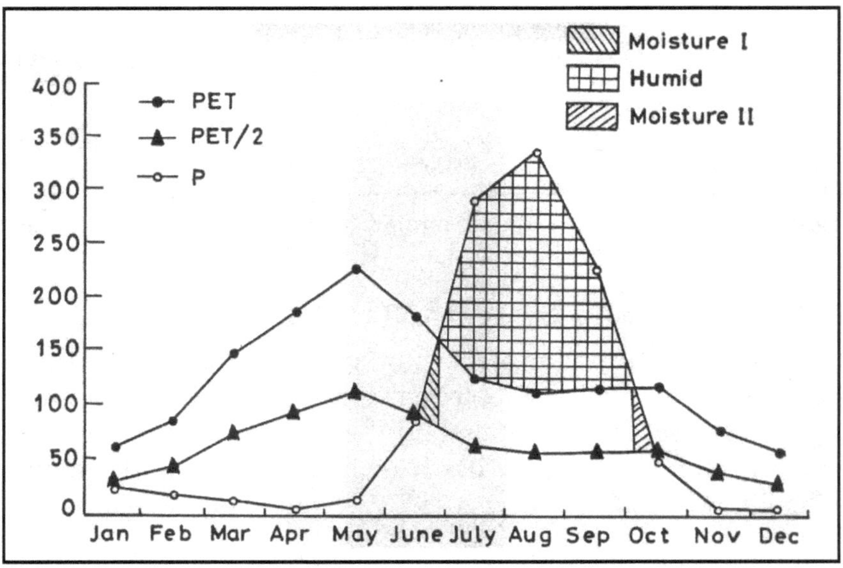

Fig. 5. Moisture availability with respect to rainfall and PET

The level and duration of moisture availability in rainfed situations depends on the rainfall.Apart from soil and site characteristics, availability of soil water in the root zone in required amount at different stages of crop growth is the major deciding factor in crop production. Quantification of such information, therefore, helps in assessing the length of cropping season in a rational manner and also provides guidelines for optimum land use.

Soil water availability periods in different categories have been worked out and presented in Fig 6. As a guide in practical farm operations the first period of very poor (VP) soil water availability and part of poor (P) availability period can be used for land preparation. This period coincides with standard week 24-25. It can also be attributed from the Fig.6 that excellent period of soil moisture availability is associated with 26 standard weeks 26-05 (June 25 to Feb. 4) indicating length of growing season in dryland areas of Eastern Uttar Pradesh as 32 weeks. Moisture Availability Index (MAI) is another indicative for preparing and planning calendar of operations. It can be interpreted from data on MAI in Table 11 that the weeks with MAI exceeding 0.50 are generally defined as having adequate moisture to sustain dryland crop production. Thus for Varanasi region, the MAI exceeds 0.50 value during July to January (approximately 07 months) and soil moisture stress to winter crops is inevitable anytime after 5th February.

RAINFALL VARIABILITY

In order to optimise use of water resources, for crop production in given region characteristics of rainfall in terms of its distribution and Variability gains urgency. Weekly distribution of rainfall and other weather parameters are presented in Table 12. About 87 per cent of the total annual rainfall is received during the monsoon season (Table 9).

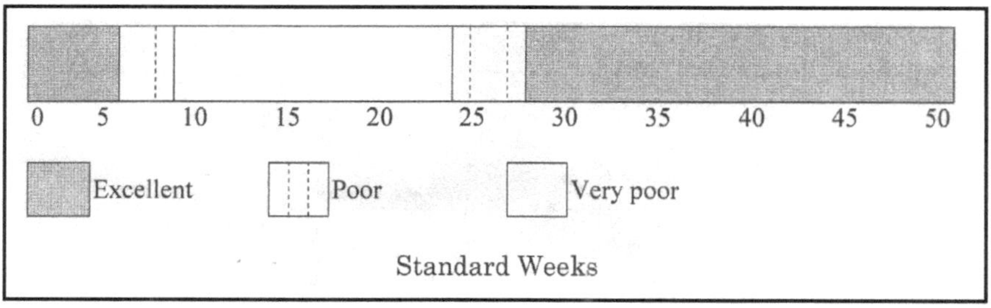

Fig. 6. Water availability calendar in dryland areas of eastern Uttar Pradesh

Table 11. Soil water availability in different categories (Varanasi)

Total Number of Weeks			
Excellent (E)	Good (G)	Poor (P)	Very poor (V.P.)
32	0	3	17
Length of effective cropping season in dryland areas of eastern Uttar Pradesh			
Week number			
Excellent (E)	Good (G)	E+G	Total duration in weeks
1 - 5, 26 - 52	-	26-5	32

Table 12. Averaged weekly values (1941-2004) of weather parameter at Varanasi

Week No.	Month	Rainfall (mm)	Temperature ⁰C)		Wind speed km/ha	Sunshine hrs	RH%		Evaporation (mm)
			Maximum	Minimum			I	II	
1	January	4.4	22.1	8.8	2.1	7.7	86	39	1.7
2		2.3	22.2	8.6	2.0	7.6	83	42	1.8
3		7.8	22.0	8.7	2.0	7.8	89	48	1.9
4		5.3	22.7	9.3	1.9	8.3	80	50	2.0
5		6.1	23.1	9.6	1.8	8.9	80	49	2.4
6	February	5.3	23.6	9.6	1.9	8.8	79	50	2.5
7		2.9	25.9	11.4	2.0	8.9	78	40	2.9
8		7.8	26.8	11.7	2.1	9.3	78	44	2.9
9		2.1	27.8	12.4	1.7	9.4	89	44	3.3
10	March	1.8	29.5	14.0	1.9	9.6	76	23	3.9
11		1.6	31.5	14.9	2.0	9.9	50	25	4.1
12		2.1	33.5	16.4	3.0	9.6	49	20	5.3
13		5.8	34.4	17.8	3.2	9.7	42	21	6.1
14	April	1.0	36.4	17.8	3.4	9.7	39	20	6.1
15		1.0	36.4	19.2	2.9	9.7	30	19	7.5
16		0.4	38.7	22.0	3.0	10.0	29	15	8.9
17		1.0	39.4	23.8	4.5	10.1	32	18	8.9
18		2.2	39.6	24.3	4.7	10.2	33	19	9.4

Contd...

Table 12. Contc...

19	May	4.9	39.8	25.1	4.9	10.2	40	22	9.7
20		2.5	40.1	25.8	4.4	10.1	44	21	9.4
21		1.0	40.2	26.2	5.0	10.4	50	23	9.5
22		5.8	40.7	29.9	5.3	9.5	50	30	9.0
23	June	8.2	39.5	27.9	6.7	9.8	56	39	9.0
24		17.6	38.6	27.7	6.3	88	63	42	8.7
25		16.3	37.6	27.3	5.9	79	69	48	8.0
26		44.7	35.7	26.9	5.7	6.5	77	59	6.7
27	July	49.3	34.4	26.6	6.8	5.6	80	63	5.5
28		79.4	33.5	26.2	6.1	5.9	83	69	5.0
29		96.7	33.0	26.1	5.9	4.8	88	72	4.6
30		49.8	32.6	26.0	5.8	4.6	86	72	3.9
31		63.1	32.2	26.0	5.6	4.6	86	71	3.8
32	August	101.1	32.4	26.0	5.4	5.5	88	73	3.3
33		83.5	32.2	26.0	5.5	6.3	87	73	3.5
34		67.5	32.1	25.8	5.9	5.7	87	73	3.2
35		60.4	32.6	25.7	5.3	6.7	86	73	3.5
36	September	54.4	31.8	25.3	5.0	6.3	86	73	3.3
37		82.9	31.9	25.0	5.9	5.6	88	75	3.0
38		37.2	32.2	24.9	5.4	6.9	84	69	3.8
39		36.0	32.2	24.2	5.6	7.2	84	66	3.7
40	October	24.8	32.4	23.4	3.1	8.1	82	63	3.6
41		9.6	32.5	22.0	2.7	9.0	82	53	3.5
42		10.4	32.4	19.0	2.6	9.2	79	53	3.5
43		1.2	31.4	18.4	2.4	9.2	77	46	3.0
44		5.1	30.9	17.2	2.1	9.5	78	44	2.8
45	November	0.3	29.8	15.3	2.2	9.6	76	42	2.5
46		2.6	29.1	14.5	2.0	9.2	78	43	2.5
47		1.9	27.6	13.6	2.3	8.7	80	43	2.5
48		0.4	26.5	11.7	2.4	8.9	79	43	2.4
49	December	2.0	25.3	10.3	2.4	8.9	82	42	2.1
50		0.2	24.7	9.8	2.6	8.7	83	42	2.1
51		1.2	23.8	9.5	2.4	8.3	88	47	2.0
52		1.2	23.7	9.4	2.8	7.9	88	50	2.0

The weekly probability of receiving rains greater than 21 mm is presented in Fig.7. It can be inferred from the figure that as the probability of receiving at least 21 mm rains during a week is increased, the number of weeks receiving this amount is decreased. In other words, at 50% level of probability 21 mm or more rains could be received between standard weeks 25 and 30.

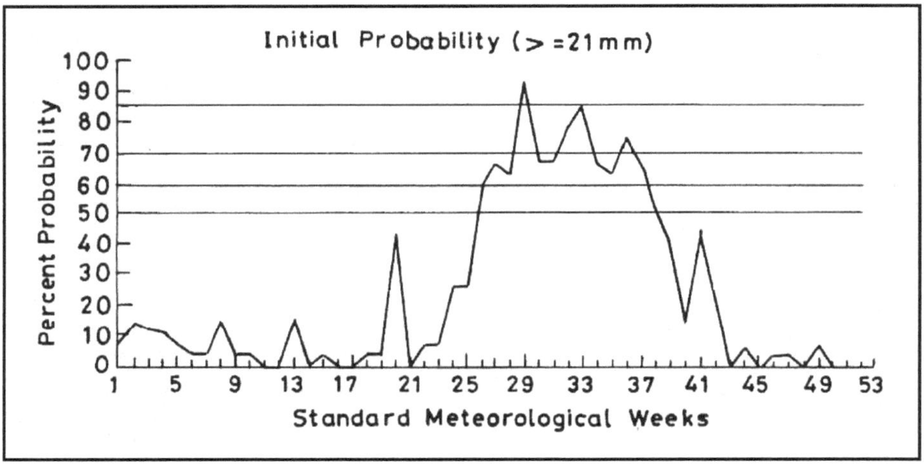

Fig. 7. Probabilities of weekly rainfall at Varanasi

The extent of variability in rainfall during different standard weeks is presented in Fig.8.

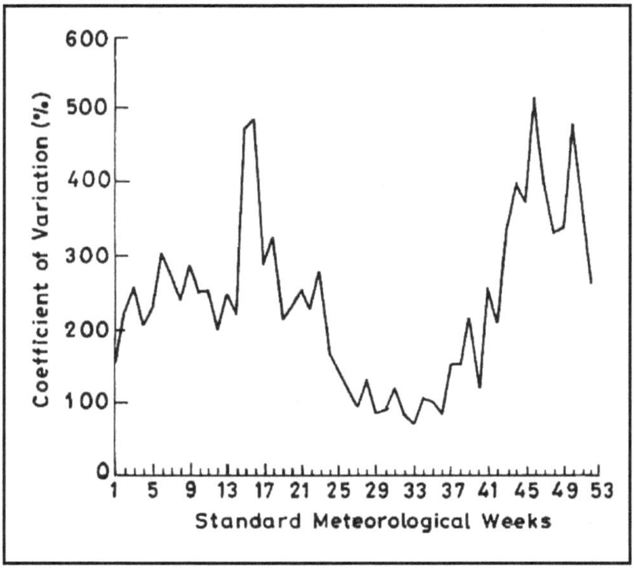

Fig. 8. Weekly rainfall variability (coefficient of variation) at Varanasi

It is evident from the figure that even during the SW monsoon period, coefficient of variability in weekly rainfall ranges from 66 to 165 percent. In general, experience has shown that the weeks with CV value upto 150% are dependable and weeks with CV value below 100% are highly dependable. As such 7 weeks (27, 29, 30. 32, 33, 35 & 36) are highly dependable and rest of the weeks during kharif season are dependable.

REFERENCES

Singh., V.P. R. K. Singh., A.S.RA.S. Sastri, S. S. Baghel and J. L. Chaudhary, (1999). *Agro-climatic Atlas of Eastern India* Indira Gandhi Agricultural University and International Rice Research Institute, Raipur.

Singh, T., Singh, S.R. and Kalyan Singh, (2006). Agroclimatic characterization of Varanasi AICRP on Dryland Agriculture., Department of Agronomy, Inst. Ag., Sc., B.H.U., Varanasi.

Climate Change and Food Security, 2008
Edited by: M.Datta, N.P.Singh & Er.D.Daschaudhuri, pp.149-162
New India Publishing Agency, New Delhi (India)
E-mail : newindiapublishingagency@gmail.com
Web: www.bookfactoryindia.com

10

Rubber Cultivation in North East India-An Agrometeorological Perspective

SHAMMI RAJ AND S.K. DEY

INTRODUCTION

Rubber cultivation in regions extending from 10°N to 10°S latitude is known as the traditional region and the areas other than its natural climatic origin is termed as the non-traditional region. Northeast India comprises majority of the non-traditional areas in India with major climatic constraints like high seasonal rainfall, excessive precipitation and to some extent dry conditions. These areas extending from 22 to 29°N latitude, holds promise in extending rubber cultivation because of its exceptional climatic make-up and abundant amount of annual precipitation (2000 mm) distributed over a fair period of time.

This perennial crop generally yields latex rubber at the 7th year after planting (with an economic life span of 20 years), although the period (immaturity period) it requires, is higher in the non-traditional region (Rubber Growers Companion, 2006). This immaturity period has also been shown to reduce with water availability in sub-humid climates (Vijayakumar *et al.*, 1998). Harvest of this crop is by tapping. Tapping is a process of controlled wounding during which thin shavings of bark are removed, thus opening the latex vessels which exude latex. Dry rubber yield is obtained after the latex is coagulated and rolled into sheets (about 500gm) which is then smoked and dried.

In the northeast, the alternate daily system of tapping is being presently followed. The general trend of rubber yield in Tripura shows a sudden rise in latex volume from early October, which reaches a peak in mid November and falls to the original level towards the end of January.

One of the important requirements of rubber cultivation is a warm equable climate with not less than 2000 mm of evenly distributed annual rainfall and temperatures ranging from 20 to 35°C. Mean daily air temperature is an important factor controlling growth and yield of rubber (Huang and Zheng, 1985; Rao *et al.,* 1993). The optimum monthly climatic requirements (Vijayakumar et al., 2000) for *Hevea brasiliensis* which are quite characteristic of its place of origin i.e. Brazil is shown in Table 1.

Table 1. Optimum monthly climatic requirement for *Hevea brasiliensis*

<table>
<tr><td>Rainfall</td><td colspan="2">≥ 2000 mm evenly distributed with 125-150 rainy days</td></tr>
<tr><td>Maximum temperature</td><td>29 - 34^{0}C</td><td rowspan="2">Monthly mean 25 - 28^{0}C</td></tr>
<tr><td>Minimum temperature</td><td>≥ 20^{0}C</td></tr>
<tr><td>Humidity</td><td colspan="2">≥ 80%</td></tr>
<tr><td>Sunshine</td><td colspan="2">≥ 2000 hours at the rate of 6 hours per day</td></tr>
</table>

None of the areas in the Northeast non-traditional region fully qualifies for the above optimum conditions. However, major areas with moderate level conditions do exist for which rubber can be successfully cultivated. This necessitates the importance of characterizing the crop-weather relations, spatial and temporal variations of the meteorological parameters for the Northeast.

MATERIALS AND METHODS

Studies carried out so far in this region have shown that the day-to-day yield is mainly being controlled by ambient as well as antecedent environmental parameters. A study was undertaken based on the relationship of environmental parameters with latex yield for five clones (RRII203, RRII118, RRIM600, RRII105 and GT1) of *Hevea brasiliensis* in the region of Agartala (Shammi Raj *et al.,* 2005). The dry rubber yield from these trees was studied with the antecedent observed and computed daily meteorological parameters. Regression equations were developed for each of the clone by the forward stepwise procedure for yield. Conducive ranges of the related parameters were identified with daily based above average yield.

On a spatial extent in any geographical region, the importance of rainfall and its variability is the most important to decide the conduciveness of rubber cultivation, more so if it is in a non-traditional region like the Northeast India. Rainfall being the most important parameter for rubber compared to other optimum requirements, its spatial variability was studied over the entire geographical area of the Northeast. Utilizing 139 raingauge stations with more than 30 year individual weekly data sets from the India Meteorological Department, the study was made using horizontal rainfall contours. The Precipitation Concentration Index (PCI) denotes uniformity of monthly distribution in a year over a particular station or the interannual variability. This index (Oliver, 1980) is described as

$$\text{PCI} = 100 \times \frac{\sum_{i=1}^{12} p_i^2}{\left(\sum_{i=1}^{12} p_i\right)^2}$$

where, pi is the rainfall amount of the ith month and annually calculated for each of the 139 stations. PCI values below 10 indicate a uniform monthly rainfall distribution in a year. Values from 11 to 20 denote seasonality and values above 20 correspond to climates with substantial monthly variability. Contours of this index were plotted in digital maps prepared for the Northeast (Raj and Dey., 2004).

Rainfall availability varies from place to place depending mainly on the topography (Raj and Dey, 2004). Information on chances of a wet week and wet spell in a season would clearly characterize the seasonal rainfall variability in a region. This could also be precisely utilized by the rubber based farming community in the Northeast for planning timely farming operations like application of fertilizers, pesticides and other agromanagement practices needed for the crop. Probability studies adopting the first order Markov Chain model were conducted in ten selected stations in the rubber growing tracts of the Northeast. Here, a week is designated as wet if the weekly total rainfall is higher than the specified amount of the cumulative weekly evaporative demand of 28 mm (Raj *et al.,* 2005) for rubber. Otherwise it is known as a dry week. The initial probabilities (%) of dry weeks P(D), wet weeks P(W), and conditional probabilities (%) of a wet week following a wet week P(W/W), a dry week following a wet week P(W/D), a wet week following a dry week P(D/W), and a dry week following a dry week P(D/D) have been calculated for all respective four seasons for every stations considered in the study using the following equations (Raj and Dey, 2005). For every season, the probability of continuation of dry/wet spells of different lengths (2 to 6 weeks) has been calculated. Probability analysis of monthly rainfall data was carried out using the Gringorten's Plotting Position method (Chow *et al.*, 1988). The minimum rainfall expected for various probability percentages have been computed, but only the important levels of 50 and 90% were discussed for brevity sake. This dependability depicts the true nature of rainfall, even when the distribution is highly skewed.

Table 2. Regression equations for yield (Y) of clones for the non-winter and winter tapping periods using environmental variables.

Clone	Non-Winter Tapping Period	Winter Tapping Period
$Y_{RRII203}$	- 7.638 + 0.147 ST-1 (8.64) * - 2.055 T_n-0 (2.93) + 1.823 T_x-2 (2.61) - 1.133 VP_{f-a}-1 (2.55) + 0.694 VD_a-1 (2.31) (R^2=0.72)	-123.353 + 0.269 ST-3 (11.65) + 5.869 T_{20a}-1 (6.84) - 2.029 T_n-1 (3.46) - 1.964 $T_{10}D$-3 (4.04) - 2.120 $T_{10\text{-}20f}D$-1 (3.36) - 1.166 VP_{f-a}-3 (2.88) (R^2=0.94)
$Y_{RRII118}$	18.378 + 0.048 ST-1 (4.14) - 2.486 V_f-1 (4.42) + 1.708 T_{10f}-0 (3.39) - 2.201 PE-2 (2.38) + 2.760 T_{10f}-1 (3.28) - 0.748 T_{5a}-3 (2.25) (R^2=0.55)	43.307 + 0.137 ST-3 (5.54) + 0.243 RD_{f-a}-1 (3.52) - 5.483 E-2 (3.50) + 0.975 VD_a-1 (3.94) - 1.531 T_5D-3 (5.31) (R^2=0.86)
$Y_{RRIM600}$	91.104 - 3.065 AT-1 (2.95) + 0.117 ST-1 (4.70) + 23.220 $T_{5\text{-}10f}D$-2 (3.86) + 3.471 $T_{10}D$-2 (2.42) + 1.810 S-1 (2.20) (R^2=0.55)	46.034 + 0.139 ST-3 (3.27) + 1.255 VD_a-2 (8.93) + 1.139 T_{5a}-3 (2.97) + 0.555 R_a-3 (3.50) + 0.396 RD_{f-a}-3 (3.08) - 6.167 E-2 (2.63) (R^2=0.86)
Y_{RRI105}	40.879 - 5.050 AT-1 (4.31) + 0.115 ST-1 (4.28) + 1.735 S-1 (2.23) + 6.214 T_{20a}-3 (3.57) - 6.082 PE-2 (2.14) - 3.000 T_n-3 (2.01) (R^2=0.62)	-95.658 + 0.417 ST-3 (10.51) + 3.518 T_{10f}-0 (3.13) - 15.069 SI-2 (3.58) + 4.516 VP_{f-a}-3 (4.09) - 3.856 V_f-1 (2.84) + 3.265 T_{20a}-1 (1.89) (R^2=0.83)
Y_{GT1}	-7.930 + 0.434 ST-1 (2.69) - 0.422 ST-2 (2.62) - 2.208 T_n-0 (2.75) + 1.891 $T_{10}D$-3 (2.06) - 0.153 PE-2 (1.43) - 0.673 T_{5a}-2 (1.93) (R^2=0.37)	23.556 - 15.758 E-2 (4.13) - 0.449 R_a-2 (4.11) + 3.154 T_n-3 (4.73) + 0.839 RD_{f-a}-3 (6.41) - 1.956 VD_{f-a}-3 (2.51) + 0.108 ST-3 (3.89) (R^2=0.83)

RESULTS AND DISCUSSION

Regression equations between latex production and weather parameters for different clones of rubber were worked out and data are presented in Table 2.The explained variance (Table 2) for the regression equations over five clones, based on parameters from the day of tapping up to three days prior to it, varied from 37 to 72% during the non-winter tapping period and 83 to 94% during the winter tapping period.

Estimated soil moisture storage values one day and three days prior to tapping were found to be the primary parameters affecting yield for these two periods respectively. RRIM600 and RRII105 being high yielding clones were found to be fairly dependent on the average temperature of the day prior to tapping. It was also observed (Table 3), that the clone RRII105, with a comparatively lower yield to that of RRIM600, was more susceptible to daily water deficit conditions during the non-winter season.

Table 3. Values of environmental parameters associated with above average yield for five different clones in *Hevea brasiliensis* during the non-winter tapping period.

Variables	RRII203	RRII118	RRIM600	RRII105	GT1
ST-1 (mm)	124 to 320	227 to 320	149 to 289	226 to 320	129 to 320
PE-2 (mm)		1.4 to 3.6		2.1 to 3.6	3.6 to 6.2
AT-1 (°C)			15 to 31	15 to 29	
T_n-0 (°C)	21 to 27				19 to 27
T_n-3 (°C)				20 to 26	
T_x-2 (°C)	26 to 35				
T_{5f}-1 (°C)		25 to 27			
T_{5a}-2 (°C)					29 to 42
T_{5a}-3 (°C)		25 to 35			
T_{10}D-0 (°C)					
T_{10}D-2 (°C)			-1.1 to 9.5		
T_{10}D-3 (°C)		2.5 to 11.0			-3.4 to 4.0
T_{20}a-3 (°C)				28 to 33	
$T_{5\text{-}10f}$D-2 (°C)			-1.8 to 1.7		
V_f-1 (mm of Hg)		19.2 to 23.3			
$VP_{f\text{-}a}$-1(mm of Hg)	-5.3 to 1.7				
VD_a-1 (mm of Hg)	0.0 to 16.0				
S-1 (Hours)			0.2 to 10.6	1.6 to 10.6	

Contribution to the above-average yield for all the five clones was found to be the highest during October-January period (40-60%). The effect of solar radiation is more pronounced in the high yielding varieties, especially during the non-winter period. Antecedent soil and air temperature also affects the daily yield of rubber. This study also confirms that varied responses of yield with the environmental factors in the non-traditional region depend on clonal characteristics (Raj *et al.,* 2005).

Studies on the spatial heterogeneity of rainfall conducted in the northeast revealed that on an average, about 62% of the annual rainfall is accounted for the monsoon season. Generally, December, January and February months comprise the lowest rainfall period. The mean annual precipitation along the entire northeastern states, ranged from >10,000 mm along the Jaintia Hills of Meghalaya to <1500mm along the east Assam, South of Nagaland and Northern Manipur. Mean annual rainfall amounts greater than 4000 mm were seen to be concentrated in large areas in the northwestern portion of Assam, and the Jaintia Hills.

Grouping the frequencies of PCI values associated with the mean distribution of the number of dry months in a year, three distinct ranges (Fig.1) were identified in the Northeast as <14.9, 15.0-19.9 and >20.0. The <14.9 category showed more than 50% of the contributions without any dry periods per year. The >20.0 PCI range showed 7 and 10 % years with 4 and 5 dry months per year respectively.

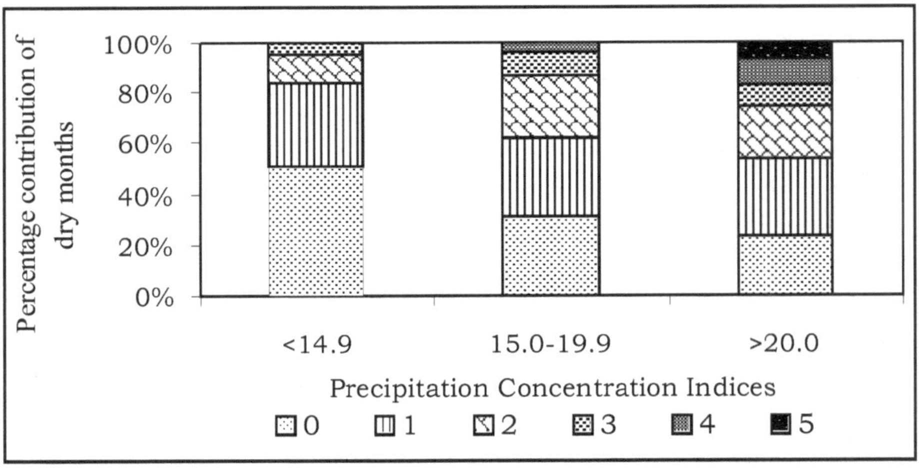

Fig. 1. Percentage contribution of dry months for the three ranges of Precipitation Concentration Indices

Superimposing the gridded base maps of the 1500-4000 mm rainfall contours with that of the PCI contours, three separate areas of suitability could be demarcated. These were designated as Optimal (< 14.9), Sub-optimal (15.0 to 19.9), and Moderate (>20.0) (Fig 2). Marginal areas in terms of rainfall suitability considered here were areas with rainfall >4000mm and <1500mm irrespective of the PCI values. The regions of Optimal, Sub-optimal, Moderate and Marginal identified comprises of 14.5, 26.9, 46.4 and 12.2 % of the total area in the Northeast.

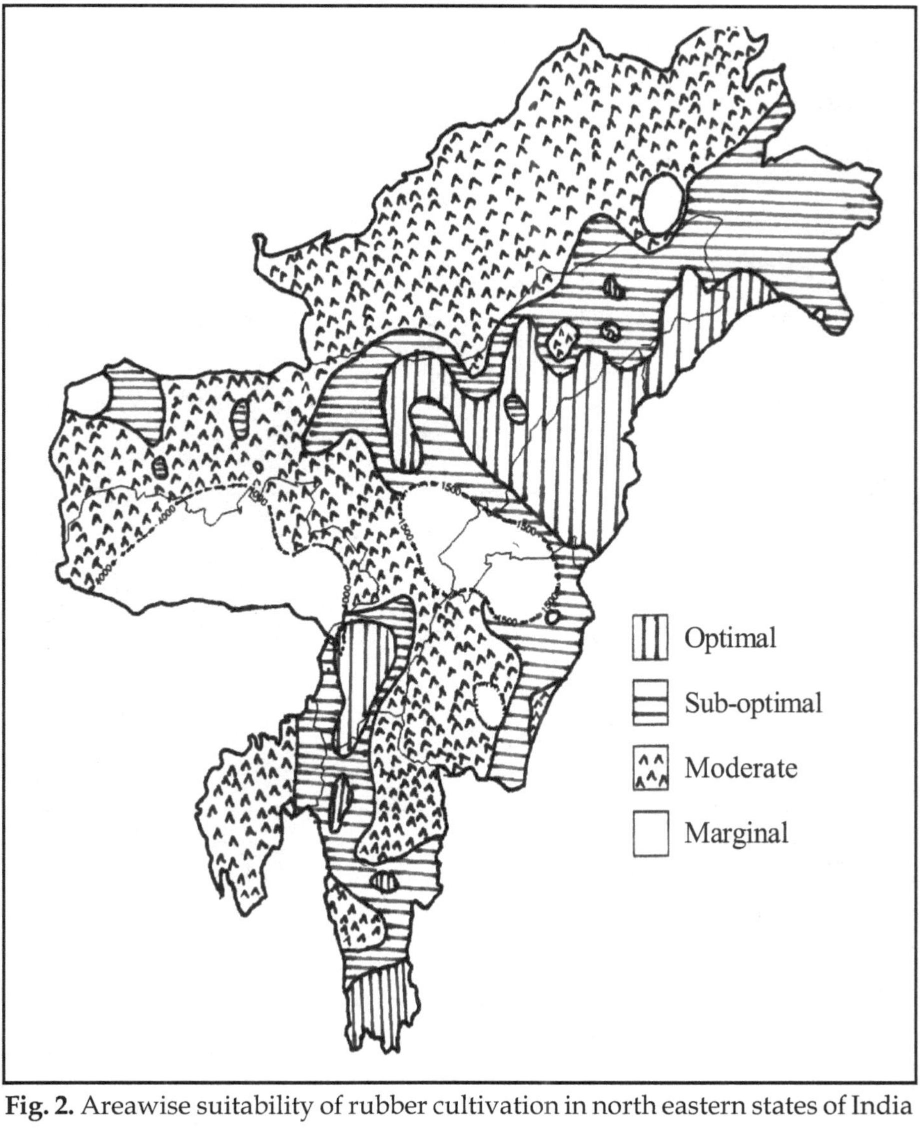

Fig. 2. Areawise suitability of rubber cultivation in north eastern states of India

Table 4. Initial and conditional probabilities of ten rubber growing places in different seasons

Rubber growing places	W	D	W/W	D/W	W/D	D/D	W/W	D/W
Winter season								
Silchar	9	91	30	70	10	90	9	91
Tezpur	2	98	0	100	3	97	2	98
Guwahati	2	98	0	100	3	97	2	98
North Lakhimpur	12	88	22	78	13	87	12	88
Lumding	2	98	10	90	3	97	2	98
Amguri	9	91	20	80	10	90	9	91
Imphal	6	94	21	79	7	93	6	94
Tura	4	96	28	72	4	96	4	96
Kolasib	9	91	22	78	10	90	9	91
Agartala	2	98	0	100	3	97	2	98
Hot-weather season								
Silchar	61	39	71	29	54	46	61	39
Tezpur	47	53	62	38	42	58	47	53
Guwahati	43	57	59	41	41	59	43	57
North Lakhimpur	57	43	75	25	44	56	57	43
Lumding	29	71	47	53	29	71	29	71
Amguri	61	39	76	24	50	50	61	39
Imphal	33	67	51	49	34	66	33	67
Tura	47	53	81	19	31	69	47	53
Kolasib	57	43	79	21	46	54	57	43
Agartala	44	56	66	34	38	62	44	56
Monsoon season								
Silchar	89	11	90	10	1	99	89	11
Tezpur	72	28	72	28	3	97	72	28
Guwahati	68	32	73	27	1	99	68	32
North Lakhimpur	87	13	90	10	1	99	87	13
Lumding	57	43	61	39	5	95	57	43
Amguri	87	13	88	12	1	99	87	13
Imphal	63	37	66	34	1	99	63	37
Tura	80	20	88	12	1	99	80	20
Kolasib	88	12	93	7	1	99	88	12
Agartala	65	35	74	26	1	99	65	35
Post-monsoon season								
Silchar	21	79	36	64	18	82	21	79
Tezpur	12	88	30	70	19	81	12	88
Guwahati	9	91	16	84	9	91	9	91
North Lakhimpur	19	81	28	72	18	82	19	81
Lumding	13	87	21	79	16	84	13	87
Amguri	18	82	26	74	15	85	18	82
Imphal	15	85	20	80	15	85	15	85
Tura	17	83	42	58	14	86	17	83
Kolasib	25	75	40	60	11	89	25	75

Probability estimates explained the nature of variability of rainfall at different places in the NE region. Probability or chances of occurrence of 76% was noted for wet weeks in general during the monsoon season. Probability of obtaining dry weeks (Table 4) in winter is more than 90%. A high value of 12% probability for a wet week was noted in the case of North Lakhimpur. The stations of Tezpur, Guwahati and Agartala showed similar characteristics of not receiving any two consecutive wet weeks during the winter season at all, while Silchar showed a high of 30%. Percentage occurrence of wet spells is higher for North Lakhimpur and Amguri during the winter season.

Table 5. Seasonwise probabilities (%) of consecutive dry and wet week spells of different lengths

	Consecutive dry spells (no of weeks)						Consecutive wet spells (no of weeks)				
	2	3	4	5	6		2	3	4	5	6
Winter season											
Silchar	79	66	53	42	33		13	7	0	0	0
Tezpur	92	86	77	68	59		0	0	0	0	0
Guwahati	94	89	82	75	66		0	0	0	0	0
North Lakhimpur	76	64	52	43	35		3	2	2	0	0
Lumding	94	89	83	76	69		0	0	0	0	0
Amguri	80	70	58	47	37		8	0	0	0	0
Imphal	87	79	69	59	50		3	0	0	0	0
Tura	92	86	78	70	62		11	0	0	0	0
Kolasib	81	70	60	50	40		4	0	0	0	0
Agartala	93	86	78	69	59		0	0	0	0	0
Hot-weather season											
Silchar	20	9	3	1	0		53	42	36	0	0
Tezpur	34	19	11	6	2		42	31	26	0	0
Guwahati	35	21	12	6	3		33	22	16	1	0
North Lakhimpur	30	16	7	2	0		61	51	46	0	0
Lumding	52	38	28	20	14		25	14	7	1	0
Amguri	27	12	6	3	0		59	51	43	0	0

Contd...

Table 5 Contd...

Monsoon season											
Silchar	8	3	0	0	0		80	70	61	0	0
Tezpur	18	11	9	7	5		50	34	24	0	0
Guwahati	22	14	10	9	9		51	37	26	0	0
North Lakhimpur	19	17	17	16	15		78	69	59	0	0
Lumding	21	11	6	4	3		33	18	10	1	0
Amguri	8	3	0	0	0		75	64	54	1	0
Imphal	19	10	8	6	5		40	24	13	0	0
Tura	34	28	24	21	20		72	61	50	0	0
Kolasib	14	8	5	5	5		81	71	63	0	0
Agartala	18	8	4	2	2		49	34	22	1	0
Post-monsoon season											
Silchar	84	80	76	74	71		14	6	1	0	0
Tezpur	89	87	85	84	83		7	1	0	0	0
Guwahati	92	90	89	88	87		0	0	0	0	0
North Lakhimpur	82	77	72	67	60		8	1	0	0	0
Lumding	88	85	83	82	81		5	1	0	0	0
Amguri	82	79	75	72	68		6	0	0	0	0
Imphal	84	79	75	72	69		4	0	0	0	0
Tura	91	88	86	84	82		16	4	0	0	0
Kolasib	83	79	76	73	70		19	7	2	0	0
Agartala	86	83	82	81	80		0	0	0	0	0

Consecutive wet weeks were obtained at more than 70% probability for Silchar, N.Lakhimpur, Amguri and Kolasib during the Hot-weather season. About 50% chance occurence was seen for a dry week following a wet week (W/D) in the case of Silchar and Amguri. Although, the nature of wet spell occurrences (Table 5) were similar for the hot-weather and monsoon seasons, relative percentages were higher during monsoon. High percentages of dry spells (75%) were noted for even the sixth week during post-monsoon followed by the winter season (51%). More than 10% dry spells of > 5 weeks were also experienced for Lumding, Tura, Kolasib and Agartala during the post-monsoon season. It was also observed that over all seasons, wet spells lasted only up to the fourth week beyond which chances were negligible, especially during the hot-weather and monsoon seasons. However during the five months comprising the winter and post-monsoon periods, very little chances for consecutive 3 wet week spells were seen. In the case of Guwahati and Agartala, no chance occurrence, were noted after the first week.

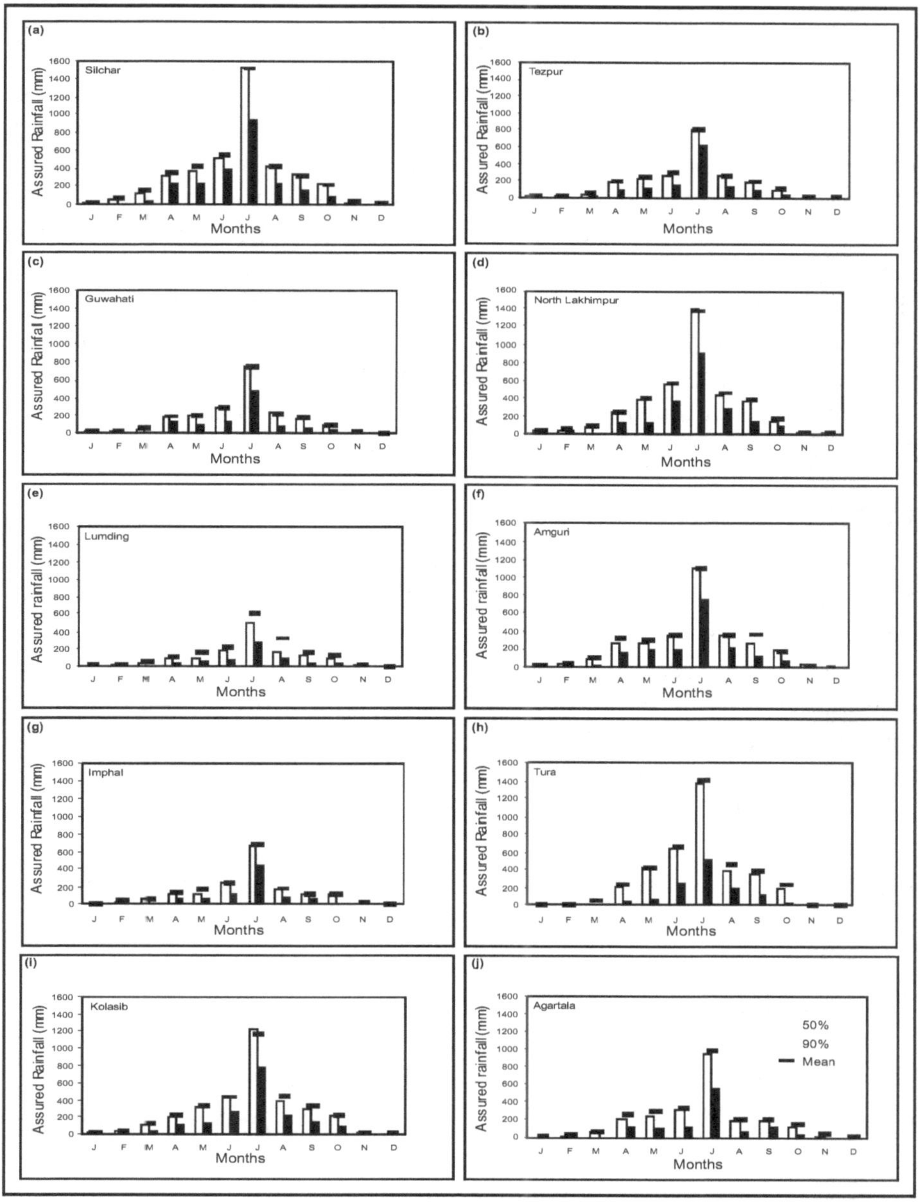

Fig. 3. Monthly assured rainfall at 50 & 90% probability levels for ten different rubber growing stations in the Northeast

The minimum amount of rainfall at a given level of probability, rather known as the dependable rainfall (mm) is expressed at the 50 and 90% probability levels (Fig.3 (a-j)). Silchar & North Lakhimpur received a more satisfactory distribution of monthly rainfall followed by Amguri and Kolasib.

Tura in Meghalaya showed the lowest assured rainfall at 90% in comparison with its 50% probability level, showing the high level of erratic behaviour of rainfall in this region. The dependable amount at 50% probability was the lowest for Lumding. It can also be seen from (Fig.3 (h)) that Tura is characterized with the highest skewness during the hot-weather period, due to high differences between the 90% level and the mean monthly rainfall. However, experiencing a high annual rainfall, the low annual value at the 90% level is quite comparable to that of Agartala. This can be corroborated by the fact that the high yielding RRIM 600 variety is more stable and adaptable to the northeastern regions of India (Meenatoor *et. al.*, 1991) and stability of this clone is comparable to Agartala in Tripura in terms of growth and yield (Reju *et al,*. 2001; Reju *et al.*, 2003).

CONCLUSION

Studies pertaining to larger proportions of yield (areawise) will help in precisely understanding the suitability of clones within an agroclimate. As a forerunner, implications of these studies can be considered for assessing development of clones for the non-traditional regions where rubber is not normally grown. Improper estimation of production of Natural Rubber (NR) within a framework of supply and demand, not only leads to import/excess stock, but also results in serious price fluctuations. Therefore, better resource management can be inducted in this non-traditional rubber growing area with identification of weather parameters having a regional influence on yield.

REFERENCES

Chow, V.T., Maidment, D.R. and Mays, L.W. (1988). *Applied Hydrology*. Mcgraw Hill, International Editions, Civil Engineering Series. New York.

Huang, Z. and Zeng, X. (1985). *Rubber cultivation in China* : Planter's Conference. Rubber Research Institute of Malaysia.

Meenatoor, J. R., Vinod, K. K., Krishnakumar, A. K., Sethuraj, M. R., Potty, S. N. and Sinha, R. R. (1991). Clone X environment interaction during early growth phase of *Hevea brasiliensis*: 1. Clonal stability on girth, *Indian Journal of Natural Rubber Research*, 4(1): 55-61.

Oliver, J.E. (1980). Monthly precipitation distribution: a comparative index. Professional Geogr. 32: 300-309.

Rao, P. S., Jayarathnam, K. and Sethuraj, M.R. (1993). An index to assess areas hydrothermally suitable for rubber cultivation. *Indian Journal of Natural Rubber Research*, 6 (1&2): 80-91.

Reju, M. J., Thapliyal, A. P., Deka, H. K., Soman T. A. and Nazeer, M. A. (2003). Growth stability of Hevea clones in a high altitude region of Meghalaya. *Indian Journal of Natural Rubber Research*, 16(1 & 2): 118-121.

Reju, M. J., Thapliyal, A. P., Deka, H. K., Nazeer, M. A. and Soman, T. A. (2001). Growth and initial yield of some Hevea clones in Meghalaya. *Indian Journal of Natural Rubber Research*, 14(2): 146-151.

Rubber Grower's Companion. (2006). : Published by The Rubber Board (Government of India, Ministry of Commerce & Industry), Kottayam-686002. India. p.94.

Shammi Raj and Dey, S. K. (2005). *Nature of rainfall variability in the rubber growing tracts of Northeast India: A study* : Preprints of Papers. International Natural Rubber Conference, Cochin, India. 6-8 November. No. 37.Oral Session: pp. 192-200.

Shammi Raj, Gitali Das, Jacob Pothen and Sushil Kumar Dey. (2005). Relationship between latex yield of *Hevea brasiliensis* and antecedent environmental parameters. 2005. *Intl. J. of Biometeorology*. Springer-Verlag GmbH: Vol.49 (3). pp. 189 – 196.

Shammi Raj and Dey, S.K. (2004). Spatial analysis of rainfall variability over the Northeast with respect to rubber cultivation. *Journal of Agrometeorology*. Vol.6 (Sp. Issue, 2004): 52-57.

Vijayakumar, K.R., Chandrashekar, T.R. and Varghese Philip. (2000). Agroclimate In: *Natural Rubber: Agromanagement and Crop Processing* (Eds): P.J.George and C.Kuruvilla Jacob. Rubber Research Institute of India, Kottayam. pp 97-116.

Vijayakumar, K.R., Dey, S.K., Chandrashekar, T.R., Devakumar, A.S,, Mohankrishna, T., Rao, S.P. and Sethuraj, M.R. (1998.) Irrigation requirement of rubber trees (*Hevea brasiliensis*) in the sub-humid tropics. Agricultural Water Management, 35(3): 245-259.

ABBREVIATIONS

Observed Parameters

T_x-Maximum Temperature; T_n-Minimum Temperature; S-Sunshine Hours; P-Rainfall; E-Evaporation; R-Relative Humidity; V-Vapour Pressure; T5, T10, T20 - Soil Temperatures at 5, 10 & 15 cm depths respectively; Subscripts 'f '-forenoon & 'a'-afternoon.

Computed Parameters

SI-Sine of the day of year; PE-Potential Evapotranspiration; ST-Soil Storage; WS-Water Surplus; WD-Water Deficit; AT-Average Temperature; DT-Difference between maximum (previous day) and minimum temperature (succeeding day), VD-Vapour Pressure deficit; $RD_{f\text{-}a}$-difference in Relative Humidity values between succeeding day morning and previous day afternoon; $VP_{f\text{-}a}$-difference in Vapour Pressure values between succeeding day morning and previous day afternoon; $T_{5\text{-}10}$D-Difference of Soil Temperature between 5cm and 10 cm depth; $T_{10\text{-}20}$D-Difference of Soil Temperature between 10cm and 20 cm depth; T_5D, T_{10}D & T_{20}D-decrease in respective Soil Temperature of 5, 10 and 20 cm depths of the previous day afternoon to that of the succeeding day morning.

Rainfall Probabilities

W-Wet week; D-Dry week; W/W-Wet week following a wet week; W/D-Dry week following a wet week; D/W-Wet week following a dry week; D/D-Dry week following a dry week.

Climate Change and Food Security, 2008
Edited by: M.Datta, N.P.Singh & Er.D.Daschaudhuri, pp.163-170
New India Publishing Agency, New Delhi (India)
E-mail : newindiapublishingagency@gmail.com
Web: www.bookfactoryindia.com

11

Influence of Weather Parameters on Plant Diseases

A.K. SINGH AND PANKAJ BAISWAR

INTRODUCTION

The weather conditions during the growing season of crops influences the onset and development of disease. The relationship between weather and plant disease development is the basis for meteopathological forecasting. On the basis of observed or forecast weather conditions, the incidence of plant disease is likely to be predicted.

Such predictions are designed to warn growers of expected significant developments in the manifestation of a particular plant disease. Such disease forecasts, if sufficiently accurate, permit efficient and competent counter measures to be adopted. A short range or medium range of disease forecast, issued at short intervals during the growing of crops, can give guidance to the farmers on the best remedial measures. Growers can decide whether a chemical control measure is needed or not, and if needed then optimum date on which to be applied and similar day to day decisions to fight against disease to minimize the losses to be caused.

The result, of adopting such a plan of weather based control measures, is to reduce the number of chemical sprays to a well timed few, thus giving an equal or even better protection at a lower cost in chemical and labour, and simultaneously reduces losses caused by phytotoxic and mechanical damage to the crop. The reduction in the use of potentially toxic chemical is a matter to be welcomed in the present world, considering the consequences of the widespread use of toxic chemicals.

PLANT DISEASES

The plant diseases for which forecasting based on weather parameters can be of practical importance to the farmers can fulfill the following four requirements.

1. The disease, which causes significant economic losses (quantity/quality) to crop

The majority of diseases cause losses by way of quantity, but some of the diseases are causing losses by way of quality like apple scab in Kameng district of Arunachal Pradesh, citrus scab in the hills of all the States of north east.

2. The disease must be variable with seasonal influence

Different parts of the disease development i.e. time of appearance, speed of disease buildup, ultimate level should be influenced by weather conditions directly or indirectly to an appreciable extent.

It is important to define that exactly what a disease forecasting system is attempting to predict. It may predict simple appearance of the disease in the current season. It may say the approximate date of appearance or it may well predict the whole course of disease development and approximate loss at the end of the season.

In majority of cases, biological observations are not just a divisible but also an essential supplement to weather information for success in prediction. The level of accuracy in prediction requires a close and continuous cooperation between Plant Pathologists and Meteorologists.

3. Availability of control measures

Prior to going for development of disease forecasting system it should be ensured that an economically acceptable disease control measure whether preventive or curative is available. There is no point in having knowledge about the appearance and disease buildup if the disease cannot be managed even after that. With the help of disease prediction, the cost of control measures can be significantly reduced and in such conditions, some of the control measures economically not acceptable will come under acceptable range. It is advisable that the farmer should get a longer notice about disease development, so that a more orderly preparation for control measures can be taken up.

4. Information on the nature of weather dependence of the disease

This is hub of the whole matter. To develop a working model for disease forecasting, information on the nature of the relationship between weather parameters and the course of disease development / the life cycle of the pathogen is required.

FACTORS FOR DISEASE DEVELOPMENT

A complex of factors determines the development of plant disease. It is the result of the interaction between the host, the pathogen (fungus, bacteria, virus or any other pathogen) and the physical environment. For successful development of disease, a susceptible host is needed. Presence of virulent strain of the pathogen is equally important. For an infection process to complete for the disease symptoms to appear, prevalence of favorable atmospheric or weather conditions is an essential requirement.

For the development of disease, the main struggle is between host and pathogen, depending on the amount of resistance possessed by host and the virulence of pathogen. The atmospheric environments may have the effect on both pathogen and host. The degree of this influence depends on the growth stage of the crop. This effect may differ from crop to crop and also between varieties of the same crop, depending on factors such as relative leafiness, erect or sprawling growth pattern, structure of plants etc. Nutrition also changes the plant growth, therefore affecting the microclimate.

The role of environment on disease development is therefore depends on the crop variety, its growth stage, plant nutrition and strain of the pathogen (virulence).

We can now discuss the three main factors: host, pathogen and environment, whose interaction results in disease development.

Host

Level of resistance possessed by a host against a particular pathogen is no doubt an important parameter of the disease development. There are some other properties of the host, which are significant for epidemiology. Density and distribution of cultivation of a particular crop is an important factor. Crops grown on a limited scale and in scattered locations may escape some of the diseases. Dense population of the crop in pure stands invites epidemics. The general conditions of the host and its parts become a relevant factor in the case of many diseases. Potato grown in poor illumination is more susceptible to many virus diseases. Soft lush crop growth is known to encourage many fungal diseases.

The resistance to disease of the host plant, which is taken as constant varietal characteristics for disease prediction, may also differ with the age or growth stage of the plant.

Pathogen

Most prediction systems are based on the assumption of a uniform inoculum load at beginning of disease build up. Primary inoculum load is therefore an important parameter, usually unknown factor in most of the forecasting models. Hyre (1964) observed that occurrence of high temperature (above 29°C) soon after a favourable infection period for downy mildew of lima beans did not allow disease development.

Different disease also interacts each other. Wallen et al (1967) concluded that a disease might flourish best, not in the environment, which is most suitable to it in isolation, but in the environment which most restrains its rivals or parasites.

Environment

The environment (E) is the microclimate in which the pathogen develops and attacks the host and ultimately results in the development of the crop disease. The problem arises about the exact location at which E can be observed. Such a crop microclimate (E), which, is often determined to a considerable extent by the surrounding plants, is basically controlled by the local climate in the open air outside the crop, i.e. farm (w). This in turn is largely determined by the macro climate (W) as observed at a standard meteorological station in the area.

There is a need of developing a relationship between E and w and again between w and W. Based on relationship, synoptic weather charts (S) were developed by many workers.

Differences between E and w depend largely on the nature and density of the crop. Level of soil moisture also determines this difference. Differences tend to be greatest on calm sunny days. Winds help in generating homogeneity. On a cloudy and wet weather situations, conditions inside and outside the crop vary little. Wind helps towards greater uniformity between local climate (w) and macroclimate (W) at standard station. It is better to depend on weather observations from a group of standard stations than from a single station for disease prediction.

Some Important Plant Diseases in the North Eastern Region

i. *Rice blast:* The disease is economically most important in the region. Its severity depends on seasonal variation. A disease-forecasting model will be practical and beneficial.

ii. *Powdery mildew and scab of citrus:* These two diseases of citrus are governed by environment and under favourable environmental conditions cause enormous losses. Both these diseases attack new leaves and shoots. During appearance of new leaves and shoots (flushes) if rain gets delayed i.e. weather will be humid but dry then powdery mildew severity will be very high. Rains before flushes favours appearance and build up of scab.

iii. *Early and late blight of potato and tomato*: A forecasting system will reduce the number of fungicidal sprays.

iv. *Northern leaf blight of maize*: This is also economically most important disease in the region. Disease forecasting model will definitely reduce the cost of maize production in the region.

Experiments conducted at Barapani, ICAR research complex for NEH region have helped in establishing the relationship between weather variables and turcicum blight incited by *Exserohilum turcicum* (Pass) Leonard & Suggs which is highly destructive disease in north eastern region. In this experiment, 24 sowings were done at monthly intervals. Data was recorded under natural epiphytotic conditions. A 0-5 scale developed by Elliot and Jenkins (1946) was used. Observations were recorded at dry silk stage. Percent Disease Index (PDI) was calculated. Data of weather variable were made average for the period between two recordings. Weather variables such as maximum and minimum temp. RH, total rainfall, rainy days, mean bright sunshine hours and wind speed were included in our study and designated as

X1, X2, X3, X4, X5, X6, and X7. PDI recorded periodically on different varieties *viz.* RCM 1-1, 1-2 (Popcorn), Hemant, Vijay composite and MLW (Meghalaya Local white) symbolized as Y1, Y2, Y3, Y4 and Y5, respectively. PDI obtained was subjected to correlation and stepwise multiple regression analysis (Gomez and Gomez, 1984).

Disease appeared in March and disease pressure was quite high from May to July (38-54%). PDI during 2003-04 was 37% whereas in 2004-05 it was 42%, which was comparatively more. Perusal of weather data revealed that during 2004-05, average rainfall, rainy days and RH were high as compared to 2003-04 whereas mean BSSH (Bright sunshine hours) was less during 2004-05 as compared to 2003-04 which signifies that disease development was favoured by high rainfall, RH, rainy days and cloudy conditions. There was not much temperature difference during both the years. Pataky *et al* (1998) also reported that NLB epidemics are related with wet weather and during favourable environment, disease severity on upper leaf canopy is high. Influence of weather variables was further established using correlation analysis between PDI and weather variables. Pooled data from both the years revealed that maximum, minimum temperature, RH, total rainfall and rainy days (independent variables) had significantly positive correlation with PDI (dependent variables) whereas mean BSSH and wind speed had negative correlation. In case of wind speed correlation was not highly significant (Table 1). Wind may have drying effect on leaves, thus reducing the disease intensity. Thakur *et al* (1989)

Table 1. Correlation matrix for Northern leaf blight development on maize varieties in relation to weather parameters (Pooled basis)

Varieties	Weather parameters						
	X1	**X2**	**X3**	**X4**	**X5**	**X6**	**X7**
V.Composit	0.688**	0.726**	0.571*	0.548*	0.723**	-0.516*	-0.199
RCM 1-1	0.706**	0.731**	0.605*	0.777**	0.815**	-0.747**	-0.583**
Hemant	0.667**	0.674**	0.546*	0.674**	0.804**	-0.686**	-0.411
MLW	0.678**	0.761**	0.750*	0.434	0.656**	-0.532*	-0.319
RCM 1-2	0.595**	0.752**	0.848*	0.638**	0.690**	-0.543*	-0.325

* Correlation is significant at 0.05 level (2- tailed);

** Correlation is significant at 0.01 level (2- tailed)

also reported that disease severity was greater at reduced light intensities. Levy and Cohen (1983) stated that light had no direct effect on spore germination but inhibited it indirectly through physiological activity of the host.

Prediction equation developed using stepwise regression analysis revealed that 86%, 76%, 63%,81% and 68% variability can be explained by minimum temperature, rainy days and Mean BSSH in all the five varieties on pooled basis. So from the models, it can be deduced that minimum temperature, rainy days and Mean BSSH were major contributing factors in disease development (Table 2).

Table 2. Multiple regression functions for prediction of northern leaf blight on different maize varieties (pooled basis)

Variety	Regression equation	R^2
RCM 1-2	$Y = 3.657 + 2.942 X^2 + 0.577 X^5 - 1.567 X^6$	0.886
RCM 1-1	$Y = 1.999 + 0.992 X^2 + 0.532 X^5 - 0.280 X^6$	0.736
Hemant	$Y = 1.286 + 3.310 X^2 + 0.369 X^5 - 0.372 X^6$	0.625
MLW	$Y = 4.898 + 1.620 X^2 + 0.499 X^5 - 0.267 X^6$	0.813
Vijay Composite	$Y = 1.643 + 1.315 X^2 + 0.778 X^5 - 1.476 X^6$	0.681

Experiment conducted on effect of leaf wetness, duration on turcicum blight development in moisture chambers as subjected to 0,3, 6,12, 18, 24 and 48 hours of continuous leaf wetness,had revealed that minimum 6 hours leaf wetness was necessary for disease development with incubation period and latent period of 8 and 12 days, respectively. Incubation period and latent period were decreased up to 18 hours to 6 and 8 days respectively and thereafter was stabilized thus indicating favourable effect of leaf wetness duration on disease progress.

From the above discussion, it is clear that models are tools that can assist us in making rational decisions, which will ultimately help growers to minimize the cost of production and increase yields. This information will also be helpful in timing the agricultural operations. Extensive efforts are needed to develop and disseminate this kind of information for betterment of agricultural community and increase farm production efficiency. Important challenges still remain in developing meaningful models under scenario of climatic changes.

REFERENCES

Broadbent, L. (1949). Potatoes and the weather. *Quart. J. Roy. Meteorol. Soc.* 75:302-309.

Elliott,C and Jenkins,M.T. (1946). *Helminthosporium turcicum* leaf blight of corn. *Phytopathology* 36:660-667.

Gomez,K.A. and Gomez,A.A. (1984).Statistical Procedure for Agricultural Research. 2nd Ed., John Willey and Sons, New York, pp. 680.

Grainger, J. (1967). Meteorology and Plant Physiology in potato blight forecasting and its mechanization. *Nature.* 171:1012-1014.

Hyre, R. A. (1964). High temperature following infection checks downey mildew of lima beans. *Phytopathology.* 54:181-184

Levy,Y. (1989). Analysis of epidemics of northern leaf blight on sweet corn in Israel. *Phytopathology* 79:1243-1245.

Levy, Y., and Cohen,Y. (1983). Differential effect of light on spore germination of *Exserohilum turcicum* on corn leaves and corn leaf impressions. *Phytopathology* 73:249-252.

Pataky,J.K., Raid,R.N., Du Toit, L.J., and Schueneman, T.J. (1998). Disease severity and yield of sweet corn hybrids with resistance to northern leaf blight. *Plant Dis.* 82:57-63.

Thakur,R.P.,Leonard,K.J., and Leath,S. (1989). Effects of temperature and light on virulence of *Exserohilum turcicum* on Corn. *Phytopathology*.79: 631-635.

Wallen, V. R., Cuddy,T. F. and Grainger, P. N. (1967). Epidemiology and control of *Ascochyta pinodes* on field peas in Canada. *Can. J. Plant Sci.* 47:395-403.

Wallin, J. R. (1962). Summary of recent progress in predicting late blight epidemics in United States and Canada. *Am. Potato. J.* 39: 306-312.

Climate Change and Food Security, 2008
Edited by: M.Datta, N.P.Singh & Er.D.Daschaudhuri, pp.171-189
New India Publishing Agency, New Delhi (India)
E-mail : newindiapublishingagency@gmail.com
Web: www.bookfactoryindia.com

12

Weather and Climatic Information for Integrated Pest Management

D.K. DAS, A. DHANDAPANI AND T.P. TRIVEDI

INTRODUCTION

Integrated Pest Management (IPM) has been defined as a "system that, in the context of the environment and the population dynamics of the pest species, utilizes all suitable techniques and methods in as compatible manner as possible that maintain the pest populations at levels below those causing economic injury" (Glass 1975). The Inter society Consortium for Plant Protection (ICPP) has defined integrated pest management as "the use of multiple tactics in a compatible manner to maintain pest population at levels below those causing economic injury while providing protection against hazards to humans, domestic animals, plants and the environment". The term 'integrated' means it is a broad interdisciplinary approach taken using scientific principles of plant protection of variety of management strategies and tactics to fuse into a single system. The terms 'pest' means all biotic agents which adversely affect plant production. This includes insects, mites, nematodes, weeds, bacteria, fungi, viruses, parasitic seed plants and vertebrates. The term 'management' indicates to keep the pest population below some level based on economic criteria. It does not mean complete eradication of entire pest population.

NEED FOR INTEGRATED PEST MANAGEMENT (IPM)

The present day crop protection is mainly relying on chemical control of pests. The dependence on chemicals cause some serious problems to the human and its ecosystems. The problems are:

a) Insects becoming resistant to insecticides,
b) Insect resurgence after becoming resistant,
c) Biomagnification of certain insecticides in food chain,
d) Persistence of some insecticides as residue,
e) Pollution of surface and ground water causing harm to aquatic ecosystems
f) Rejection of our agricultural products at foreign market.

The adoption of IPM strategies has received the most attention to minimize the above problems.

The IPM strategy suggests

a) Use of safe pesticides and botanical pesticides.
b) Use of pesticides in appropriate time based on pest phenology and weather information to minimize pesticide use.
c) Use of biological control agents like parasites and predators.
d) Use of resistant crop cultivars.
e) Use of cultural methods to have minimum pest problems.
f) Development of transgenic plants resistant to insect pest and disease.

To understand the IPM techniques, it is necessary to understand the interacting factors in a crop or pest ecosystem first. Fig.1 depicted the various factors to be considered in a general pest management programme. Proper understanding of each factor, their interrelationship and techniques of suitable manipulation will lead to a successful pest management programme.

USE OF WEATHER AND CLIMATIC INFORMATIONS IN IPM

Meteorology is the science of atmosphere which deals with weather and climate with a wide range of branches from Aviation Meteorology to Agrometeorology. The informations which are useful can be classified into two broad categories: weather forecasting and agrometeorological informations.

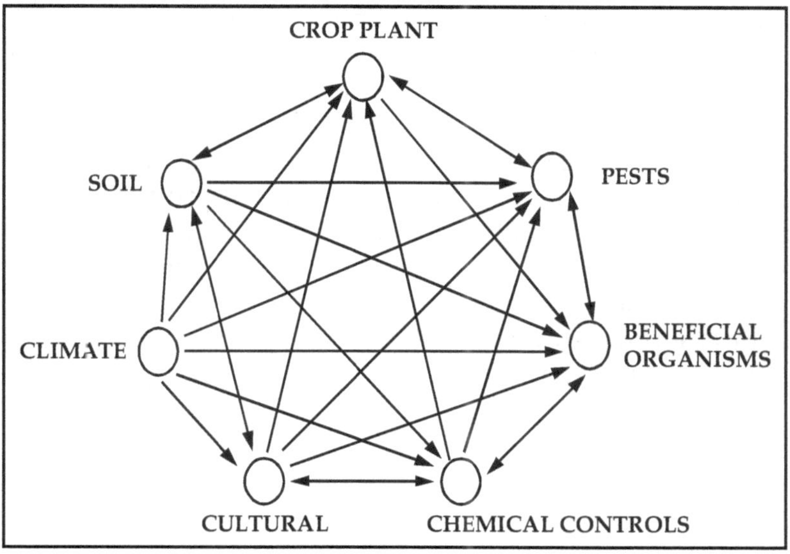

Fig. 1. Typical Interacting factors that need to be considered as a part of a general pest management program (after Steiner, 1966)

Weather forecasting is an important branch of meteorology. Weather forecasting can be divided into 3 types.

a) Short range - up to 48 hours or 2 days.
b) Medium range - 3-10 days
c) Long range - more than 10 days.

Short range weather forecasting is used to provide very short time gap to prepare for pest management practices. Medium range weather forecasting provides good scope for planning, preparation, purchasing of pest management inputs and also provides adequate time for pest management as well as farm management practices. Medium range forecasting of different weather parameters can be utilized for different pest management measures. Long range forecasting is also useful in IPM particularly pest forecasting as several workers have noticed the post-monsoon pest population depending on amount and distribution of monsoon rainfall. India Meteorological Department (IMD) makes forecast about monsoon in April. Therefore, in April itself we could have an idea about the likely pest scenario of September, October, November, December.

Apart from weather forecasting, another branch of meteorology, *i.e.* agrometeorology has significant role to play in IPM.

Agrometeorology is an interdisciplinary science in which the main scientific disciplines involved are atmospheric sciences, soil sciences, plant and animal sciences including their pathology, entomology and parasitology. Its subject matter links together the physical environment and biological responses under natural conditions. The subject of agrometeorology has a great practical utility in protection against or avoidance of adverse climatic risks. The dangers to crops and livestocks which have meterological contents include the incidence and extent of pest and disease. Therefore, the linkages between weather forecasting, agrometeorology and pest management can work together successfully for the benefit of farmers in the following areas.

1. Forecasting of pest and disease outbreak.
2. Pest phenological information based on temperature based degree day concept.
3. Scheduling insecticidal, fungicidal spray using weather information, pest phenology and crop phenology to minimize the use of chemicals.
4. Selection of biocontrol agents and appropriate timing of applying biocontrol agents.
5. Determination of crop stress due to insect pest and diseases by remote sensing techniques and assessment of yield loss.
6. GIS for pest mapping and forecasting.
7. Modification of microclimate for pest suppression.
8. Soil environment and relation with soil borne pathogens and nematodes.
9. Pest risk analysis
10. Assessment of future pest scenario with changing climate.
11. Strengthening agromet advisory services.

FORECASTING OF INSECT PEST AND DISEASE OUTBREAK

There is a close relationship between plant disease and weather. Therefore, in terms of prevailing weather conditions, the incidence of several diseases can be forecasted. Viral and bacterial disease are more weather-dependant than insects due to the fact that viral and bacterial pathogens remain at fixed location, consistently exposed to a particular type of weather conditions for a sufficiently long period. Insects, however, are mobile and move from place to place at different times and hence are less affected by uniformity of weather. It is certain that most disease are more dependent on weather than are insects; however, spread of certain insect pests is directly weather dependent.

For forecasting of insect or disease attack, sometimes, only one weather parameter is sufficient. For example, in USA, if the sum of average temperature of December, January and February remains below 38 ^{0}C, there is no chance of the occurance of bacterial wilt of corn in the coming season. But if the sum of average temperature as above 29.5 ^{0}C leaf blight of corn is likely to occur (Mather, 1974). Coming to the insect, no incidence, low incidence and high incidence of corn flea beetle are expected if the above sum of temperature remains < 26 ^{0}C, 26-32 ^{0}C and > 32 ^{0}C, respectively.

In most of the cases, single weather parameters can not explain the pest population to a desired level of accuracy. So, other parameters like rainfall, humidity, solar radiation, photo period, dew period used to be incorporated. For forecasting of wheat rusts (yellow, brown and black rusts) of Northwest India, mean temperature, relative humidity, sky condition and dew information are required. Similarly, ranges of several weather parameters (temperatures, dew, rainfall, cloudiness, relative humidity) combine together to forecast potato blight by Dutch rule, Cook's rule, Beanmont's rule, Irish rule, Wallim's rule etc. For example.

Potato blight by Dutch rule:

1. Dew during night >= 4 hours
2. Minimum temperature (night) >= 10 ^{0}C
3. Mean cloudiness on following day >= 8/10
4. Rainfall in 24 hours following the night of dew >= 0.01 cm

As a result of this combination, blight will appear within a fortnight.

Weather parameters have been modelled by multiple regression techniques to assess the per cent incidence of 'Karnal bunt' disease of wheat by Srinivasan and Mavi (1983).

The equation developed was

$$Y = 18.475 - 0.544X1 + 0.108X2 - 0.059X3 - 0.118X4 + 1.192X5 \quad (R^2 = 0.9066)$$

Where, y = Karnal bunt disease incidence (%)
x1 = Maximum temperature during 11th week (^{0}C)
x2 = Minimum temperature during 11th week (^{0}C)
x3 = Relative humidity during 11th week (%)
x4 = Cumulative rainfall during 11th week(mm)
x5 = Number of rainy days during 11th week

A computer based simulation model MOTHZV has been developed by Witz *et al* (1985) for predicting the population dynamics of Heliothis species. The model uses early season number of eggs, larvae or adults to forecast the timing and size of later, potentially damaging population. Pheromone trap has provided the means for measuring early season number of Heliothis adults. This pheromone trap data along with climatic variable and crop phenology were used as input to the MOTHZV model to predict the timing of future Heliothis generations for a pilot test on the management of Heliothis species.

Requirement of a Pest Forecasting model

To have a successful pest forecasting programme the most important need is accurate historical data of three kinds mainly.

a) Weather Data

Weather data include temperatures, wind, atmospheric pressure, humidity, precipitation, soil moisture and radiation. The data should be accurate. Inaccurate data will lead to development of an inaccurate model which ultimately would lead to loss of crores of rupees.

Linking weather forecasting to pest forecasting

In all pest forecasting programme, weather data is an essential component. Forecasting of future pest population is done from the past/present weather. But emergence of population and size of population are determined by the future weather in which the pests are going to get exposed. In forecasting models, it is assumed that the weather of next few days or weeks would be at par with those of previous years. This may not happen in reality. Therefore, there is a need to foresee the future weather parameters. This need will be fulfilled by availing forecasted accurate weather data of the coming future.

Apart from regular weather data, some micrometeorology data are required for development of an accurate model as the environment within canopy, in which the pests thrieve,differs significantly from outside environment. The micromet data which are of major significance are temperature, humidity and radiation profile within canopy and duration of wetness period. Soil temperature data are also required as some stages of insects are completed within soil. This area needs more attention in future studies.

b) Pest population data

Accurate monitoring of pest population on daily basis is very much necessary for simulation models and daily to weekly basis for medium range forecasting. For monitoring the population, the devices are light trap and pheromore trap. Manual counting of eggs, larvae and pupae also serves the purpose.

Data on natural enemies can also be incorporated as pest population reducing factor.

c) Crop data

This is one of the most important data. Particularly crop phenological information is required for simulation models. With similar type of weather condition and initial population the future population can vary depending on crop area, crop phenology, and associated crops. For example, larval survival of Helicoverpa was 78%, 55% and 85% when fed with leaves, flowers and pods of Bengal Gram respectively (Tripathi et al, 1987) and the adult emergence was in the order of 89%, 84% and 44%. For identification of hosts, their area coverage and phenological stage, remote sensing techniques can be suitable tools for rapid and wider area information for forecasting purpose.

Expert Systems

This is an extension of pest forecasting models. After getting a forecast, the expert system will tell what to do. This expert system is a knowledge based system where the knowledge of expert pest managers' (i.e. plant protection persons, IPM specialist) are incorporated. A pest forecasting model will give output interms of

1. Time of pest initiation
2. Time of peak pest population

3. Time of its declination
4. Size of pest population

The knowledge of expert will be comprised of

1. Whether to take any preventive measure or not based on ETL values or size or population.
2. Selection of pest management techniques (chemical/ biological/ cultural).
3. Selection of proper doses
4. Selection of proper time.

The 'expert system' will use data on pest population, pest phenology, crop phenology, weather parameters and will give recommendation for proper management practice. With the increased use of computers in agriculture, this computer based expert system will be used for very rapid decision making in pest management and other areas.

Different expert systems are in practice for pest management in developed countries, e.g. NERISK is an expert system to enhance the integration of pesticides with arthropods biological control (Messsing and Croft, 1990). COMAX is an expert system for cotton crop management (Lemmon 1986).

In India, monthly pheromone trap moth data of 1983-90 along with weather data of corresponding period collected from PAU, Ludhiana have been analysed at NCIPM by Das *et al* (2000) and it was found that there were major two peaks of moth catches in a year. The main peak was during March-April and the second peak was during October every year. During March-April peak, chickpea was the main crop and during October, cotton was grown almost over entire area. The peak population of *Helicoverpa* moth during March-April and October can be predicted in advance by multiple regression models using different weather parameters along with previous season's population of *Helicoverpa*. The population density during March-April (PM-A) has been regressed with different weather parameters as well as pest population density of previous five months separately and cumulatively. It has been found to depend on total moth catch per trap during previous October to February (PO-F), mean monthly relative humidity recorded in the afternoon during previous

February (RHEF) and mean monthly minimum temperature of previous February (TminF) as follows:

$$PM\text{-}A = -1032.65 + 2.06\ PO\text{-}F - 34.26RHEF + 516.74TminF \quad (R^2 = 0.75)$$

Apart from weather parameters, the total population cumulated from October to February is another important parameter. The population, which uses to undergo diapause from October due to severe winter is emerged again after February.

Similarly, the population density during October (PO) can be predicted in advance with multiple regression using weather parameters as well as pest population of preceding months as follows:

$$PO = -5.39 + .12\ PM\text{-}A + 1.28\ PJ - 0.51\ RJ\text{-}J \quad (R^2 = 0.87)$$

PM-A = Pest population density of March and April,

PJ = Pest population density of June,

RJ-J = Total amount of rainfall during June and July.

The data collected from ICRISAT, Hyderabad were analysed and a thumb rule has been found out to predict the level of attack during any year using surplus/deficit rainfall of different months and that thumb rule has justified the severity of pod borer attack in different districts of Andhra Pradesh during the winter of 1997-98 (Das et al 2001).

The surplus or deficit of monsoon rainfall (*i.e.*, the excess or less rainfall of June, July, August and September) which is referred here as 'A' and surplus or deficit rainfall during November which is referred here as 'B' of any place in this region could play the crucial role. A and B can be calculated as:

A = (Actual total rainfall during four monsoon months in a year – Long term average rainfall of that period of that place)

B = (Actual total rainfall during November month in a year - Long term average rainfall of that month of that place)

Positive (+) value of A or B indicates surplus rainfall and negative (-) value indicates rainfall deficit.

A+B- LOW	A+B+ MODERATE
A-B- MODERATE	A-B+ SEVERE

The constructed rule has been tested to predict the disastrous situation in crop damage at different districts of Andhra Pradesh during 1997-98 winter. The actual as well as long term average rainfall data of 1997 of Warrangal, Khammam, Karimnagar, Rangareddy and Mehboobnagar were obtained. Although, the population density data of pod borer were not available, data on the extent of damage were available. The A and B values of the above districts during the winter of 1997-98 showed A-B+ combination (Table 1) indicating severe attack or damage of *Helicoverpa armigera* and this was true as cotton, redgram, chilles and sunflower were devastated during this year in the above districts.

Table 1. A and B parameters, crops affected and level of damage by *Helicoverpa armigera in* different districts of Andhra Pradesh

Districts	A	B	Level of damage	Crops affected
Warrangal	-305.5	+17.1	SEVERE	Cotton, pigeon pea, chilles, sunflower
Khammam	-142.0	+9.2	SEVERE	Cotton
Karimnagar	-321.0	+29.5	SEVERE	Cotton, pigeon pea,, chilles
Rangareddy	-213.5	+17.1	SEVERE	Cotton, pigeon pea,
Mehboobnagar	-250.1	+18.5	SEVERE	Cotton, pigeon pea,

The thumb rule developed earlier to predict the level of population of *Helicoverpa armigera* at Andhra Pradesh was put into validation in chickpea-pigeon pea based ecosystem at Gulbarga, Karnataka using the historical as well as current data. After studying the historical data on monthly rainfall and per cent pod damage of Gulbarga, the thumb rule was fine tuned with one modification. To compute B parameter, the rainfall values of October were taken instead of those of November (Das *et al,* 2001a). Out of 10 years, the thumb rule predicted infestation level accurately for eight years (Table 2). The rule is adopted by Zila Parishad & Gulbarga District Agricultural Office to forecast the level of H. armigera infestation in pigeon pea.

PEST PHENOLOGICAL INFORMATION BASED ON TEMPERATURE BASED DEGREE DAY CONCEPT

The effect of temperature on growth and development of insect as well as diseases has long been established. During winter (low temperature) insects take longer period for completion of different phenological stages as well as total life cycle and during summer a shorter period is required to complete each stage of its life cycle. Several workers tried to have consistent time unit through out the season for insect developmental stages. The degree day (DD) or growing degree day (GDD) or day degree (DD) or heat day or heat sums which are nothing but summation of departure of daily mean temperature from a base temperature over a period of time found to be better unit for consistency than real time unit.

Mathematically

Degree day = å (Tmean - Tb)

where, Tmean = (Tmax + Tmin)/2 and

Tb is the base temperature below which pest does not grow.

Table 2. A and B parameters, percent of pod damage by *Helicoverpa armigera in* different years at Gulbarga, Karnataka

Year	A	B	Predicted level of damage	Percent pod damage	Actual level of damage
1989	+140.9	-40.5	LOW	25.8	LOW
1990	+42.6	+88.8	MODERATE	90.7	SEVERE
1991	-232.7	-63.6	MODERATE	43.1	SEVERE
1992	-180.5	+12.6	SEVERE	100	SEVERE
1993	-156.7	+96.0	SEVERE	71.8	SEVERE
1994	-354.7	+145.4	SEVERE	65.3	SEVERE
1995	-197.5	+70.6	SEVERE	80.2	SEVERE
1996	+9.9	+36.0	MODERATE	54.6	MODERATE
1997	-126.0	+23.2	SEVERE	99.6	SEVERE
1998	+266.6	+178.4	MODERATE	32.5	MODERATE

Assumption: LOW=<30%, MODERATE = 30-60% and SEVERE =>60% pod damage.

The degree day concept is widely used in USA and European countries for predicting the level of attack of pests, relating degree days with calendar dates for egg hatching and forecasting the optimum moment of destroy adult moths so that they will not lay eggs which will produce fruit burrowing larvae with fixed number of degree days. Another interesting application with degree-day model is the prediction of the levels of attack of pests. Development of model with 5-6 years database should facilitate the right timing of insecticide application based on the predicting levels with (say 2%, 20%, 50%, or 100%) levels of pest attack and crop phenology (Nathan, 1997).

This degree day concept has been used to forecast the timing of attack of cabbage root fly (Delia radium) by Collier *et al* (1991) and Collier and Finch (1988).

SCHEDULING INSECTICIDAL/ FUNGICIDAL SPRAY

The spray operations in agricultural systems are recommended based on time interval. For example - Three fortnightly sprayings with 0.04% Endosulphan or two application of Phorate (10%) grannules at 2nd and 6th week after transplanting is recommended against rice steam borer. This type of recommendation does not consider weather parameters as well as pest phenology into account. During the time of recommended applicaion rainfall may occur and every thing would be washed out. High wind speed may hamper spraying operation. The pest may not be at proper stage which is succeptible to the insecticide. To over come this problem it is now needed to develop and follow weather information based spraying schedule. The schedule should take into account the phenological information of the pest based on degree-days. Application of chemical spray against Colorado Beetle and Potato blight in Soviet Russia based on meteorological analysis and forecasting reduced crop losses of about 1.5 million Rubles per year. In Canada, the IPM is based on the meteorological information to regulate or economize the intensity and amount of spraying. The day-degree concept is widely used in USA and European countries for predicting the levels of attack of pests, relating degree-day with calender dates for egg hatching, forecasting the optimum moment to destroy adult moths. Collier and Finch (1988) given a preliminary forecasting system of carrot fly based on degree-days.

In our country, National Centre for Medium Range Weather Forecasting (NCMRWF) is giving medium range forecasting of temperatures, rainfall and wind speed. These parameters will be utilized to calculate degree-days for predicting insect stage and appropriate time for spraying.

SELECTION OF BIOCONTROL AGENTS AND APPROPRIATE TIMING FOR APPLICATION OF BIOCONTROL AGENTS

Biological control of pest is a very important tool in IPM. Use of predators and parasites for managing insect pests is gaining momentum at this stage. In this system, three kinds of living organisms are acting together- host plant, pest and parasite/predator to make the system more complex. Prevailing weather has different effects on these three interacters. Decisions of controlling pest are to be taken based on ETL and crop stage . Similarly, selection of parasite/predators will be done observing the efficacy of each predators under the prevailing weather. For example, both Telemonus ullyetti and Trichogrammatidea lutea are used to parasitize eggs of cotton bollworm, Helicoverpa armigera. But T.leutea is most effective at 28 ^{0}C, 25% RH and LD 14:10 where as T.ullyetti dominates at 22 ^{0}C, 70% RH and LD 10:14 (Kfir et al 1988). Therefore, slection of parasites should be made according to prevailing weather.

Another important point is the selection of timing of application of biocontrol agents. Most virulent stage of parasite should attack the most susceptible stage of pest. A degree-day based calculation or model for phenological stages of both parasite and pest should be developed and followed so that no mismatching should occur. This will reduce the amount of biocontrol agents as well as the cost of crop protection.

DETERMININATION OF CROP STRESS DUE TO PEST AND DISEASES BY REMOTE SENSING

Remote sensing can provide the basis for improving the predictive capabilities of pest population, plant disease and plant growth models used in IPM. For many years, both ground and aerial-based photography have been the main stay of remote sensing. Conventional color films cover the visible region of the spectrum (0.4 to 0.7 um). Color infra red (CIR) films are sensitized to record not only visible light but also a portion of near infra red wavelengths (0.7 to 0.9 um). The number of reports on CIR in the literature attests to widespread use for agricultural purposes. These range from monitoring soil moisture and drought stress to detection of plant disease. Although the individual insects are not normally detectable because of small size, often their metabolic by products or defoliating habits are, and this provides the basis for their detection.

The nature of many fundamental meteorological process can be characterized by the spatial and temporal variability of earth's surface temperature. Infrared thermometry provides a non-contact method for documenting these patterns over geographical regions in a cost-effective manner. It is based on the principle that the thermal radiation emanating from a surface is proportional to the fourth power of its absolute temperature. Any stress of physical or biological origin which interferes with the movement of water through the soil-plant-atmospheric continuum should be detectable by using this thermal approach. Canopy temperatures are useful for discriminating between certain diseased and healthy plants. Kumar et al (1978) reported that radiant temperatures of corn canopies increased with the severity of southern corn blight disease.

The disease which cause hampering of water movement in the soil plant-atmospheric continuum can be detected as the diseased plant, use to show few degrees higher canopy temperature than a healthy plant. The temperature difference between crop canopy and ambient air is also very useful for detection of stress. Stress degree day (SDD), and index is a good indicator of final yield when calculated over reproductive phage. Therefore, yield loss due to stress by insects and diseases can be estimated based on temperature condition of the crop canopy.

GEOGRAPHIC INFORMATION SYSTEM (GIS) FOR PEST MAPPING AND FORECASTING

GIS is a potential tool to prepare pest affected zones. Preparation of pest maps using GIS on different rice and pests at district level for different years is completed at NCIPM. Super imposition of different years pest data along with weather and soil data will help to delineate 'hotspot', 'moderately affected' and 'less affected' zones of different pests through out India. Key weather parameters of hotspot zone can be used as indicators for pest forecasting.

MODIFICIATION OF MICROCLIMATE FOR PEST SUSPPRESSION

The microclimate within the crop canopy is most important from pest development point of view. Outside temperature may be 40 ^{0}C in which pests can survive, but within canopy it may be 35 ^{0}C (i.e. within tolerable range for a pest). So, study of microclimate, its relationship with meso climate and simulation of microclimate needs proper attention. If the input parameters, daily maximum and minimum temperatures can be replaced by leaf or canopy temperatures in forecasting/simulation models, the predictability is expected to increase.

Another important microclimatic parameter, wetness duration have definite relationship with development of pathogens. Weiss *et al* (1980) found white mold disease (Sclerotiana sclerotiorum) of dry edible beens (Phaseolus vulgaris) is dependent on leaf temperature and wetness duration. Wetness duration is single most microenvironmental factor which is most difficult to measure and predict. Fortunately, simulation of wetness duration has been started in India by Sanjeeva Rao and other at NCMRWF using temperature, relative humidity, wind, cloud cover and some plant parameters. This will help the future workers to predict plant diseases in advance.

The crop microclimate changes during the process of natural plant development. It becomes more humid and the upper and lower limits of temperature become less extreme. Water deposited by rain, irrigation or dew persists for a longer time. Modern agricultural practices have pushed us to dense plantings and high canopy densities, sometimes leading to an increase in foliage over and above its natural development. The too much foliage facilitates pathogen development. One of the objectives of IPM is to make the habitat less favourable for pest. This can be achieved by modification of microclimate through modification of canopy. This may be achieved by proper rate of seeding, pruning, fertilization or irrigation, by proper training or trellising or by breeding cultivars with spare canopy.

MODIFICATION OF SOIL ENVIRONMENT FOR CONTROLING SOIL BORNE PATHOGENS AND NEMATODES

Soil borne pathogens and nematodes would appear to be well shielded from the direct effect of solar radiation under normal field conditions. The findings of Israeli investigations have demonstrated that solar radiation can be utilized to control soil-borne diseases. When field soil is mulched with polyethylene sheets, the soil temperature is raised to the lethal level for pathogenic fungi.

Polyethylene mulching increased soil temperature by 7-13 degree Celsius at a study conducted by Nematology Division of IARI. Mulching for 3 weeks significantly increased the growth of rice seedlings. This was attributed to reduction of nematode population at mulched plots. Soil solarization at the same place reduced *Meloidogyne, Helicotylenchus* and *Aphelenchus* sps by 87%, 58% and 77% respectively.

Besides soil temperature, another important parameter to affect growth of soil pathogens is soil moisture. Relatiavely wet conditions enhance the development of white mold (*Sclerotiana*) epidemics on the aerial parts of bean plants in the field. A number of diseases incited by fungi in the genus *Phytophthora* have historically been associated with very wet conditions. For example, the late blight of potato caused by *Phytophthora infestans* only infects large numbers of potato tubers during crop growth if soil moisture remains well above field capacity for at last 24 hours (Lapwood, 1977). Not only wet conditions but moisture stress also causes development of diseases. For example, *Macrophomina phaseoli* is usually observed to cause charcoal rot of grain sorghum in the field when there is the proper combination of high temperature, low soil moisture and fairly matured plants (Edmund, 1964).

Apart from diseases, soil environment has some significance influence on insects' life cycle as some part of its is passed through soil. For example, *Helicoverpa* spp. use to spend pupal stage in the soil.

Therefore, the proper study of soil environment with a view to control the soil borne pests will lead to develop the successful IPM technology.

PEST RISK ANALYSIS

Pest forecasting deals with the foreseeing of outbreak of any pest which is already present in the location or surrounding area; where as pest risk analysis (PRA) deals with the outbreak of exotic pests which are not present in the country. If any exotic pest gets an entry through any means, by chance, the possibility of survival and spread of that pest in new location is the study area of PRA. Climatic matching of different locations of the country with the country of origin of the pest is the main technique. For import of any agricultural produce PRA is done compulsorily by plant protection and quarantine department to check the spread of potentially dangerous insect, pathogen, weeds associated with it. For export also, PRA is done for the pests present in our country (e.g. fruit fly in mango) and have potential to be spread in the importing country.

ASSESSMENT OF FUTURE PEST SCENARIO WITH CHANGING CLIMATE

Climate change and global warming are the hot topics of discussion now a days. Workers are assessing the probable impact of global warming on insects' life cycle. Collier et al (1991) simulated the future attach of Cabbage rootfly (*Delia radicum*) in UK increasing daily mean temperature by 3°C, 5°C, 10°C. An increase of 3°C in mean daily temperature would cause the cabbage root fly to become active about a month earlier in the year than at present. Under such conditions, the emergence of flies from the overwintering population would be less synchronised, as the completion of iapause and post-diapause development would occur at the same time in different individuals within the population. However, these would continue to be 3 generations in a year. But with temperature increases of 5°C to 10 °C, the fly would complete four generations each year. New strategy would be required to manage this pest.

Similarly all insect pest will be expected to have more generations in a year which indicates a greater amount of IPM inputs requirement in future to manage the pests. On the other hand, in our country 2-3°C temperature increment may cause some reduction of pest population in summer as it will be lethal to some soil borne diseases or pests.

CONCLUSION

National Centre for medium Range Weather Forecasting (NCMRWF) is issuing weather forecast 5 days in advance. With the help of this forecasts, farmers are advised for different farm operations like, sowing, transplanting, irrigating, harvesting of crops. This Agro advisory service could be improved, if it can forecast the time of pest attack, its ETL(level of population causes economic loss) value for taking pest management practices. Pest forecasting can be linked to weather forecasting. Degree days based model will be predicting accurate timing of larval or adult actively based on forecasted temperature. Spraying operation should be done or not could be predicted from forecasted rainfall and wind information. What will be optimum time for application of parasite or predator can be fore seen. Agrimet Department of IMD, Pune has published available pest forecasting models through its website.

Indian Council of Agricultural Research (ICAR) has also started advisory service on plant protection through its website. Weather information is also used to some extent for this purpose. The plant protection advisory in the country is in right direction and this can be further strengthened through effective collaboration work weathermen, agrometeorologist and IPM specialist.

REFERENCES

Collier, R.H., Finch, S, Phlps, K, Thompson, A.R. (1991). *Ann. appl. Biol.* 118: 261-271.

Collier, R.H., Finch, S.(1988). *Brighton Crop Protection Conference*-(1988). 3C-17.

Das, D. K., Trivedi, T.P., Singh Joginder and Dhandapani, A. (2000). *International Conference on Managing Natural Resources for Sustainable Agricultural Production in the 21st Century. Voluntary Papers – Natural resources,* Vol.2: 621-622.

Das, D.K., Trivedi, T.P. and Srivastava, C.P. (2001). *Indian J. Agr. Sci.* 71(6) 421-423.

Das, D. K., Yelshetty, Suhas, Trivedi, T. P. and Dhandapani, A. (2001a). *National Symposium on Plant Protection Strategies for Sustainable Agri-Horticulture. S K University of Agricultural Sciences and Technology. October 12-13, 2001. R S Pura Jammu.*

Edmunds, L.K.(1964). *Phytopathol*. 54:514-517.

Glass, E.H.(1975). *Entomol. Soc. Am. Special Publication*. 75-2 pp 141.

Kfir, R., Hamburg, H. van; Van Hamburg, H.(1988). *Environ. Entomol.* 4: 664-670.

Kumar,R, Silva, L.F.; Baur, M.E.(1978). Publication N78-29537 (National Technical Information Service, Springfield, Virginia).

Lapwood, D.H.(1977). *Ann. Appl. Biol.* 85:23-42.

Lemmon, H.E.(1986). *Science.* 233(7):29-33.

Mather, J.R.(1974). *Climatology. Fundamentals and Applications*. McGraw Hill Company: 181-214.

Messing, R.H., Croft, B.A.(1990). *Acta Horticulturae.* 276:15-19

Nathan, K.K.(1997). *Course Material- Master Trainers' Training on IPM,* March 17-22,1997, NCIPM, LBS Building, Pusa Campus, New Delhi 110 012.

Rotem, J., Ben-Joseph, J.(1970). *Plant Dis. Reptr.* 54:768-771.

Srinivasan, G.S., Mavi, H.S.(1983). *Proc. Int. Conf. on Biometeorology*, New Delhi.

Steiner, H.(1966). Proc. *FAO Symp. Integr. Pest Control. 3:13-20.*

Tripathi, S.R, Rajesh Singh; Singh, R.(1987). *Giornale-Italiano di-Entomologia.* 3(17) : 405-408.

Weiss, A., Kerr, E.D.; Steadman, J.R.(1980). *Plant Disease.* 64:757-759.

Witz J.A., Hartstack, A.W., King, E.G., Dickerson, W.A., Phillips, J.R.(1985). *Southwestern Entomologist*. Suppl. No. 8:56-70.

Climate Change and Food Security, 2008
Edited by: M.Datta, N.P.Singh & Er.D.Daschaudhuri, pp.191-198
New India Publishing Agency, New Delhi (India)
E-mail : newindiapublishingagency@gmail.com
Web: www.bookfactoryindia.com

13

Role of Agro-meteorological Advisory Services in Mizoram

K.A. PATHAK, BRAJENDRA AND A.K. VISHWAKARMA

INTRODUCTION

Mizoram is a hilly terrain lying in the north eastern part of the country. The total geographical area of the state is 21,087 sq.km and lies between 92° 15′ to 93° 29′ E longitude and 21° 58′ to 24° 35/ N latitude. Out of six agro climatic zones of the NEH region, Mizoram comes under sub tropical hill zone and posses a sub tropical and humid climate due to high rainfall (Fig.1). These climatic conditions are very conducive for the cultivation of a large number of agri-horticultural crops. The physiography of the state is divided into hills and valley lands. About 90 % area of the state comes under hills and only 10 % area is under valley. The valley lands are sporadically distributed in small patches except in Champhai district. The slope of the hills in Mizoram is much steep as compared to the other states in the North Eastern region. The land use statistics of the state is presented in Table1.

The state posses a wide range of flora and fauna. The forest vegetation mostly comprises of tree species, bamboo species, shrubs etc. in abandoned *Jhum* lands. The state is mostly inhabited by Mizo tribe and the farmers are mostly practicing *Jhum* cultivation in hill slopes except transplanted rice cultivation in valley lands.

Table 1. Land use statistics of Mizoram from 2000-01 to 2003-04 (Area' 000 ha).

Particulars	2000-01	2001-02	2002-03	2003-04
Total Geographical area	2108.7	2108.7	2108.7	2108.7
Area under forests	1626.475	1626.475	1541.608	1593.500
Not available for cultivation	130.390	130.490	132.390	133.930
Other uncultivated land excluding fallow land	58.410	34.863	69.853	26.723
Fallow lands	192.094	199.060	247.069	232.825
Total cropped area	104.689	117.812	122.307	126.932
Area sown more than once	3.358	3.891	4.527	5.210
Net sown area	101.331	113.921	117.780	121.722
Net irrigated area	9.686	10.219	10.66 9	11.352
Gross irrigated area	11.629	12.162	12.612	15.749

AGRO-CLIMATIC ZONES OF NEH REGION

- Alpine zone
- Temperate and Sub- alpine zone
- Sub-tropical hill zone
- Sub-tropical plain zone
- Mild tropical hill zone
- Mild tropical plain zone

Crop production is a complex phenomenon and depends on abiotic factors (soil, weather, plant, etc.) and biotic factors (insect pest, disease, *etc.*). These factors varied frequently with time during different crop growth stages and any sudden change in these factors can bring drastic change in crop productivity. Weather is one of the most important factor playing an major role in crop production and thus crop weather relationship is of immense practical importance. All the agriculture operations like land preparation, sowing, harvesting, etc. are directly dependent on weather parameters. At the same time, these factors also indirectly influence the biotic factors. Therefore, forecasting of the weather will be of great importance and will help in escaping the vagaries of erratic weather conditions.

WEATHER PARAMETERS

The weather parameters recorded in Kolasib District of Mizoram is presented below.

Rainfall

The average annual rainfall of the state is 2648mm and it varied between 2102 mm in Champhai district to 3203mm in Lawngtlai district during the year 2002-03. It is evident from Fig.2 that maximum annual rainfall of 3583mm was recorded in the year 2004 and the minimum rainfall of 1762mm was recorded during the year 1994. The mean monthly rainfall varied from 9mm - 450.6mm in the month of January and August respectively Fig.3. The average percent rainfall distribution pattern based on 25 years data is presented in the Fig.4. This indicates that the state receives maximum rainfall during the months of July and August followed by June, May and September respectively.

Based on the existing pattern, the rainfall of the state may be classified into three categories.

1. Pre- monsoonal rain (March to May) - Twenty seven per cent of the total rainfall.
2. Monsoonal rain (June to October) - Seventy per cent of the total rainfall.
3. Post- monsoonal rain (November to February) - Three per cent of the total rainfall.

The rainfall is generally erratic during the monsoon season which is mainly received from June to October. Heavy to very heavy rains during monsoon affect the kharif crop growth and restrict agricultural operations while at the same time, it also restricts the cultivation and growth of rabi crops due to lack of moisture. All the farm activities are dependent on rainfall and it is very essential for the farmers to have prior information on the rainfall pattern.

Temperature

The data presented in Fig. 5 showing mean monthly temperature based on average of data from 1986-2004 shows that the minimum temperature varied between 13.29^{0}C in the month of January to 23.65^{0}C in the month of July and the maximum temperature varied between 21.18^{0}C in the month of January to 29.12^{0}C during the month of July. The temperature of the Kolasib, Mizoram is favourable for growing a wide range of crops during kharif and rabi seasons. Besides this, the temperature along with other climatic factors is very conducive for incidence of various insect pests and diseases also.

Relative Humidity

Due to high rainfall, the state is having high relative humidity. The data presented in Fig. 6 showing mean monthly relative humidity based on average of data from 1986-2004 shows that the minimum relative humidity varied between 73 percent in the month of March to 78.1 percent in the month of July and the maximum relative humidity varied between 73 percent in the month of March to 88.1 percent during the month of July. The relative humidity of Kolasib, Mizoram is favourable for growing a wide range of crops during kharif and rabi seasons. Besides this, the relative humidity along with other agro climatic factors is very conducive for incidence of various insect pests and diseases also.

Wind Velocity

The data on wind velocity recorded during the year 2004 is presented in Fig.8. It is evident from the figure that the wind velocity in the state varied between 0.04 km/h in the month of November to 4.41 km/h during the month of March. During March ,the Kolasib, Mizoram usually receives thunder hail storms and after this germination of the crop seeds sown in the *Jhum* lands, begins.

CONCLUSION

It can be concluded from the above descriptions of weather parameters that the Kolasib District, Mizoram possesses very heavy and erratic type of rainfall, moderate temperature and high humidity. These conditions are favorable for crop growth and also very conducive for incidence of various insect pests and diseases. Therefore, it is very essential for the farmers to have prior information on the weather forecasts and based on that the farm operations can be planned. This can be achieved only through effective timely agro meteorological advisory services to the farmers for increasing crop productivity and to minimize weather related avoidable losses.

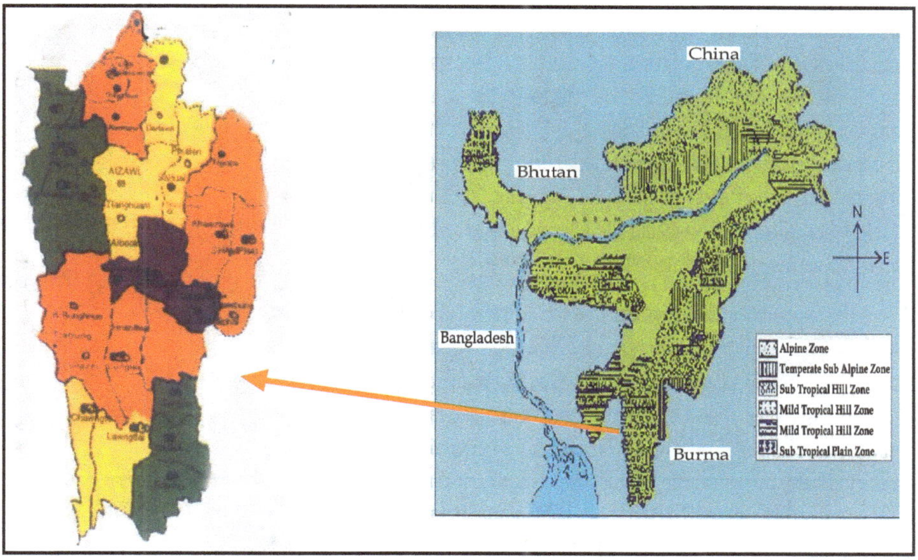

Fig. 1. Agro Climatic Zones of NEH region

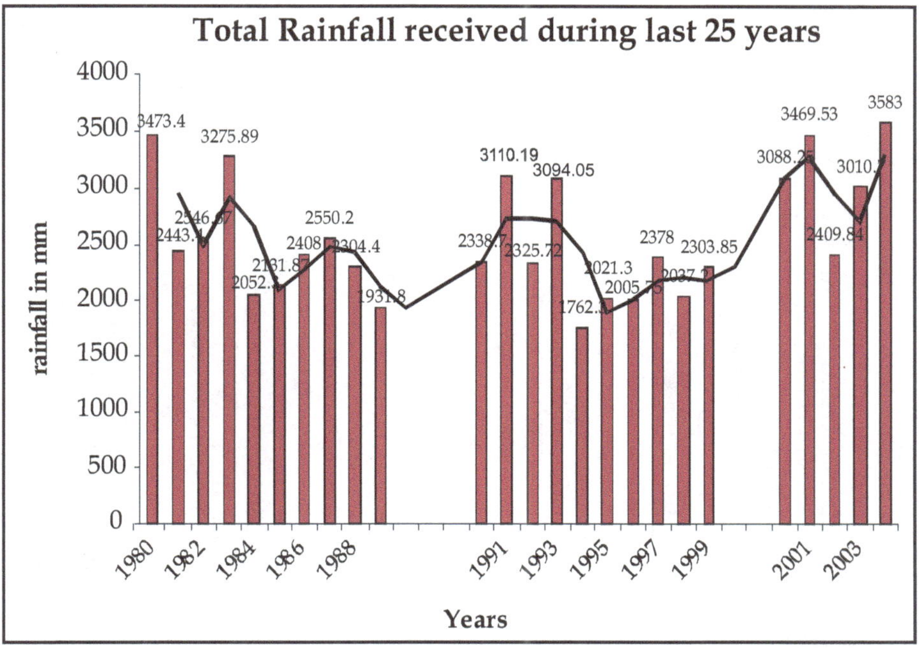

Fig. 2. Total annual rainfall in Mizoram

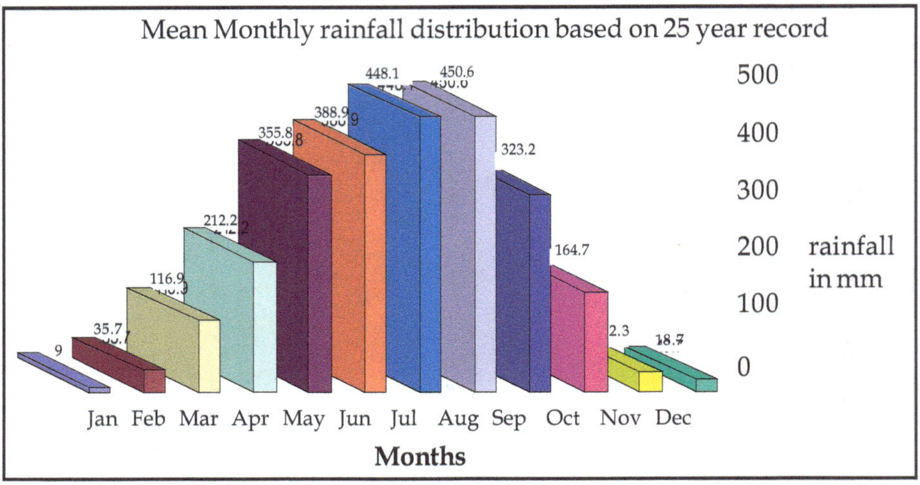

Fig. 3. Monthly rainfall distribution of Kolasib in Mizoram

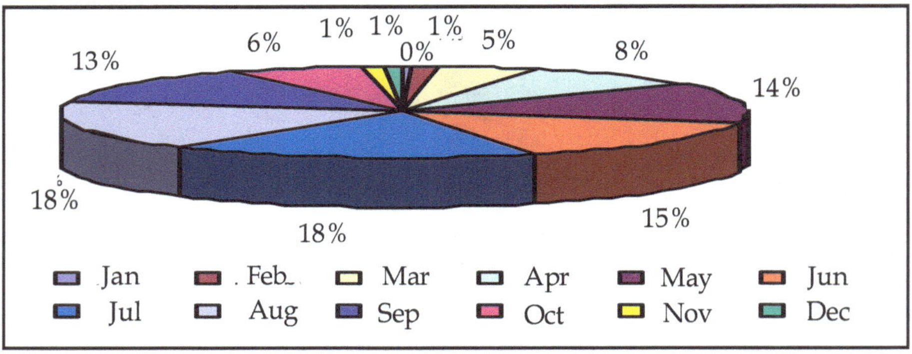

Fig. 4. Percent rainfall distribution pattern of Kolasib based on 25 years data

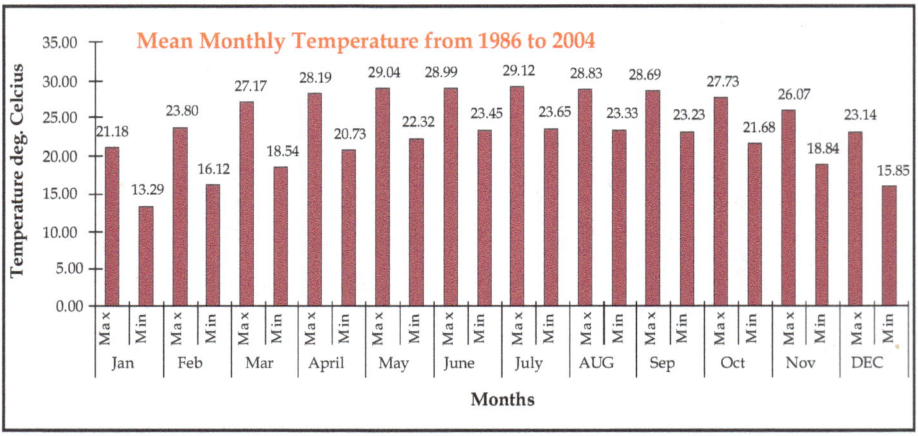

Fig. 5. Average mean monthly minimum and maximum temperature

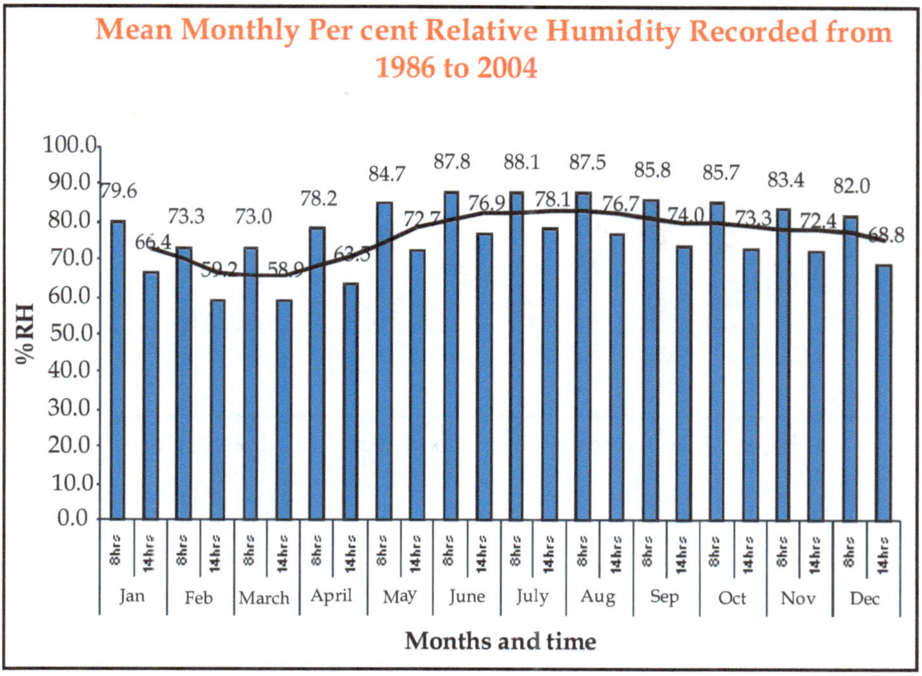

Fig. 6. Average mean monthly minimum and maximum relative humidity

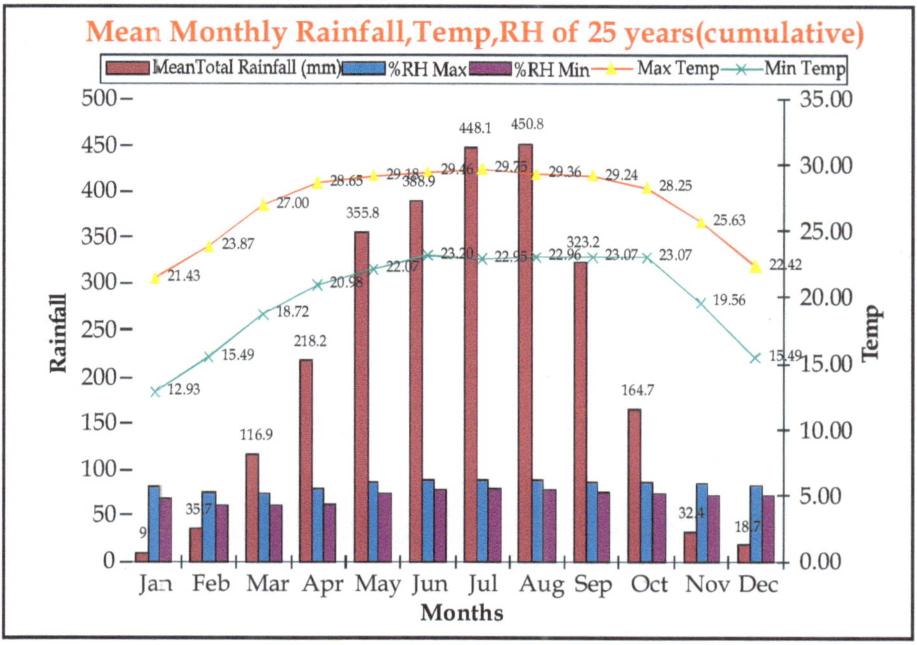

Fig. 7. Cumulative Mean monthly rainfall, temperature and relative

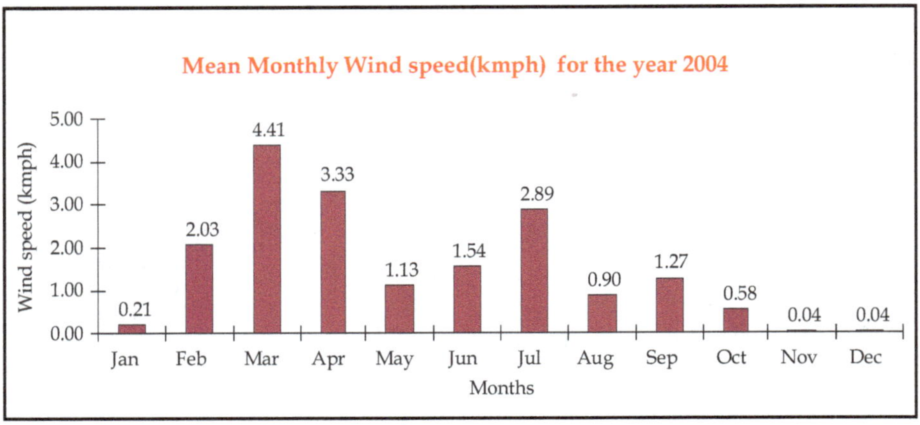

Fig. 8. Mean monthly wind velocity Km/h for the year 2004

Climate Change and Food Security, 2008
Edited by: M.Datta, N.P.Singh & Er.D.Daschaudhuri, pp.199-213
New India Publishing Agency, New Delhi (India)
E-mail : newindiapublishingagency@gmail.com
Web: www.bookfactoryindia.com

14

Analysis of Rainfall Characteristics at Lembucherra (Tripura) in North East India

D. DASCHAUDHURI, M. DATTA AND N.P. SINGH

INTRODUCTION

Several biophysical factors govern the agricultural productivity of a region. Among the key biophysical factors, weather and more particularly rainfall has the over riding impact on agricultural productivity since rainfall is a primary source of water and the first requirement for raising any crop. This element influences various dynamics of crop growth and development and accounts from nearly two thirds of variation in agricultural productivity. The amount of rainfall at a particular place is important; an equal important factor is its temporal distribution, which is uneven, uncertain and erratic in most part of the country. Breaks in monsoon are common phenomenon during the monsoon period. The analysis of rainfall characteristics such as occurrence and distribution of dry or wet spell is essentially required for successful crop planning, development of befitting agro techniques, recommendation of suitable crop variety, designing of water harvesting ponds, earthen dams and other soil conservation structures. It is also an established fact that the crop development will be affected if the dry spell coincides with sensitive phonological stage of the crop and sometimes beneficial if it coincides with the ripening stage. Hence for the purpose of crop planning and to carry out the agricultural practices, it is important to know the sequence of dry and wet periods. An attempt has been made in this paper to analyse the rainfall in standard meteorological week as well as monthly basis at different probability levels for Lembucherra by using computer based forecasting technique.

A lot of work has been carried out in past by various investigators on rainfall analysis. The criterion set by Raman (1979) for rainfall of 1mm for defining a rainy day is not suitable for agricultural purpose. Ashokeraj (1979) used the criteria fixed by IMD for defining the rainy day i.e. the day with at least 2.5mm rain is called rainy day. The same technique was also used in determining the onset and offset of effective monsoon in Umiam, Meghalaya by Satapathy et al. (1998) and found very effective and logistic. The same technique is also used in present paper. When probability of occurrence of dry spell of different length in a week bonded by wet week is known; adequate steps may be taken by shifting the sowing time or arranging minimum irrigation to get optimum yield.

LOCATION AND METHODOLOGY

The climate is humid subtropical characterised by high rainfall. Based on climate and altitude, the vegetation in Tripura may be classed with the tropical forest type. Evergreen and moist deciduous forest generally predominates in Tripura. The meteorological data are recorded at Lembucherra (Tripura), which is located at $25^0 53'$ N latitude and $91^0 55'$E longitude with an altitude of 16.2m above mean sea level. Lembucherra is located west part of Tripura near capital town, Agartala.

Data Used

The rainfall and evaporation data used in this study have been collected from Agrometorological Observatory of ICAR Research Complex for NEH Region, Tripura Center, Lembucherra (Tripura). All the parameters are observed from 1992 - 2005. Due to remoteness of the area, the existence of class A meteorological observatory (approved by IMD) are very few, so availability of data length is also less.

ANNUAL, SEASONAL, MONTHLY AND WEEKLY VARIATION OF RAINFALL.

The rainfall data so collected are analysed in annual, seasonal, monthly and weekly basis. Standard week defined by the India

Meteorological Department (IMD) are followed to define the weekly rainfall. Seasonal distribution is defined as per the following guidelines

Summer or Pre monsoon	:	April to May
Monsoon	:	June to September
Post monsoon	:	October to December
Winter	:	January to March.

For analysis on monthly basis, a calendar month is considered as a monthly rainfall. Similarly for analysis of distribution of rainy days , a threshold limit of 2.5mm per day is decided to consider a day as rainy day.

Onset and Cessation of Wet Season

One aspect of crop planning is forward and backward accumulation of rainfall to determine the onset and termination of wet season based on precipitation data. Cropping strategies are greatly influenced by the variability of the onset and end of the rainy season. Generally, farmers opt for planning or broadcasting their seed when a certain amount of rainfall has sufficiently moistened the top soil. However 75 mm accumulation of rainfall has been considered as the onset time for the growing season for dry seeded crops and 250 mm accumulated rainfall for initiation of puddling i.e. wet land preparation of rice fields. Similarly, the end of the wet season is determined by backward summing of rainfall data. It is considered that 500 mm accumulated values represent the week after which sufficient rain would be expected to sustain a second crop or other short duration field crops, assuming a fully charged soil profile at planting (WMO, 1982).

For calculation of forward and backward accumulation, the weekly rainfall is added from 1st week onwards to find out the corresponding week number in which the cumulative rainfall has reached a sum total of 75 mm and 250 mm. The forward accumulation of 75 mm from a critical week, which was considered as 22nd week, represents the week in which monsoon rain starts. Also the forward accumulation of 250 mm from the critical week represent the week in which puddling is to be done for transplanting of rice. Dash *et al* (1992), Rath *et al* (1996), Pandarinath (1991) used forward accumulation of 200 mm as the week for transplanting of paddy. But Satapathy *et al* (1999) considered 250 mm forward accumulated rain for Meghalaya because in Meghalaya, paddy is grown on bench terrace on the foot hill and in valleys where there is high seepage loss due to slope and soil type. But in Tripura, paddy is grown mostly in lowland with few exceptions. Thus we consider forward accumulation of 200 mm as the week for identifying of wet period.

Similarly, the weekly rainfall of 52nd week is added backwards, (i.e. 52nd week+ 51st week+ ....) to get 200 mm, 300 mm and 500 mm total rainfall and corresponding week numbers are noted. The probability of each week is calculated by the simple statistical technique.

For backward accumulation, the rank order and probability levels are arranged in ascending order and the corresponding week numbers are arranged in the same manner. Similarly for backward accumulation, the rank order and probability levels are arranged in descending order and corresponding week numbers for 500mm, 300 mm, 200 mm respectively are arranged in descending order.

Two computer programs were used to calculate the onset and endset of wet period. In first computer program, the inputs are weekly rainfall (1992-2005); critical week from which the forward accumulation will take place (i.e. 22nd week), values for forward accumulation (i.e. 75 mm, 200 mm); value for backward accumulation (i.e. 200 mm, 300 mm and 500 mm). The result gives the rank number, probability and standard week numbers for different values for forward and backward accumulation. In the second program, the inputs are the same, but outputs are the different. It gives the yearwise rainfall characteristics of the place indicating the week in which monsoon starts and ends and the duration of the monsoon in that year.

Weibull's Method of Probability Analysis

The weekly rainfall data have been analysed at different levels of probability by using Weibull's method. In this method, the weekly rainfall is arranged in descending order of magnitude and the maximum magnitude was given rank1, and next magnitude was given rank 2 and so on. The probability 'p' of the week having rainfall exceeding or equaling normal value was calculated by using Weibull's formula $p = m/(n+1)$, where p is probability of occurrence, m is rank number and n is the number of years of data used.

RESULTS AND DISCUSSION

Based on the analysis of 14years (1992 to 2005), annual rainfall data at Lembucherra (Tripura) varied between 1457.0 mm (1992) to

3421.7mm (2001) with annual average rainfall of 2409.6 mm. Yearly total rainfall showed the alternate increasing and decreasing trend with exception from 2000 to 2003, where a consecutive two years (2000 to 2001) were found an increasing trend followed by two years (2002 to 2003) a decreasing trend. Maximum rain occurred during the month of July, which contributes about 18 % of total yearly rain followed by June and May. The month November, December, January and February almost remained dry. These four months jointly contributed only 4 % of total yearly rainfall. The variability of annual and seasonal rainfall are shown in Figs. 1 to 10.

Annual Rainfall and its Variability

1. Mean Annual rainfall: 2409.6mm
2. Highest annual rainfall recorded: 3421.7mm (2001)
3. Lowest annual rainfall recorded: 1457mm (1992)
4. Highest Rainfall in a single day: 286mm (6th June 2001)
5. Distribution of Rainfall more than 100 mm in a single day

1992 - nil	1993 - 3 days	1994 - 1 day
1995 - 1 day	1996 - 2 days	1997 - nil
1998 - 3 days	1999 - 3 days	2000 - 3 days
2001 - 7 days	2002 - 1 day	2003 - 2 days
2004 - 6 days	2005 - 2 days	

6. Mean Annual Rainy Days: 101 days
7. Highest rainy days in a year: 117 days (1993 & 2002)
8. Lowest rainy days in a year: 81 days (1995)

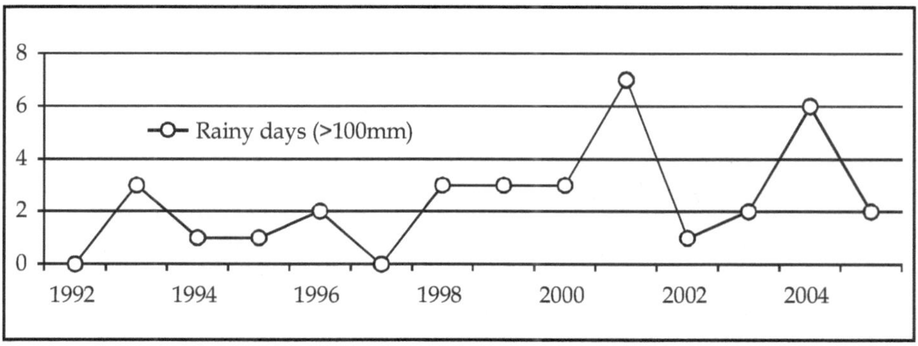

Fig. 1. Distribution of rainy days (>100mm) Database: 1992 - 2005

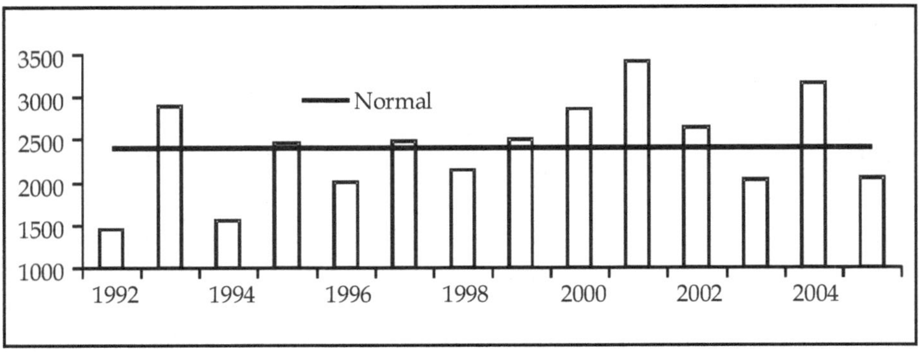

Fig. 2. Distribution of Yearly Rainfall (mm) Database: 1992 - 2005

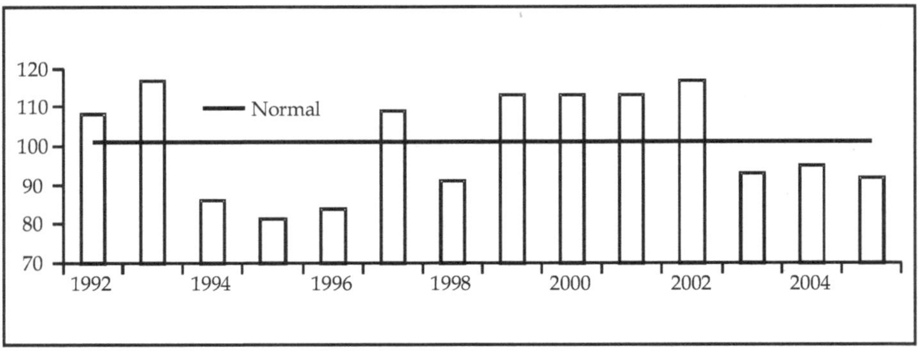

Fig. 3. Distribution of Rainy Days Database: 1992 - 2005

Seasonal Rainfall and Its Variability

Winter Rainfall (January to March)

1. Mean rainfall: 143.2 mm
2. Highest rainfall recorded: 351.8mm (1994)
3. Lowest rainfall recorded: 27mm (1999)
4. Highest Rainfall in a single day: 123.2mm (27th March 1994)
5. Mean Annual Rainy Days: 7 days
6. Highest rainy days in a year: 15 days (1994)
7. Lowest rainy days in a year: 1day (1999)

Pre- Monsoon Rainfall (April to May)

1. Mean rainfall: 567.0 mm
2. Highest rainfall recorded: 1051.8 mm (2000)
3. Lowest rainfall recorded: 301.9 mm (1994)
4. Highest Rainfall in a single day: 163.0 mm (8th May 2001)
5. Mean Annual Rainy Days: 22 days
6. Highest rainy days in a year: 43 days (1992)
7. Lowest rainy days in a year: 9 days (1995).

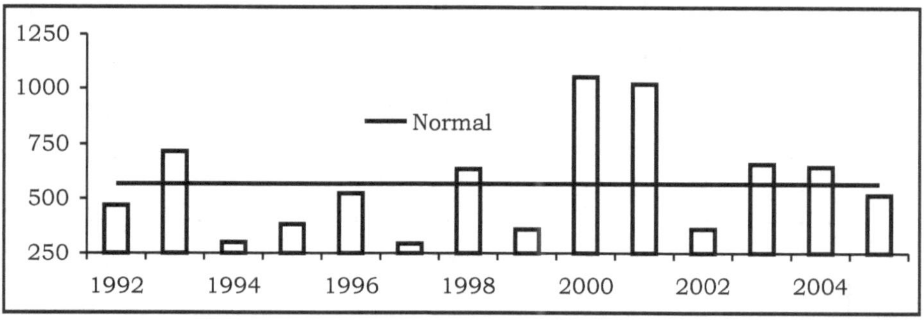

Fig. 4. Distribution of Pre Monsoon Rainfall (mm) Database: 1992 - 2005

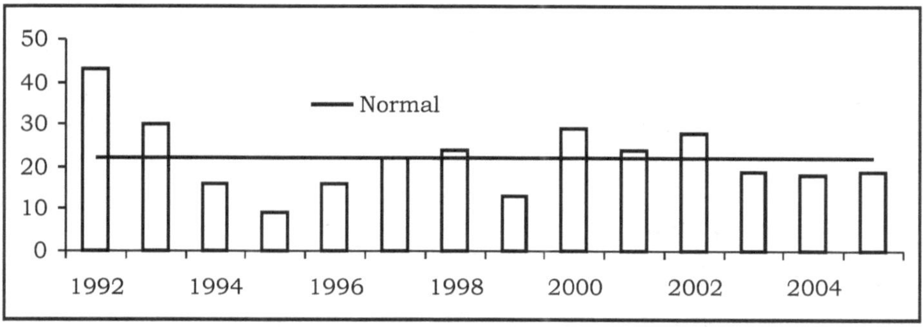

Fig. 5. Distribution of Rainy Days during Pre Monsoon Database: 1992 - 2005

Monsoon Rainfall (June to September)

1. Mean rainfall: 1470.1 mm
2. Highest rainfall recorded: 2318.3 mm (2004)
3. Lowest rainfall recorded: 737.9mm (1994)

4. Highest Rainfall in a single day: 286.0mm (6th June 2001)
5. Mean Annual Rainy Days: 63 days
6. Highest rainy days in a year: 86 days (1999)
7. Lowest rainy days in a year: 48 days (1994 & 1998).

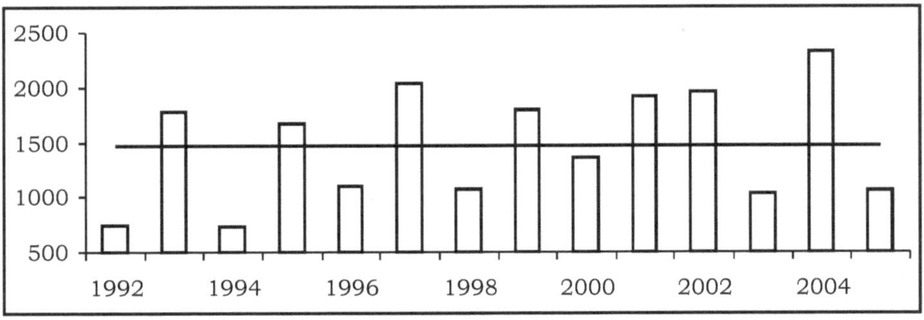

Fig. 6. Distribution of Monsoon Rainfall (mm) Database: 1992 - 2005

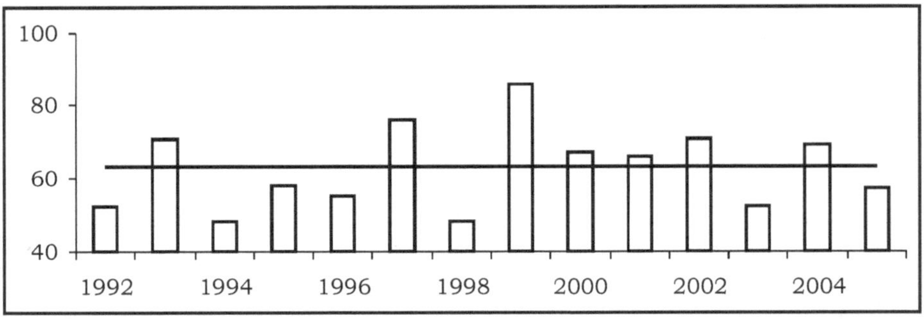

Fig. 7. Distribution of Rainy Days during Monsoon period Database: 1992 - 2005

Post Monsoon Rainfall (October to December)

1. Mean rainfall: 229.4 mm
2. Highest rainfall recorded: 345.4 mm (1995)
3. Lowest rainfall recorded: 83.6 mm (1997)
4. Highest Rainfall in a single day: 121.0 (19th October 1999)
5. Mean Annual Rainy Days: 10 days
6. Highest rainy days in a year: 18 days (2001)
7. Lowest rainy days in a year: 5 days (1997).

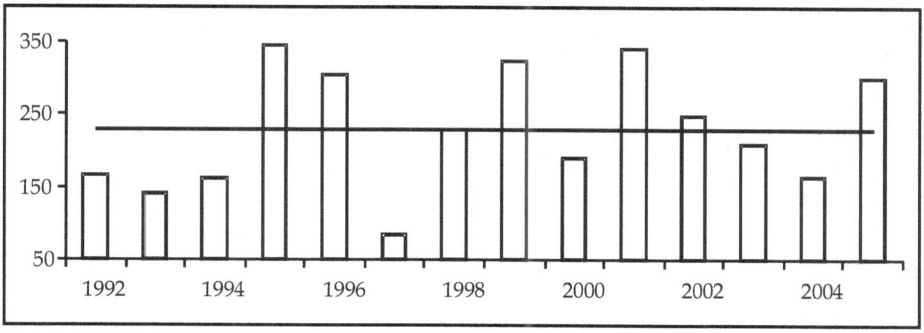

Fig. 8. Distribution of Rainfall during Post Monsoon Rainfall (mm) : Database: 1992 - 2005

Seasonal Distribution of Annual Rainfall

The distributions of rainfall in different seasons are shown in fig 11. It is noted from the figure that about 60 % rain occur during monsoon i.e. from June to September followed by post monsoon season contributing about 10 % of total yearly rain. It is thus indicated that there are good chances of cultivation of rabi crop including vegetables with the help of soil moisture recharged during monsoon. But dry winter indicates soil moisture stress particularly in upland. Artificial recharge of ground water and water harvesting structure may provide a better opportunity of growing of crops particularly in upland. Pre - monsoon rain, which is about 24 % of total yearly rain, also helps farmers to grow rice (both upland and lowland) in early kharif season.

Weekly Rainfall Probability at Lembucherra

The weekly observed minimum and maximum rainfall and computed weekly normal rainfall and the probability of the weekly

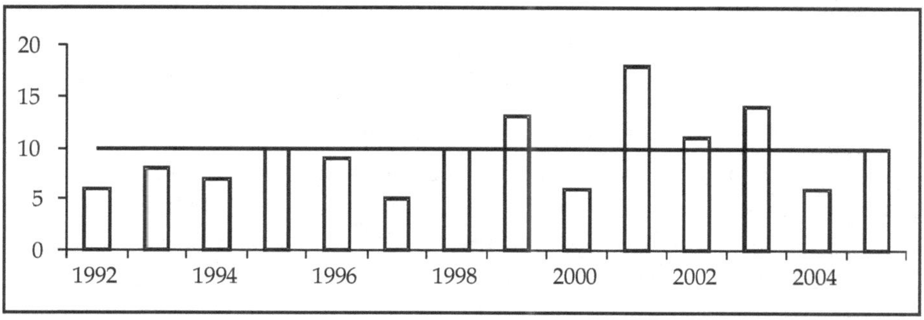

Fig. 9. Distribution of Rainy Days during Post Monsoon period : Database: 1992 - 2005

normal rainfall equaling or exceeding the normal in any year are presented in table 1. It can be assumed that major part of rainfall is received during 20th to 42nd week. The probability of normal rainfall during these weeks is almost 50 % except 37th week where probability goes below 30 %. The probability of weekly normal rain during 4th to 8th week and 49th to 52nd week is very less. This is perhaps due to the fact that these weeks throughout the study period remain dry except few irregular rains, which has been used in computing the probabilities.

Weibull's distribution has been found suitable for Lembucherra (Tripura) for the rainfall analysis. The weekly rainfall at different probability levels of 50, 60, 70, 80 and 90 percent has been computed by using Weibull's method and presented in Fig 10.

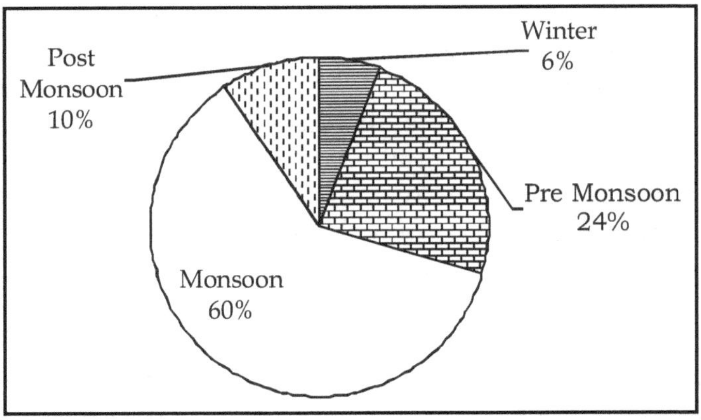

Fig. 10. Seasonal distribution of Annual Rainfall

ONSET AND ENDSET OF WET PERIOD

The results of forward and backward accumulation are presented in table 2. It gives the week number in which different accumulated rainfall amounts occur at different probability level. For forward accumulation, the amount considered are 75 mm and 200 mm and for backward accumulation, it is 200 mm, 300 mm and 500 mm.

The result shows that at 80 % probability level , 75 mm rain occurs in 24th week and 200 mm rainfall occurs in 26th week taking forward accumulation into account. In backward accumulation, 200mm rainfall occurs in 39th week, 300 mm in 36th week and 500 mm rain in 31st week at 80 % probability level.

The further analysis of the above results indicates that pre-monsoon shower in May helps in seedbed preparation. Farmers should

Table 1. Weekly observed minimum, maximum and normal rainfall and the probability of the weekly rainfall equaling or exceeding the normal in any year.

Week no	Minimum (mm)	Maximum (mm)	Normal	Probability (%)	Week no	Minimum (mm)	Maximum (mm)	Normal	Probability (%)
1	0.0	0.0	0.0	93.3	27	22.4	152.0	69.1	50.0
2	0.0	0.4	0.0	93.7	28	19.5	275.2	103.7	31.3
3	0.0	28.2	4.0	37.5	29	15.1	577.0	136.4	31.3
4	0.0	13.0	1.8	25.0	30	4.0	183.2	83.9	50.0
5	0.0	14.0	1.7	25.0	31	16.2	284.7	97.3	31.3
6	0.0	27.2	5.3	25.0	32	9.4	140.7	64.6	37.5
7	0.0	35.2	4.4	25.0	33	0.0	165.4	65.3	31.3
8	0.0	217.0	22.2	25.0	34	0.0	231.9	70.3	43.8
9	0.0	104.4	12.0	31.3	35	12.0	179.8	60.9	37.5
10	0.0	12.0	1.4	25.0	36	8.6	145.8	45.0	43.8
11	0.0	190.4	27.2	37.5	37	7.2	110.4	67.5	25.0
12	0.0	157.2	21.2	31.3	38	0.0	243.6	86.0	43.8
13	0.0	246.2	48.4	31.3	39	0.0	202.8	60.1	37.5
14	0.0	98.0	28.1	50.0	40	12.4	129.8	63.4	50.0
15	0.0	120.2	29.6	50.0	41	0.0	108.4	31.3	50.0
16	0.0	278.2	59.9	37.5	42	0.0	174.2	40.7	50.0
17	0.0	268.6	46.9	37.5	43	0.0	133.9	31.4	25.0
18	0.0	405.8	106.8	31.3	44	0.0	47.6	12.8	31.3
19	0.0	344.0	75.8	37.5	45	0.0	1285.8	19.8	31.3
20	0.2	200.6	95.2	50.0	46	0.0	63.4	5.2	18.8
21	0.0	173.0	76.2	50.0	47	0.0	117.4	9.9	18.8
22	0.0	184.6	84.2	56.3	48	0.0	0.5	0.0	93.8
23	0.0	205.2	105.3	43.8	49	0.0	6.2	0.4	12.5
24	0.0	335.0	96.3	50.0	50	0.0	20.2	1.4	12.5
25	0.0	198.6	126.9	31.3	51	0.0	46.6	3.3	12.5
26	19.2	306.2	98.6	31.3	52	0.0	7.8	0.6	12.5

broadcast their seed before 11th June as 75 mm of rainfall is expected at 80 % probability level and considering 22nd week as critical week. If farmers want to puddle the field and transplant paddy, then they should do it just after 25th June, as 200 mm rainfall is expected at 80 % probability level in 26th week. The farmers of Tripura can grow rabi

Table 2. Start and End of Rainy Season at Lembucherra by forward and backward accumulation

Rank No	Probability	Forward Accumulation		Backward Accumulation		
		75mm	200mm	200mm	300mm	500mm
1	6.67	22	23	41	40	38
2	13.33	22	23	41	40	38
3	20.00	22	23	41	40	37
4	26.67	22	23	40	40	37
5	33.33	22	23	40	39	36
6	40.00	22	24	40	39	36
7	46.67	22	24	40	38	36
8	53.33	22	24	40	38	35
9	60.00	23	24	39	38	34
10	66.67	23	25	39	38	33
11	73.33	23	26	39	36	31
12	80.00	24	26	39	36	31
13	86.67	25	26	38	34	29
14	93.3	25	28	37	33	29

crop during early October to end of December as 200mm rainfall is expected during 39th to 52nd week at 80 % probability level. So, there is a good potential for vegetables during this period.

Yearwise onset and endset of rainy season are presented in Table 3. The result reveals that the earliest date of onset of monsoon at Lembucherra by forward and backward accumulation of weekly rainfall is 28th May while latest is 18th June with mean date of onset is 2nd June. The earliest withdrawal of monsoon is on 10th September while latest is 8th October with mean date of withdrawal is 27th September. Mean duration of monsoon is 117 days and maximum and minimum duration of rainy season is 133 and 84 days, respectively.

Table 3. Start and end of rainy season at Lembucherra by forward and backward accumulation of weekly rainfall (Database 1992 - 2005)

YEAR	Onset week	Onset date	Receding week	Receding date	Duration (week)
1992	23	4th June	38	23rd September	15
1993	22	28th May	39	30th September	17
1994	25	18th June	37	10th September	12
1995	24	11th June	41	14th October	17
1996	22	28th May	40	1st October	18
1997	22	28th May	39	24th September	17
1998	22	28th May	40	1st October	18
1999	22	28th May	41	14th October	19
2000	22	28th May	39	24th September	17
2001	22	28th May	41	14th October	19
2002	23	4th June	40	1st October	17
2003	23	4th June	40	1st October	17
2004	22	28th May	39	24th September	17
2005	25	18th June	40	1st October	15

Mean probable Onset : 22.8th Week (2nd June)
Earliest probable Onset : 22nd Week (28th May)
Latest probable Onset : 25th Week (18th June)
Mean probable Endset : 39.6th Week (27th September)
Earliest probable Endset : 37th Week (10th September)
Latest probable Endset : 41st Week (8th October)
Mean duration : 117 days
Minimum duration : 84 days
Maximum duration : 133 days

So critical analysis of the above factor concludes that, farmers can grow paddy twice in a year (June to September and September to December) as during September to December ,the area gets a rainfall of 300mm at 80 % probability. Moreover, the farmers can also grow winter vegetable without any moisture stress during September to December.

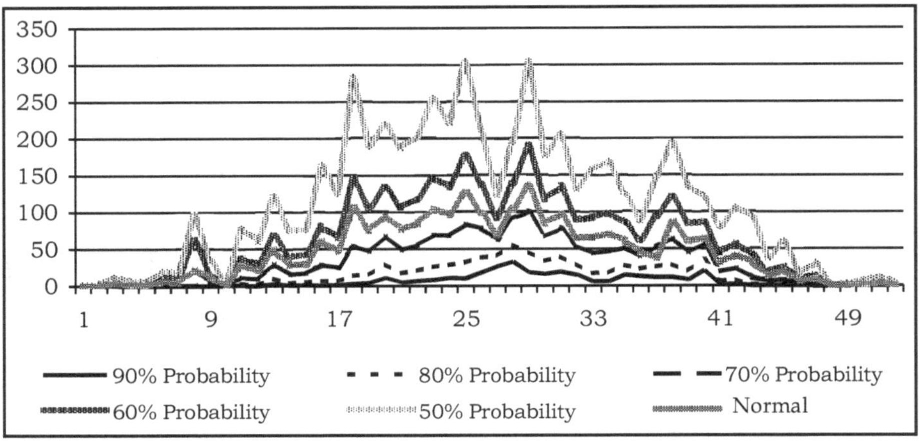

Fig. 11. Distribution of Weekly Rainfall at Different Probability Level

CONCLUSION

From the study as conducted, it is concluded that the annual rainfall of Lembucherra is 2409.6mm distributed by 101 rainy days. About 60 percent of the annual rainfall occurs during south west monsoon. Pre monsoon period contributes about 24 percent of annual rainfall. Winter is almost dry. The monsoon starts on first week of June and continues upto end of September with a mean duration of 117days. During monsoon period, the area gets about 1470 mm rainfall. Farmers should broadcast their seed before 11th June and transplant paddy after 25th June, as 200 mm rainfall is expected at 80 % probability level in 26th week. The farmers of Tripura can grow rabi crop during early October to end of December. They can grow for two short duration rice crops in a year i.e. from June to September and from September to December. Crops like mustard, groundnut and other oilseed crop can also be grown during October to December. Vegetable and floriculture should be undertaken during these months. Probability of getting dry week is very high during 1st to 13th week and 43rd to 52nd week.

REFERENCES

Agnihotri, Y.(1993). A study on occurrence of wet/dry spells and weather cycles at Chandigarh using Markov chain model.

Ashokraj, P. C. (1979). Onset of effective monsoon and critical dry spell. IARI Research bulletin no. 11, WTC, IARI, New Delhi, pp: 6-18

Chaurasia, R., Minakshi and Sharma, P.K. (1995). Climate and its Variation in Ludhiana, *Indian J. Soil Cons.* 23(1) :1-5, 1995

Dash, M. K. and Senapati, P. C. (1992). Forecasting of dry and wet spells at Bhubaneswar for agricultural planning, *Indian J. Soil Cons.*, 20; 1&2, pp: 75 - 82

Khambete, N. N. *et al.* (1978). Characteristics of short period rainfall in Gujurat. *Indian Journal of Metrology, Hydrology and Geophysics.* 23(3), pp : 521 - 527

Raman, C. R. V. (1979). Analysis of commencement of monsoon rains over Maharashtra state for agricultural planning. *Scientific report No* - 216, IMD, Pune.

Rath, H et-al. (1996). Forecasting of Dry and wet spells at Boudh for agricultural planning, *Indian J. Soil. Cons.*, 24(1):28-36.

Satapathy, K.K.; Jena, S.K. and D. Daschaudhuri, (1998). Characteristic of Monsoon & rainfall pattern at Umiam, *Journal of Soil & Water Conservation;* 42; 154-161.

Satapathy, K.K.; Jena, S.K. and D. Daschaudhuri, (1999).Climate and its variation in Umiam, Meghalaya; *Indian J. Soil Cons.*; 27(2), 166-170.

Satapathy, K.K.; Jena, S.K. and D. Daschaudhuri, (1999). Climate of Umiam; *Research Bulletin,* ICAR Research Complex for NEH Region,Umiam , Meghalaya 793 103.

Tiwari, K. N. and Saxena, A. K. (1987). Dry spell and wet spell analysis with reference to water harvesting - A case study- IV. *Indian Journal of Power & River Valley Development,* pp: 295 - 301.

W.M.O. (1982). *Technical Note No.* 179:149-158.

Climate Change and Food Security, 2008
Edited by: M.Datta, N.P.Singh & Er.D.Daschaudhuri, pp.215-227
New India Publishing Agency, New Delhi (India)
E-mail : newindiapublishingagency@gmail.com
Web: www.bookfactoryindia.com

15

Scope of Agro-meteorological Advisory Services in Manipur

S.V. NGACHAN AND P.T. SHARMA

INTRODUCTION

The state of Manipur is located at the easternmost border of India and is characterized by flat valleys comprising of four valley districts (Bishnupur, Imphal East, Imphal West and Thoubal) surrounded by five mild to steep slope hill districts (Chandel, Churachanpur, Senapati, Tamenglong, and Ukhrul). The state is geographically bordered by Nagaland in the north, Burma in the east, Mizoram in the south and Assam in the west. The total geographical area of the state is 22,327sq.km divided into 20,089sq.km in the hills and 2,238sq.km. in the valley. More than 98% of the population in the hill is in rural areas whereas around 60% of the population in the valley is in rural and around 40% are in the urban areas.

The AAS Unit, Imphal is located at $24^{0}45'$ N latitude, $93^{0}54'$ E longitude at an altitude of 774m above msl. The agro-climatic zone of the state covers sub tropical plain zone, sub tropical hill zone, temperate hill zone and sub temperate zone.

PRESENT STATUS OF AGRO-METEOROLOGICAL SERVICE

The weather condition is localised due to undulated hilly topography and varies at short distances. Variation of weather condition at a short distance of 3-4 km. is also observed at many places especially rainfall and temperature which are important meteorological parameters in agriculture. For example, there was report of hailstorm at different places in the valley on 24th December, 2005 but no rain was ever recorded at our observatory at Lamphelpat itself. In January, 2004 there was frost occurrence for a period of two weeks in low to high altitude areas as well as in plain sub tropical zones. Normally, the hill zones such as Tamenglong, Senapati and Ukhrul receives maximum annual average rainfall of 2000-2300mm with an exceptional record of 4000mm in Tamenglong. Part of Thoubal and Bishnupur districts, majority hill areas of Churachandpur and Chandel districts receive less than an average of 1200mm rainfall annually. While the Imphal East and Imphal West districts plain zone receive an average of 1500mm rainfall annually.

At present, the agro advisory service in the state is located only at ICAR Research Complex, Manipur Centre, Imphal. Hence, the accuracy of weather forecast and advisory is limited in Imphal East and West only. Majority of the hill districts cannot be covered and vagaries of monsoon in such areas require expansion of automatic weather stations for accurate analysis of meteorology in the agriculture of Manipur.

Current meteorological data are at present not available for different zones/districts and hence various weather information and weather forecast are not effective for far places and hilly areas. However, six more numbers of automatic weather stations are in the process to be installed at different places of the state in collaboration with North Eastern Space Application Centre, Meghalaya.

Temperature

The normal maximum temperature varies from 21.5^{0}C during January to 29.1^{0}C during June and August. The variation of maximum temperature during the last few years is indicated below (Figs. 2-6). The maximum temperature in most of the months (Fig. 2) of the years 2001, 2002 and 2003 is observed more than the mean average maximum temperature whereas the maximum temperature for the years 2004 and 2005 is observed lower than the mean average.

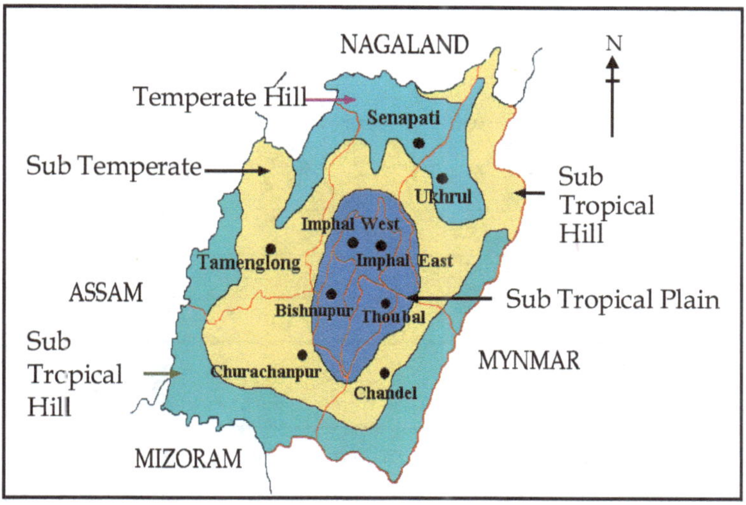

Fig. 1. Agro Climatic Zones of Manipur

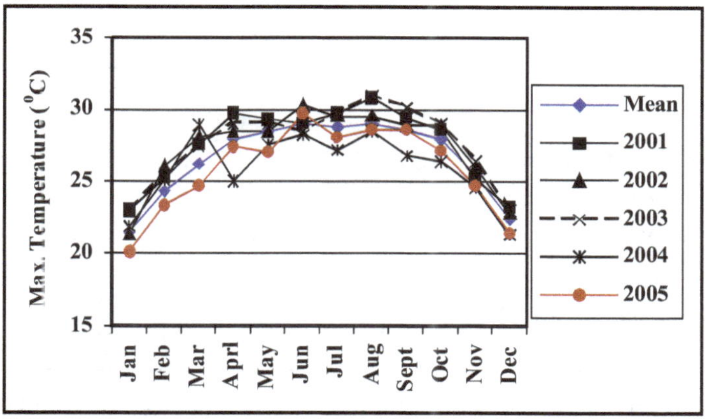

Fig. 2. Variation of Maximum Temperature during 2001-2005

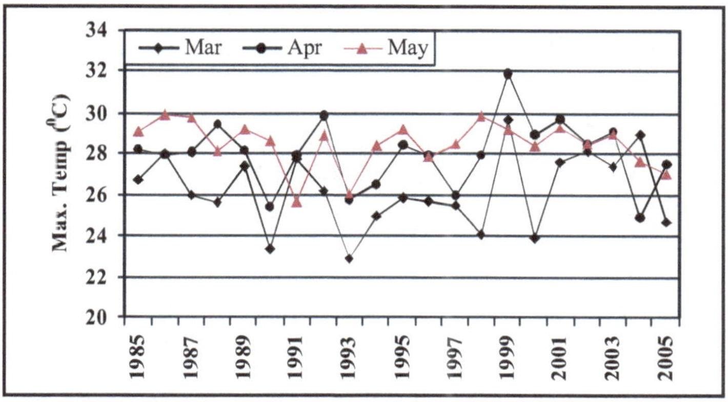

Fig. 3. Variation of Maximum Temperature during Summer

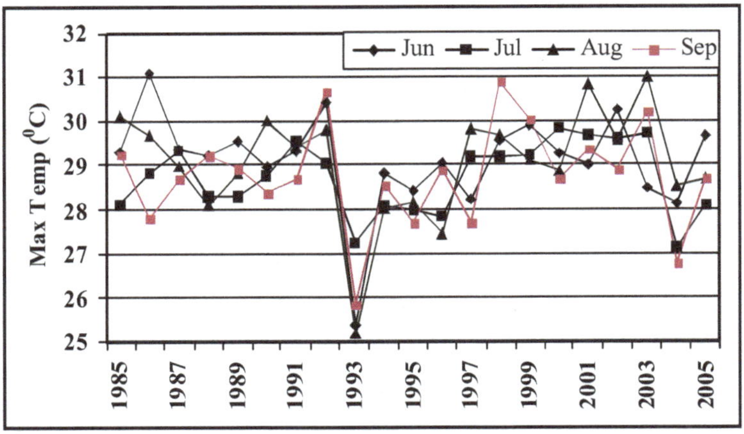

Fig. 4. Variation of Maximum Temperature during Monsoon

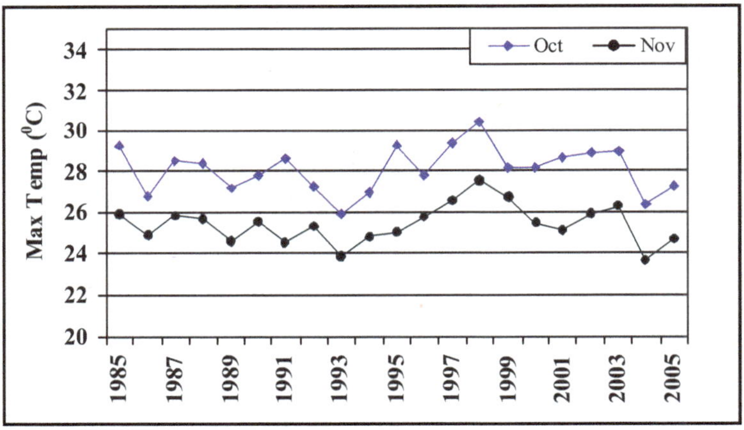

Fig. 5. Variation of Maximum Temperature during Post Monsoon

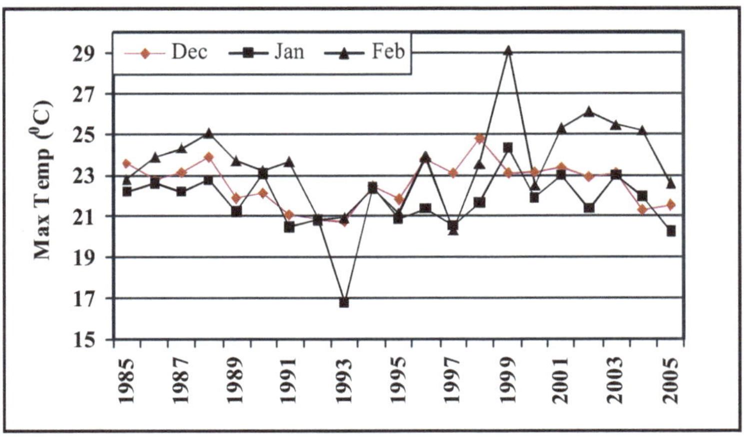

Fig. 6. Variation of Maximum Temperature during Winter

The monthly average minimum temperature recorded during the past few years is presented in Figs. 7-11. The difference in minimum temperature during 2001-2005 as compared to the normal temperature is shown in Fig. 7. It is observed that the minimum temperature for most of the months during the years 2001-2005 is more than the normal temperature.

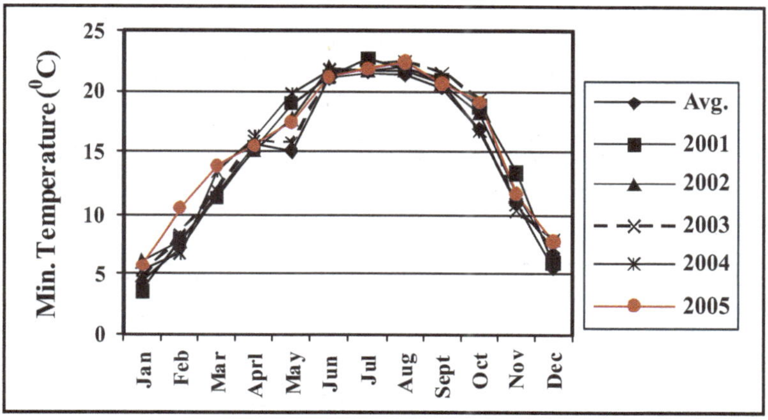

Fig. 7. Variation of Minimum Temperature during 2001-2005

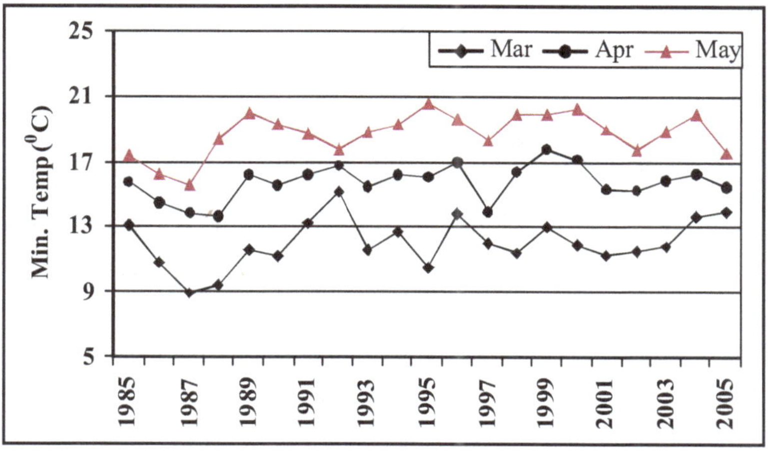

Fig. 8. Variation of Minimum Temperature during Summer

During 2005, the monthly average minimum temperature for the months of January, February and December are 5.6^{0}C, 10.5^{0}C and 7.7^{0}C respectively as compared to the normal minimum temperatures of 4.2^{0}C, 7.1^{0}C and 5.5^{0}C, respectively. It is ,therefore observed that the minimum temperature is increased to a considerable rate during the last few years.

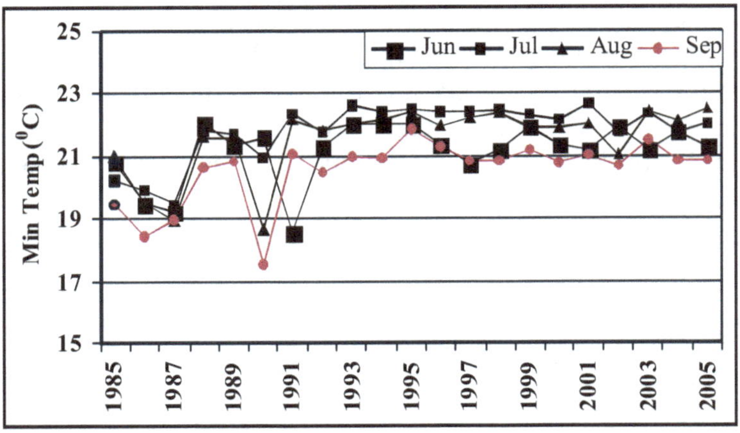

Fig. 9. Variation of Minimum Temperature during Monsoon

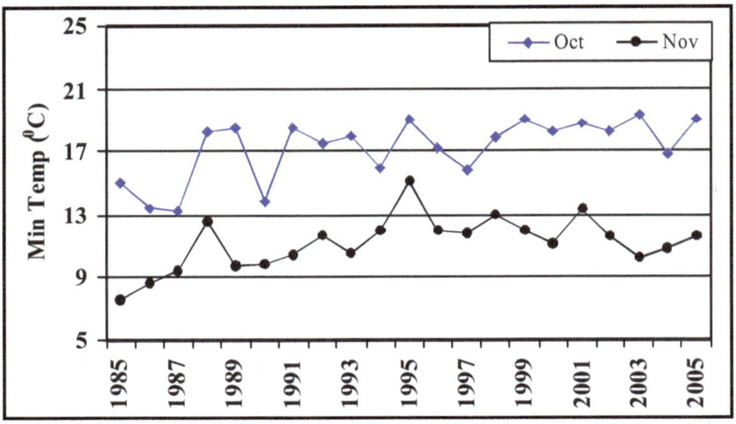

Fig. 10. Variation of Minimum Temperature during Post Monsoon

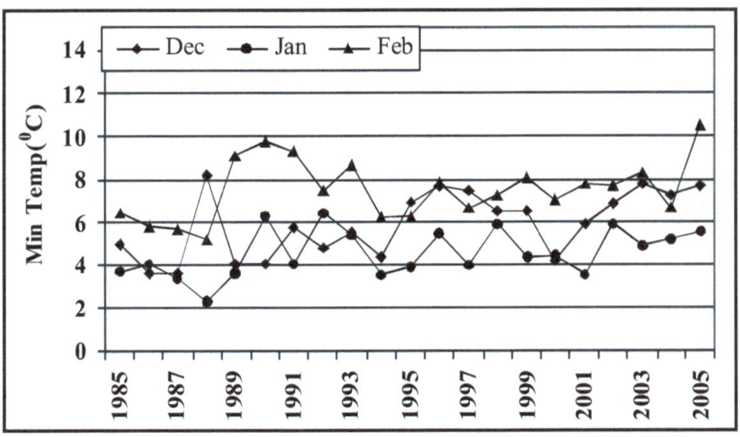

Fig. 11. Variation of Minimum Temperature during Winter

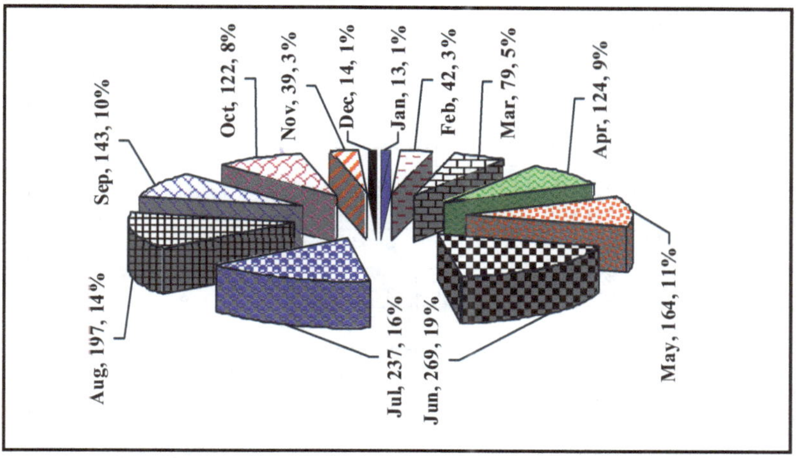

Fig. 12. Monthly Average Rainfall for The Period of 1954-2005

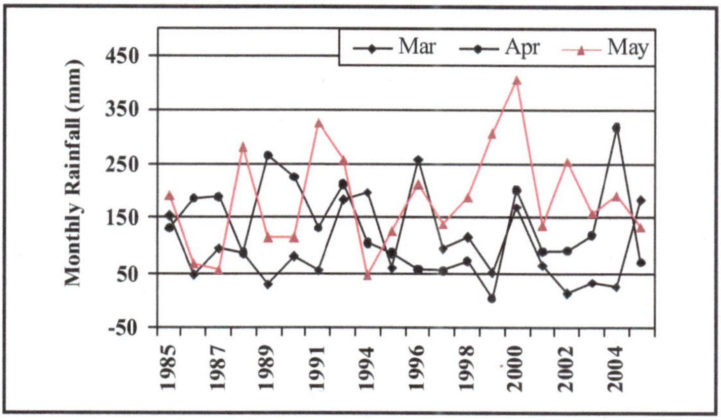

Fig. 13. Variation of Monthly Rainfall during Summer

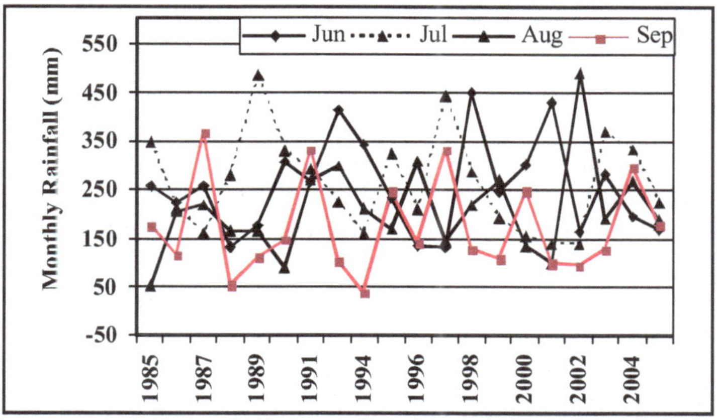

Fig. 14. Variation of Monthly Rainfall during Monsoon

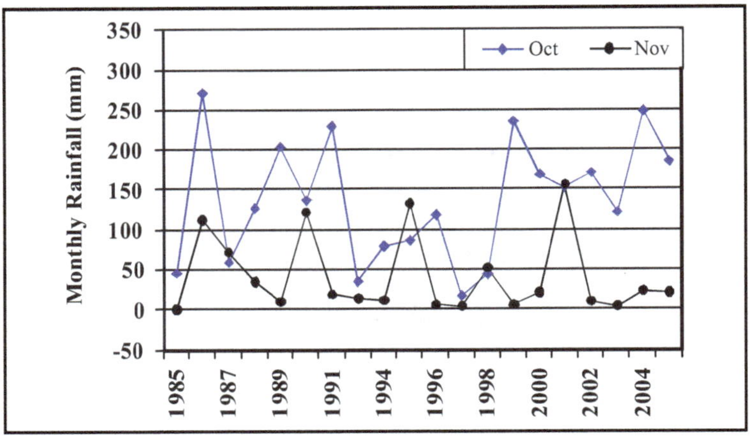

Fig. 15 Variation of Monthly Rainfall during Post Monsoon

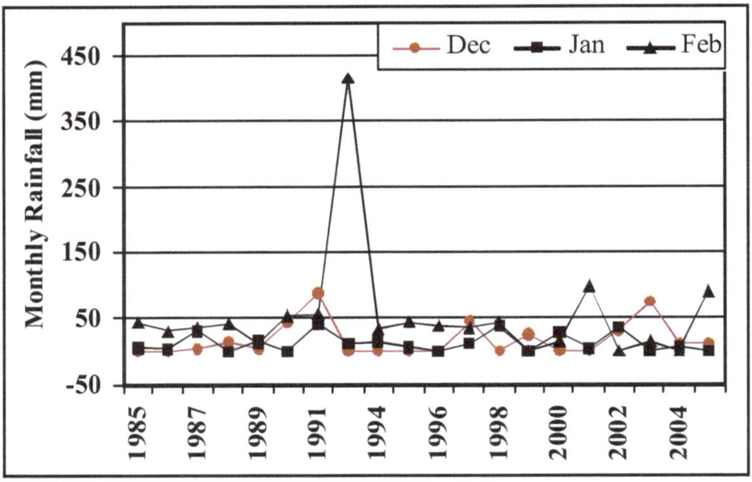

Fig. 16 Variation of Monthly Rainfall during Winter

Rainfall

The annual average rainfall recorded at Imphal area is around 1450mm varying from 929 mm during 1979 to 2172 mm during 1993. However, the annual rainfall sometimes exceeds 4000 mm in the Tamenglong district of Manipur. The normal monthly rainfall distribution pattern is presented in Fig. 12 and variation of monthly rainfall for different years is given in Fig. 13-16. Maximum monthly rainfall of about 271mm occurs during June followed by around 238mm during July. The average rainfall during the months of January, February and December are 12.7mm, 42.3mm and 13.8mm, respectively. However, no rain was recorded during December 2001, February 2004, January 2005 and January 2006 and only 1.5 mm during February 2002 was recorded.

The amount of rainfall and onset of monsoon is different in different parts of the state. The amounts of rainfall as well as distribution pattern in these areas are not yet fully known.

Relative Humidity

High relative humidity prevails during the morning as well as afternoon hours throughout the year. The average relative humidity during the morning hours varies from 67.8% in March to 90.1% in December. However, it is more than 70% in most of the months. The average relative humidity during the afternoon hours ranges from 49.6% in February to 77.0% in August.

MAJOR CROPS GROWN

The major cropping season in the plain areas of the state is classified as pre-kharif (Feb-Mar to Jun-Jul), Kharif (Jul-Aug to Nov-Dec) and Rabi (Oct-Nov to Feb-Mar). However, the cropping season in the hilly areas varies according to the altitude and two crops mainly kharif and rabi/ pre-kharif are grown. In the plain areas (sub tropical plain zones), two crops of paddy viz. pre-kharif and kharif crops are grown. In hill areas, monocrop paddy (kharif) or monocrop vegetables (rabi) are grown. In mid hill zones, maize followed by soybean/ mustard/ ricebeans are grown. Pulses, vegetables (cabbage, cauliflower, potato, tomato, Brinjal etc.), pineapple, banana, citrus, papaya, mango etc. are also grown as single crop.

Although production and productivity variation of food grains is observed from year to year due to vagaries of monsoon, agro-meteorological advisory services have greatly helped in management of irrigation, pests and disease control, management of floods and untimely distribution of rain etc. Most of the crops in the state are rainfed including rice. The yield of paddy during the past few years is correlated to the amount of rainfall except in cases of flood and untimely distribution of rain whereas the yield of maize does not seem directly correlated to the total annual rainfall (Fig. 18). In maize, intermittent rain followed by rainless days (of 7-10 days) attracts maximum weevils and cob borers in June/ July. Hence, the critical period for maize cultivation during the cob formation has been brought under agro-meteorological services. For example, in 2003-2004, total failure in maize cultivation was observed due to such pests in sub tropical hill zones where maximum population of weevils and cob borers multiply at the temperature ranging from 280C to 340C and high humidity of 80% to 95%.

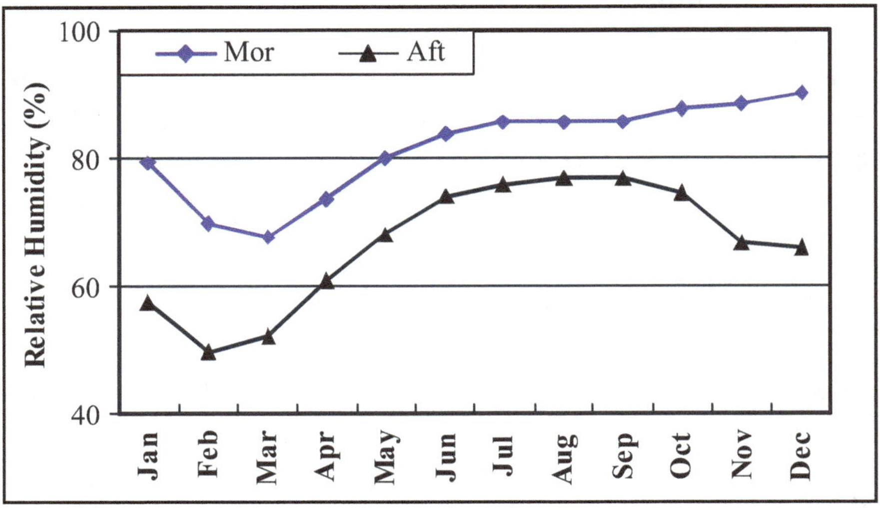

Fig. 17. Variation of relative Humidity at Imphal

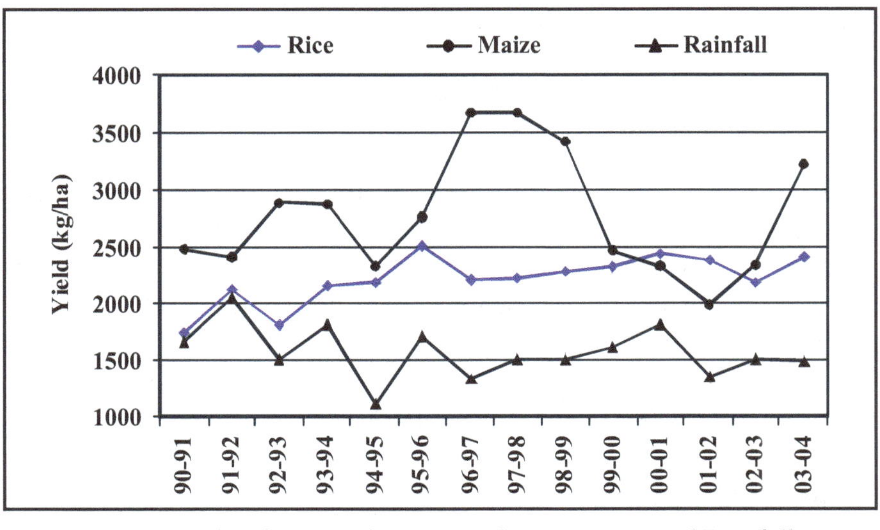

Fig. 18. Yield of Rice and Maize with respect to Total Rainfall

Rice

Rice is the major cereal crop of Manipur cultivated in 157.9 thousand hectares producing 381.2 thousand tonnes during the year 2003-04. About 58% of the total cultivated areas in the plain are occupied under high yielding varieties. Pests and disease control by cultivating resistant varieties and advancing the date of sowing from middle June to first week of July is advocated. These practices escape many pests and diseases including stem borer, blast disease of rice. In less rainfall areas (average annual rainfall 1000-1200mm), e.g. part of Thoubal district, Churachandpur and Chandel districts suitable variety - RC Maniphou 7 (photosensitive variety) records maximum potential yield of 8.2t/ha (average of 5.6 t/ha) against state average of 2.1t/ha. Similarly, in upland jhum fields, Bhalum-1 and Bhalum-2 are found resistant to many pests and diseases with a potential yield of 4.2 t/ha against state average of 0.9t/ha. Timely weather forecast through bulletins, radio and daily newspapers renders major services to the farmers in Manipur. A critical weather period in September end and first week of October for blast infestation has to be taken care for susceptible varieties by forewarning of meteorological data through agro-meteorological advisory services.

Rice is cultivated mainly as monocrop during the rainy season. Most of the crop in the state is dependent on rain and irrigation facility is very low. Damages of crops due to pests and diseases are also a major problem faced by the farmers. ICAR Research Complex, Manipur Centre has developed some of the high yielding and disease resistant varieties of rice suitable for different season as in Table1.

Table 1. High yielding varieties of rice developed at ICAR, Manipur

Variety	Location	Season	Maturity Period (d)	Yield (t/ha)	Optimum sowing season
RCM-5	Upland Rainfed condition	Pre-kharif & Kharif	115-120	4.5-5.0	May to June
RCM-7	Valley	Pre-kharif	110-115	4.5-4.8	Feb to March
RCM-8	Valley	Pre-kharif	110-115	5.0-5.5	Feb to March
RCM-9	Valley	Kharif	135	5.5	June to 1st week of July
RCM-10	Valley	Kharif	128	6.4	June to July

Source : Agricultural Technologies for Manipur, ICAR, Manipur Centre, 2000

Gall midge is also frequently observed at many parts of the state reducing the yield of rice significantly. Other major pests of rice in the state are rice stem borer which can destroy upto 95% of the grain yield. Brown plant hoppers in severe cases destroy the crop considerably. The most favourable condition for development and survival of the nymphs is at around 29^{0}C and relative humidity of 60-80% which is in the range of the weather of the state. It is reported that rice leaf rollers have capacity in destroying the crops upto 80% of the grain yield.

Vegetables

As the weather in the state is mostly humid and wet throughout the year, it is favourable for occurrence of various pests and diseases. Damping of disease is very common in nurseries of different vegetables. The loss due to this disease is considerable due to inadequate control measures. The loss may be minimized by giving suitable and timely advisories to the growers. The state produces about 4,632 thousand tones of tomato from around 579 ha of land. The crop is frequently damaged due to powdery mildew and late blight thus reducing the yield to a great extent. Potato is also commonly grown as a winter crop in a large area of the state. There are frequent frost occurrences during the winter season. Cold weather forecast in January and February is highly essential in saving tomato and potato crops by irrigating and sprinkling water in early morning before sunrises as the crops are highly susceptible to frost.

There are such other numerous pests and diseases which destroy other crops. As the farmers are not aware of the possible extent of losses, dissemination of timely precautions and effective control measures are highly essential.

CONCLUSION

There are many crop losses due to effect of weather at different parts of the state. Heavy loss of paddy yield is commonly occurred during flood and drought years. Alternatives like disease resistant, short duration and high yielding varieties are essential in such conditions. The weather of this state is favourable for many pests and diseases in crops and livestocks. Blast in rice, powdery mildew in pea, fruit rot, damping off in nurseries etc. are some of the major diseases. Monitoring and analysis of weather report to correlate pests and disease occurrence, farming systems, diversification of crops as per the weather condition etc. are required to be covered in all the areas of the state rather than a single advisory centre at ICAR, Imphal west. There is a need of agro advisory services to cover the hill regions where maximum climatic variation is observed. The selection of weather based seasonal crop, disease and pests resistant varieties, scientific methods of sowing, planting, nursery preparation, irrigation and water harvesting methods, plant protection measures, harvesting and storage of tuber, seed and grains to avoid from field and post harvest losses along with other methods of farming like crop diversification, organic farming, paddy cum fish culture etc. will largely depend on successful implementation and dissemination of agro advisory services to the farmers of the state.

Climate Change and Food Security, 2008
Edited by: M.Datta, N.P.Singh & Er.D.Daschaudhuri, pp.229-241
New India Publishing Agency, New Delhi (India)
E-mail : newindiapublishingagency@gmail.com
Web: www.bookfactoryindia.com

16

Late Quaternary Environmental Changes and their Impact upon Coastal and Paleo-coastal Bengal Basin in India A Climato-Geomorphic Study

SUDIP DEY

INTRODUCTION

It is well accepted that the climatic changes also influence the geomorphic process efficiency and landform development (Ahamad, 1985). Glaciation, sea level changes, flooding and many other geo-environmental events are related with global and regional climatic changes along with tectonic and other non-tectonic factors like natural vegetation, soil, organic development etc. Considering the role of climate on landform development and dynamics of the Late Quaternary climate, a climato-geomorphic study was conducted on the coastal and paleo-coastal areas of the Bengal Basin namely-lower Western (West Bengal coast) and Eastern (Tripura Paleo-coastal Fold Belt Province) Geo-provences. The main objectives of the study were to investigate (i) Development of the landforms during Late Quaternary period with sea level changes and (ii) Climatological influences on the landform modifications.

PHYSICAL BACKGROUND OF THE STUDY AREA

Coastal parts of Western Geo-province

This is the coastal stretch of Indo-Gangetic plain (G.S.I., 1995). This area has been nourished by alluvial depositions of river Subarnarekha, Ganga and their tributaries within last 6000 years (Niyogi, 1970; Bandyopadhyay, 2000; Paul 2002). Some early scientific studies (Niyogi, 1970; Chatterjee,1972; Umitsu, 1987; Merh, 1987; Banerjee and Sen, 1987; Chanda, S and Hait, A.K, 1996; Roy and Chattopadhyay, 1997, Banerjee,.1998; Hazra *et al.* 2002) also show that the shoreline of the Bay of Bengal remained very dynamic throughout the Holocene epoch (Dey, 1999) shifting some 90 km to 120 km northwards from the present position (Dey and De, 2002; Dey *et al.*, 2002).

Paleo-coastal parts of Eastern Geo-province

The lower geological formation of the eastern geo-province or Tripura Fold Belt Province shows that this area started to develop during the Upper Tertiary period under marine environment. The geological succession of this area is shown in Table 1.

Table 1. Geological succession of Rudrasagar Lake Area

Geological period	Epoch	Time (Ma years)	Group	Lithology
Quaternary	Holocene	0.01	Alluvium	River born clayey silt, clay with silt and sands.
	Pleistocene	1.6		
Tertiary	Pliocene	5.3	Dupi Tilla	Estuarine brown clayey sand, mottled silty clay, white to grey sandrock with silt bands and white clay.
			Tipam	Marine coastal and estuarine sand rocks with pockets of shale, clay, brown sand shale and fossil wood.
	Miocene	23.3		
			Bokabil	Marine coastal shale with minor sandstone.

Source: Geological Survey of India, (2004)

On the basis of the literature review and field investigation, it can be assessed that during the Miocene epoch, the Tripura Fold Belt Province was dominated by the marine and coastal environment. Evidences of marine and coastal sediments such as the Bokabil Formation and Tipam Group and depleted marine and coastal vegetation roots, which are preserved beneath the surface, strongly support the geological history of 25 million years since the Miocene period. The bedding and ripples in the Bokabil Formation and Tipam Group strongly support the fact that the environmental conditions were different at this Lake during the late Tertiary and early Quaternary periods. Sediment deposition during the Miocene epoch in the Bengal Basin plays a vital role for the shifting of the shoreline. Tectonic uplift and climatic changes might play additional roles with regards to the relative fall of sea level. Research by Nandi (1985) proves that sediment deposition occurred under warmer climatic conditions than recent times.

METHODOLOGY

The existing literature such as reports of Geological survey of India, Digha Development Authority, Department of Environment of Government of West Bengal etc and recent research papers published in different journals and presented in different seminars, congresses etc were considered to prepare a complete fieldwork programme.

Historical literatures often preserve some valuable evidences of physical environmental change as historical events are influenced by environmental conditions. To assess the early shoreline position and impact of environmental changes upon this coastal area through historical period, some existing relevant historical records of this area available from different literatures were consulted. There are also some available antic cartographic works and charts, which were used as strong evidence of early shoreline changes.

FIELD WORK

The field work was conducted from January 1999 to July 2005 in different places of West Bengal coastal areas and Southern Tripura. Different techniques have been applied and instrumental survey works have been done during fieldwork to find out the geomorphologic evolution of this area. The aims of the field work were- i) To find out the geological and geomorphic evidences of long-term evolution of landforms and ii) To detect the climatological variations of the different geomorphic units and its impact upon long-term development. Geomorphic units of these places were surveyed on the basis of process efficiency. Various field maps like geomorphic maps, dune extension map, present shoreline map etc were prepared during field work.

A wide area, upto 40 km inland from the present shoreline in the Eastern Geo-Provence was surveyed in search of the evidences of early shoreline positions and paleo-environmental conditions of this coastal tract. Marine shells and mangrove roots were collected from various deep tube-well holes and bore-holes of thermal power project of Kolaghat-Tamluk region (21°31′ N to 21°55′ N latitudes and 88°11′ E to 87°38′ E longitudes). Remains of depleted roots of mangroves were also collected beneath the sand of the present coastal area for laboratory testing.

LABORATORY TESTING AND MAPPING

The samples collected during field survey were tested in laboratory. Chemical tests and detection of ages of the collected materials were done in the chemical and geological laboratory of the Department of Ground Water Survey of Government of West Bengal, Mednipur branch. Geomorphic data collected through field investigation during various seasons have used for preparing various maps. Pre-Holocene geological development was drawn on the basis of field data and some relevant literature. Some antic maps and literature were consulted along with field data for detection of coastal development during last 6000 years. Geological and geomorphological map of Digha area (Scale-1:50000), current landuse pattern and disaster management plan of Digha coast (Scale-1:50000), present landuse map for Digha planning area (Scale-1:50000) published by the Geological Survey of India (1995). landuse plan for 2011along Digha planning area (Scale-1:50000, published by Digha Development Authority (1996), Survey of India toposheets no.-73 O/06, 73 O /10 and 73 O/ 14 (Scales: 1:63360 of 1931-'32 and 1:50.000 of 1968-'69), and Satellite imageries IRS-IC, LISS-3 Geocoded FCC Nos-7306, 73010 and 73014 (Scale:1:50000, dated-6.1.1997) published by National Remote Sensing Agency were also consulted for the preparation of final maps. Early coastline position map, existing geomorphic map, extension of sand dunes and vegetation cover map and shoreline shifting map were prepared by using sophisticated Remote Sensing and GIS software for this purpose.

RESULTS AND DISCUSSION

Late Quaternary Sea Level Changes along the Bay of Bengal Coast

Sea level changes obviously have a direct relationship with climatic changes. Evidence indicates that a gradual sea level rise accompanied accelerated coastal erosion along the most of the coastal tract throughout the world. Relative sea level rise, product of combined eustatic and isostatic changes contributes to the process of submergence and erosion in the time scale of centuries to millennia. Besides the phenomenon of long-term sea level change, the seasonal nature of climatic conditions in this part of tropical area exercise substantial impacts upon the sea level change. Seasonal sea level changes can accelerate erosion over a time scale of 3 to 4 months.

Table 2. Coast land development of Western Geo-province with climatic change.

Geological period	Years before present	Global climatic significance	Sea level change	Shoreline position (respective to present shoreline)	Geomorphic and geological significance
Holocene	30	Rapid rise of temperature	Rise	1000m to 1500m seaward	Present coastal geological and geomorphic units developed during this period. Present beach, dunes, older dunes, spill basin, active inter tidal mudflats and paleo-intertidal mudflats have been developed during this period.
	200	Temperature rise after little ice age.	Rise	6000m seaward	
	6000	Cooling after Climatic optimum	Fall / Stable	15000m to 2000m inland	
	8000	Climatic optimum	Rise	Inland position (data not available)	Early coastal areas lost marine environmental condition, as presently that is situated 15-40 km inland. Geomorphologically those are now inland fluvial plain though evidences of marine environmental are found 20-90 m under the surface
Pleistocene	82000	Ice	Fall	40000 m to 45000m inland	
	215000	Ice	Fall	80000m to 90000m inland	

Date source: Niyogi (1968), Goswami (1998) and Field data

Due to early fall of sea level in the early Pleistocene ice age , the coastline of Bengal advanced nearly 80-90 km seaward. Evidence of marine coastal sediments, mangrove roots and marine shells are found below 26.6m depth under the surface in Kolaghat-Tamluk region (21°55′ N to 21°31′ N latitudes and 87°38′ E to 88°11′ E longitudes), which indicate the previous position of shoreline during the Pleistocene era (Chanda and Hait, 1996, pp-117-124). The present geomorphic divisions like the beach, active dunes, mudflats etc of Digha-Shankarpur coastal tract has developed within last 6000 years (Paul, 2002, p-78) with the last sea level fall after Holocene climatic optimum through sand deposit by Subarnarekha river (Bandyopadhyay '2000, P-17).

Late Holocene Shifting of Shoreline in Western Geo-province

The Geological Survey of India has classified the present geological units of Mednipur coastal area into three classes, namely, I). Ancient Digha-Junput coastal deposits (developed during 5870 ± 140 Yrs. B.P), II.) Older Digha-Junput coastal deposits (developed during 2920 ± 60 Yrs. B.P.), and III.) Recent Digha-Junput coastal tract (Geological Survey of India,1995, p-3). Their position and period of development is presented in Table 3.

Table 3. Present coastal geological Units and their development through time and space

Geological Units Period of Development	Position
Recent Digha -Junput Coastal Deposit	Around 2km. From the sea
Older Digha-Junput Coastal Deposit	Contemporary Beyond 2km from the sea to 5km. From the sea.
2920±60 Y.B.P. Ancient Digha-Junput Coastal deposit. –5870± 140 Y.B.P.	Beyond 5km from the sea to 15 km From the sea.

Data Source: GSI Report 1995

Remains of depleted mangroves under the present sandy beach in this area are marked evidence of sea level change during late Holocene period. Early fall of sea level of this area about 6000 years B.P can be attributed to the gradual disappearance of mangroves as with the fall of sea level the mangroves of this coastal tract lost contact with the tidal environment. Evidences of mangrove roots under sand or mud (1-20m below the surface) along the beach of Digha-Shankarpur coastal tract strongly suggests that some mangrove species existed till 150-200 years B.P (Chowdhury, 1999, P-18-20) and macrotidal environment existed in this mesotidal coast.

LAND MODIFICATION IN EASTERN GEO-PROVINCE DURING LATE QUATERNARY PERIOD

The late Quaternary period is marked by several climatic changes, which might play an important role for the development of the geomorphology of the present study area. During the late Pleistocene epoch, additional discharge in the rivers occurred everywhere and many lakes and wetlands were formed in the lower parts of this area (Dey *et al.,* 2002). Since 9000 years BP to 8000 years BP (a period corresponding to the Holocene Climatic Optimum) ,the average temperature was nearly 2°C higher than at present and there was also higher precipitation (Siddhartha, 2001), which might have caused flooding in lowlands. The additional discharge of the rivers of this area in the eastern geo-province discharging in the Lakes modified the landform and the natural lake basins and lowlands (locally known as "Lunga").

The Rudrasagar Lake is one of the main evidences of this event. Rudrasagar Lake preserves significant evidence of early climatic changes in Tripura region. Global climatic changes and regional tectonic changes during the Tertiary and Quaternary periods can be considered as the main factors regarding the origin of this Lake and Basin. In this regard ,it can be said that this Lake area is one of the younger geomorphic units of the eastern province of the Bengal Basin and presents a problem regarding its development. It is evident that the Lake developed until very recently without any human interventions. The evolution of this area can be divided into three phases of sedimentation namely: (i) Early Miocene: Bokabil Formation (ii) Mid-Miocene to Pliocene: Tipam and Dupitila Formation (iii) Quaternary: Recent fluvio-alluvium. The additional discharge of the River Gumti and other streams discharging in the Rudrasagar lake during the Late Quaternary Pluvial or the Holocene Climatic optimum Lake modified the Lake basin and the lake which had developed.

Presently existing lowlands or "*Lunga*" lands provide strong evidence of river erosion on comparatively less consolidated sediment beds during the late Pleistocene and early Holocene. These are then formed by the erosion and further deposition of recent alluvium. Tectonic activities are believed to have resumed even in recent times (Nandi, 1977) and have exerted significant influence on landform evolution.

CHANGING OF RIVERS IN LOWER WESTERN GEO-PROVINCE

Holocene climatic changes along with tectonic changes also influence to change the paths of many rivers of Bengal basin. It is commonly known that with the increase of Kinetic energy[1] (K), which is related with climatic conditions like rainfall and temperature,the value of meander factor[2] (M_f) is decreased. In these cases ,the numbers of meanders are decreased and the river channel shifts towards its source. Here, two rivers of the western geo-provinces- i) Subarnarekha and ii) Ganga were studied.

The river, Subarnarekha is originated from Chhatanagpur Plateau and developed its estuary along Bay of Bengal coast in the Western Geo Province. According to Pal (2002), this river shifted westward over the last 8000 years and left many mouths. That process might start during the Holocene Optimum with extra discharge. Pal{2002} also accorded that Digha inlet might be the one of the relict mouths of Subarnarekha. The Digha Mohona was examined and the following points emerged.

i) Digha river is only one Km long but its basin is almost 47 km^2 which is unmatched with its length.

ii) The width and the depth near the inlet is also unmatched with the length and average width of the river.

iii) From LISS-III data, it has also been observed that the amount of sediment deposition is very high for a small tidal inlet like Digha river. This is really not matching with this tidal river.

Lower part of Ganga is also marked by westward shifting of its channels. It is well accepted that that Adi Ganga (Tolly Nala) was the main flow of this river only 600-500 YBP. This can be a good evidence of temperature rise and increase of river energy by melting of glaciers in the Himalayan highlands. Pahtick (1982) and Banerjee (1998) also supported the increase of river discharge over the last 500 years in Ganga-Brahmaputra valley. Both of them accorded that due to heavy sediment deposition on the Ganga-Brahmaputra valley over the last 500 years ,this part is subsiding eastward and as a result of that, there is a relative rise of sea level along the eastern parts of the coastal Bengal Basin while a tilting occurred to the west which causes a relative fall of sea level to the western parts of the coastal Bengal basin.

SUPPORTING EVIDENCE OF TEMPERATURE RISE OVER LAST 500 YEARS

Mednipur Coast

The historical records show that the shore line along western geo-province is shifting landward due to gradual rise of sea level over the last 400 years. According to a government report in the 18th century, a small village named Birkul (Bircool) along the Mednipur coast became a seaside resort for the European officers. They constructed a bungalow, which was used by Lord Warren Hastings. This bungalow was later washed away by the transgression of the sea. However, today a very common myth exists among the fishermen of this area that the bungalow of Hastings often emerges from the water at the time of low tide at a distance about 5 km away from the shore. Further the Public Works Department constructed an inspection bungalow, about half of kilometre inland from that place, within mouza Digha, but with transgression of the sea this entire area has gone under sea. In 1852 ,Bayly described the natural beauty and fresh environment of this area (O'Malley 1911, P-5-6). But the eroding nature of this area was also found in his description as he says"......of the three bungalows upon them, two are gone and one all but gone."

Evidences of landward shifting of shoreline is found along the coast of West Bengal, which indicates the recent rise of sea level due to climatic change. Here the present authors have studied three parts of West Bengal coast namely - a) Mednipur coastal tract, b) Southern Sagar island coast of South 24 Parganas and c) Bakkhali coast of South 24 Parganas to understand the tendency of landward shifting of shoreline and sea level change. The results are as follows:

The rate of landward shifting of the shoreline of western part of West Bengal coast or Mednipur coast has been estimated by studying beach profiles and by the interpretation of the Survey of India toposheets (73 O/06, 73 O/10, 73 O/14 of 1930-31, 1:63360 and of

1968-69, 1:50000) and satellite images (National Remote Sensing Agency IRS-IC, LISS-3, GEOCODED FCC Numbers-7306, 73010, 73014, 1:50000, Date of publication-6.1.1997). From the recorded data and comparative study with topo-sheets, it is found that during 1931-32 to 1968-'69 the westward part of Mednipur coastal tract (from Subarnarekha river estuary to Jatranala inlet) was under prominent accretion. Rest of the area, from Jatranala to Pichhabani inlet was under erosion. Though from Jalda inlet to Pichhabani inlet some little accretion was found during this period, net positive change of shoreline is negative during that time. During 1968-'69 to 1997, a remarkable change is found along the coast. During this time, a new zone of accretion has emerged towards the east, from Jaladha inlet to Pichhabani inlet. From Jatranala to Jaladha inlet, the morphological change shows maximum erosion.

Southern Sagar Island Coast

Evidence of landward shifting of shoreline is found along the coast of southern Sagar Island. Most apparent erosion is observed at the south-eastern part of this island (Fig-4 (b). From the comparative study of Survey of India , toposheets (No 79 C/2, Scale 1:63,360 of 1922-'23, 1942, and scale 1:50,000 of 1968-'69) satellite image (IRS-1C LISS-3 Geocoded Scene no. 79C02 Scale 1:50,000, published on 6th January 1997) and present beach profile ,it can be estimated that a sharp loss of landmass due to erosion occurs along this coastal part during 1922-'23 to 2002 (Fig-4/b) at a remarkable rate of about 0.16 km^2 Year $^{-1}$.

Bakkhali Coast

Coastal erosion due to landward shifting of shoreline is a remarkable feature of the Bakkhali coastal region (Dutta and Chattopadhyay, 2003). It has been observed during the fieldwork the as much as 40% of the land area of Laxmipur mouza of Bakkhali region recorded in 1968 settlement map no. 70 is lying under the sea (Fig-4(c). The annual rate of high water line shifting has been observed about 9m to 12m in recent years. This has caused loss of valuable agricultural land and human dwelling.

CONCLUSION

Climatic change influences upon not only physical environment like landform but also human civilisation. The geo-environmental history of the Bengal Basin shows that the climatic condition remained very dynamic in this area which also influences the civilisation. One remarkable historical evidence of climatic stability of little temperature fall is the shoreline change due to sea level fall along the Bengal Basin coast during last 2000-3000 years and the deterioration of historically famous Tamralipta port (situated within 21°31′ N to 21°55′ N latitudes and 88°11′ E to 87°38′ E longitudes), which is presently known as Tamluk (O'Malley, 1911,Pp-265-268). This place is now situated more than 40 km inland (north-east ward) from the present day coastline. Tamralipta developed before the birth of Lord Buddha (544 BC) and till 8th century, it was actually the getway and a major trade centre of Eastern India. The name of this port was first described in the great epic of India, "*The Mahabharata*" (in "Adiparba"-18th chapter and "Sabhaparba"-29th chapter). It is well accepted that *The Mahabharata* was written nearly 1000-500 BC (Bari 2000,p-15). Ptoemy (*circa* 150 AD) noticed Tamralipta (*Tamalites* in his "India Intra Gangem Map") as a vital coastal port of eastern India (O'Malley, 1911,p-265). Chinese pilgrim Fa-Hian (405-411 AD) described it as being on the sea mouth. He described Tamralipta was a coastal state. During 7th century, Hiun-Tsiang wrote that Tamralipta port was situated near an inlet and only 2 km inland from the sea. The difference in observation between these two authors about the site and situation of Tamralipta port can be used as an important historical evidence of southward shoreline shifting (Dey and De, 2002;p-69-72). Now this part is existing as fluvial agro-based inland region and no coastal character in the society of this place remained. Under these circumstances, the history of climatic change during Holocene period is very important to assess the environmental condition of this younger landmass for proper planning.

REFERENCES

Acharyya, S.K. (1998). Break-up of the greater Indo-Australian continent and accretion of blocks framing south and east *Asia. J. Geodyn*. 26, 149–170.

Ahamad, E.1988. *Geomorphology*, Kalyani Publishers, New Delhi, pp.88.

Alam, M et. al. (2003). An overview of the sedimentary geology of the Bengal Basin in Relation to the regional tectonic framework and basin-fill history; *Sedimentary Geology*, 155; Elsevier, pp.179–208

Banerjee, M. (1998). Appraisal of Trend of Coastal Zone Evolution in The Western Geo-Province of Bengal Basin in Relation to Present Day Coastal Zone Problem

Chatterjee, S.N.(1972). Some aspects of batch erosion and littoral drift at Digha, West Bengal. *Journal River Resarch Institute*. 3:1-15.

Chanda, S and Hait, A.K. (1996). Aspects and appraisal of Late Quaternary vegetation of lower Bewngal basin, *Palaeobotanist*, 45, pp.117-124.

Curray, J.R., Moore, D.G., (1974). Sedimentary and tectonic processes in Bengal deep-sea fan and geosyncline. In: Burk, C.A., Drake, C.L. (Eds.), *The Geology of Continental Margins*. Springer-Verlag, New York, pp. 617- 628.

Curray, J.R., Emmel, F.J., Moore, D.G., Raitt, R.W., (1982). Structure, tectonics and geological history of the northeastern Indian Ocean. In: Nairn, A.E.M., Stehli, F.G. (Eds.), The Ocean Basins and Margins. *The Indian Ocean*, vol. 6. Plenum, New York, pp.399- 450.

Dey, S. (1999). Natural Environmental Control Through Seasons on Morphology of Digha-Shankarpur Coastal Tract of Mednipur, West Bengal; *Indian Journal of Geography and Environment*; Vol-4; pp.16-23

Dey, S .(2002). Impact of Sea Level Change upon Landform Development along Digha Coastal Tract and Risk Assessment by Process Efficiency Process; *Indian Journal of Geomorphology*; Academic and Law Serials, N. Delhi; 7, pp.111-117.

Dey, S and De, R. (2002). Evidences of Environmental Changes at Ancient Coastal Civilisation of Tamralipta; *Indian Journal of Geography and Environment*; 7; pp.80-85.

Dey, S and Haque, M .(2003). Natural Environmental Influences upon the Coastal Dune Development around Digha-Shankarpur Area, West Bengal; *Journal of Scientific Development and Environmental Research*; Mid-Waves; 2; pp.6-10.

Dey, S and De, R. (2002). Evidences of Environmental Changes at Ancient Coastal Civilisation of Tamralipta; *Indian Journal of Geography and Environment*; 7; pp.80-85.

Dutta, G.N. (1974). Some Geomorphological Aspects of Tripura State, *Abstracts of Proceedings of Indian Science Congress*, pp.130.

Falvey, D.A. (1974). The development of continental margins in plate tectonic theory. *J. Aust. Pet. Explor. Assoc.* 14, 95- 106.

Ghosh, M.M. (2002). The natural lakes of Tripura-An ecological study, *The Cosmic Zone*, 5, pp.71-79.

Hazra, S Ghosh T, Baksi, A and Ray, N. (2001). Sea Level Change: It's Impact on West Bengal Coast, *Indian Journal of Geography and Environment;* 6: pp.25-37.

Hutchison, C.S. (1989). Geological evolution of South-East Asia. Clarendon Press, London, 368 pp.

Lee, T.T., Lawver, L.A. (1995). Cenozoic plate reconstruction of Southeast Asia. *Tectonophysics* 251, 85- 138.

Mckenzie, D., Sclater, J.G. (1971). The evolution of the Indian Ocean since the Late Cretaceous. R. *Astron. Soc., Geophys. J.* 24, 437- 528.

Merh, S.S.(1987). Quaternary sea level changes: The present status *vis-à-vis* records along coasts of India, *Indian Journal of Earth Science*, 14, pp.235-251.

Menon, K.D. (1975). Tripura District Gazetteer; Government of Tripura; Agartala; pp.1-27.

Mamud, A. (2004). Ecological changes in the wet lands of North-East India, *The Scientific Journal of the SSIRA*, 3, pp.4-9.

Nandi, M .(1974). The Relation of Agartala Floods to Haora River Catchments, Tripura State; *Records of Geological Survey of India*; 108; Pt-2; pp. 149-152.

Nandi, M .(1977). Physiography of Tripura State; *Geog. Rev, Ind. Geog. Soc. Ind*, 39, No-4, pp.367-373.

Nandi, M .(1985). Geographical Account of Tripura; An Anthology of Tripura (Edited By J. Gan-Chowdhui); Inter-Indian Publication; New Delhi; pp.1-9.

Niyogi, D. (1970). Geological background of beach erosional Digha, West Bengal, Bulletin of the Geological Mining and Metallurgical Society of India, 43: pp.1-36.

Climate Change and Food Security, 2008
Edited by: M.Datta, N.P.Singh & Er.D.Daschaudhuri, pp.243-265
New India Publishing Agency, New Delhi (India)
E-mail : newindiapublishingagency@gmail.com
Web: www.bookfactoryindia.com

17

Climate and Horticultural Crops

M.SANKARAN, JAI PRAKASH, N.P.SINGH AND M.DATTA

INTRODUCTION

India has varied agroclimatic conditions, which favour the cultivation of various horticultural crops. Based on the adaptability and system of cultivation, the horticulture can be grouped into four major class such as Temperate, Sub-tropical, Tropical and Arid. India has become the second largest producer of both fruits and vegetables with a total production of 46.75 million tonnes of fruits from an area of 4.56 million ha and 93.16 million tonnes of vegetables from 6.30 million ha (2003-2004). Besides, there is a tremendous scope for increasing the production of flowers, spices and condiments, plantation crops and medicinal and aromatic plants. Floriculture has emerged as a viable agribusiness. The commercial cultivation of flowers in open field conditions was much popular than protected cultivation in India. But, now- a-days, the protected cultivation has brought a great revolution in Indian floriculture and it helps for the quality production and export to earn sizeable amount of foreign exchange. India has approximately 350 ha under green house cultivation with export value of approximately Rs.330 crores during 2003-04(APEDA).On the other hand, the prevailing climatic conditions in India are highly suitable for cultivation of the almost all spices from tropical black pepper to temperate saffron. The major spices are mainly grown in humid tropics like Western Ghats and NE States. Out of 21 agro-ecological sub regions in India, spices are grown in 14 different agro ecological sub regions. Fruits, vegetables, flowers spices and condiments, plantation crops and medicinal plants are widely distributed across the country, while plantation crops are restricted to a few regions.

FRUIT CROPS

Most of the fruits crops are grown in tropical and subtropical zones of India either as rainfed crops or with supplemental irrigation. The effect of climate is well known in fruit crops and each cultivar requires a specific climatic condition for their survival. The cultivar and climatic factors are well studied in several fruit crops. It is evident that, within the fairly broad parameters of adaptable soil and reasonably good cultural practices, temperature is probably the most important component of the climate and soil culture complex which causes difference in growth and fruit characteristics among the fruit-producing areas in the country. Among climatic factors (temperature, humidity, sunlight and rainfall), the temperature is an important parameter , which significantly affects the quality and production. Humidity also plays a pivotal role in determining the suitability of crop in the region. Limitation of rainfall can be overcome by supplemental irrigation but the effect of temperature can be rarely changed.

Table 1. State wise area and production of fruits (2003-04)

S. No.	State	Area ('000 ha)	Production ('000 MT)	Productivity (t/ha)
1.	Andaman & Nicobar Islands	3.90	22.1	5.6
2.	Andhra Pradesh	656.0	7580.10	11.55
3.	Arunachal Pradesh	48.40	101.20	2.09
4.	Assam	94.30	1181.20	12.50
5.	Bihar	307.20	3454.80	11.24
6.	Chandigarh	0.1	1.1	11.00
7.	Chattisgarh	17.0	401.10	23.59
8.	Dadra & Nagar Haveli	0.7	7.1	10.14
9.	Daman & Diu	0.0	0.0	0.00
10.	Delhi	0.1	1.0	10.00
11.	Goa	10.30	81.60	7.92

Contd...

12.	Gujarat	201.30	2957.40	14.70
13.	Haryana	31.60	257.20	8.13
14.	Himachal Pradesh	201.80	588.20	2.91
15.	Jammu & Kashmir	157.60	1180.40	7.50
16.	Jharkhand	31.80	296.30	9.31
17.	Karnataka	242.50	3833.40	15.80
18.	Kerala	170.50	1002.20	5.87
19.	Lakshadweep	0.30	1.10	3.66
20.	Madhya Pradesh	47.20	910.20	19.28
21.	Maharashtra	126.50	9769.70	7.72
22.	Manipur	51.20	320.90	6.26
23.	Meghalaya	23.80	199.60	8.38
24.	Mizoram	21.20	42.50	2.00
25.	Nagaland	13.3	48.90	3.67
26.	Orissa	227.2	1353.40	5.96
27.	Pondicherry	1.0	21.20	21.20
28.	Punjab	43.60	628.20	14.40
29.	Rajasthan	23.40	221.00	9.44
30.	Sikkim	10.0	8.10	0.81
31.	Tamilnadu	221.0	3609.40	16.33
32.	Tripura	30.80	487.40	15.82
33.	Uttar Pradesh	269.40	2320.40	8.61
34.	Uttaranchal	79.0	644.60	8.15
35.	West Bengal	172.90	2111.60	12.21
Total		**4675.40**	**4564.60**	**9.76**

Source: National Horticulture Board, Gurgaon

Banana (*Musa* sp)

Among the fruit crops, banana ranks first in production and is widely grown in Andhra Pradesh, Assam, Bihar, Gujarat, Kerala, Karnataka, Maharashtra, Madhya Pradesh, Orissa, Tamil Nadu and north-eastern region . The highest productivity is noted in Maharashtra due to use of tissue cultured seedlings, high density planting and drip irrigation but the lowest productivity is in north-eastern regions, because of lack of systematic cultivation in the region. The cultivar and production systems vary with agro- ecological region and consumer preference. Among the cultivars, Cavendish group (AAA) and Monthan (ABB) are very well adopted in dry regions where sigatoka leaf spot is not a constraint. Growth of banana is influenced by temperature. It grows well with a mean monthly temperature of 26.5°C. If it is grown in region experiencing the temperature of < 15°C, then the plant will show the "*choke throat*" symptom. Harvesting is delayed in the region experiencing temperature below the mean temperature owing to reduced rate of leaf production. Resultantly , the first crop cycle varies from 11-12 months in Kerala or coastal Karnataka to 13.5 or 18.5 months in semi-arid subtropics (Maharashtra). On the contrary, bananas in semi-arid subtropics are least affected by diseases, which are more prevalent in humid tropics. Low temperature also affects the shooting and finger filling and causes lower production if shooting coincides with winter. This adverse effect of low temperature can be avoided by adjusting the time of planting and modified cultural practices. The fruits produced in regions experiencing mild temperature have better quality.

Mango (*Mangifera indica* L.)

Mango is widely adapted in tropical and subtropical regions except temperate region although suitability of cultivars varies in different agroclimatic zones. Uttar Pradesh is having the largest area under mango but the productivity is much lower as compared to Andhra Pradesh. Phenomenon of double cropping (Off season bearing) is prevalent in Kanyakumari district of Tamil Nadu. Mangoes from coastal Maharashtra, Andhra Pradesh and Tamil Nadu are earliest to come in the market, while late crop of mango is produced in north-western India. Alphonso, a popular variety from Konkan region of Maharashtra is of export quality and this variety produces sponge tissue if the soil temperature is high. Similarly, the coastal region is highly successful in production of mango grafts through epicotyl grafting because of prevailing high humidity in the region. Dashehari, a popular mango of Uttar Pradesh when grown in Andhra Pradesh, produces 40-50 days early crop. This influence of temperature is gainfully utilized for early marketing of this variety. Kanyakumari and parts of Madurai districts of Tamil Nadu produce 2 crops of mango in a year. The mango malformation is a common problem in the region experiencing severe winter (Northern India)

Citrus

The citrus fruits (mandarin, sweet orange, acid lime and lemon) are grown widely in India and their productivity depends on climate, cultivars and type of production system. Mandarin (*C.reticulata*) is predominantly grown in eastern region of Maharashtra, parts of Madhya Pradesh, Punjab (Kinnow), Meghalaya (Khasi Mandarin), Tripura (Jampui orange) and other north eastern region. Shevroy and Palani hills (Kamala orange) in Tamil Nadu also grow substantial amount of mandarins. Sweet oranges (*C.sinensis*) are largely grown in Andhra Pradesh, Maharashtra and parts of Tamil Nadu, and Punjab, Haryana and Rajasthan in north western region. Acid lime is common in Andhra Pradesh, Maharashtra, Karnataka, Tamil Nadu, Bihar and West Bengal. Although it is difficult to delineate the influence of agroclimatic conditions on productivity, fruit quality is invariably influenced under similar agroclimatic regions having marked seasonal variation for physico-chemical quality. In Vidharbha region of Maharashtra, water requirement is much higher for *ambia* crop (October-November) than the *mrig*. On the other hand, *mrig* crop (February-March) has excellent attractive colour contrary to *ambia* crop. Soil moisture stress coinciding with maturation improves TSS and reduces acidity. Growth pattern, flowering, fruiting, maturity and fruit quality of citrus are influenced by temperature. Shoots produced in cooler and warmer months differ in size of leaves and internodal length. Trees attain larger canopy in tropical climate than in subtropical climate. In subtropical regions, flowering reaches its peak in March, while in tropical region, it is adjusted with soil moisture stress. Cool weather in subtropics and moisture stress in tropics are known to condition a major part of the shoots to flower at one time during the year. However, lime and lemon continue to flower and produce fruits all the year round in tropics. While in the region with dry summer and cool winter, these bloom mainly in spring. Regulated soil moisture stress in several citrus crops results in profuse bloom. Seasonal variability in seed content of citrus fruits indicates indirect effect of temperature on pollination and fertilization. More pronounced influence of temperature is seen on maturation of fruits. Fruits produced in cooler climate require longer period for maturation than those in warmer climates, which is true for all the cultivars of citrus.

Growth rate of the citrus fruit increases with temperature between 10 and 30°C if other factors are constant. Even under the same agro-ecological conditions, fruits produced in different seasons have varying time taken for maturity. Mandarins produced in northeastern India, especially in Darjeeling, Sikkim and Meghalaya, are flatter, and smaller having no neck, while fruits produced in central Maharashtra are larger with neck. Fruit colouration is also influenced by temperature; cool nights with hot dry days favour best colour development.

Kinnow mandarins, which have attractive colour, remain green or yellowish when grown in tropical environment. Similarly, Sathgudi, which does not attain orange colour in south India, develops attractive orange colour when grown in northwestern or north-eastern India. Seasonal variation in colour development in Nagpur mandarins is evidently due to temperature prevailing during the fruit development and maturation. Evidence also suggests for the influence of temperature on internal quality of fruits. Dry hot weather favours degradation of acid contrary to the fruits grown in humid cool climate.

Grape (*Vitis vinefera L.*)

Grape is grown largely in Maharashtra, Karnataka, Tamil Nadu and Andhra Pradesh as tropical crop and in Punjab, Haryana and parts of Uttar Pradesh as subtropical crop. It is also grown in Kinnaur districts of Himachal Pradesh as a temperate crop. This crop is cultivated from temperate to arid region but widely adapted in arid climate of both tropics and subtropics. Cultivar and production systems largely vary with the highest productivity recorded in tropical region. However, fruits produced in arid tropics have better quality. High humidity and rain during fruit development and maturation adversely affect the quality and production. In sub topical region, the fruit maturity is coinciding with the rainy season, which severely affects the quality and productivity of grapes. In Theni and Madurai districts of Tamil Nadu, the farmers are harvesting 5 crops in 2 years period due to suitable climatic condition.

Pineapple (*Ananas comosus L*)

Pineapple is adapted to humid tropics and subtropics of the country. The optimum day and night temperatures are reported to be 30°C and 20°C, respectively. The high density planting (Double row system, 90x60x30cm) has become popular in this crop. Highest productivity of pineapple is reported from West Bengal and lowest from northeastern regions. Varietal adaptability has become the key factor e.g. 'Queen' variety is grown in Tripura on commercial scale, 'Kew' variety is grown in Assam, Meghalaya and West Bengal and 'Mauritious' variety grown in Southern India.

Sapota (*Achras sapota*)

Sapota is grown both in humid and dry conditions with high productivity in Karnataka and Gujarat, and lowest in West Bengal. This is grown as a rainfed crop in Karnataka and in Gujarat, it is grown with supplementary irrigation and is widely adopted across the country even in marginal and degraded lands.

Papaya (Carica *papaya*)

The productivity of papaya is high in Karnataka and low in northeastern region. The temperature has profound influence on the sex in papaya. The optimum temperature for papaya is reported to be 21°C to 33°C (Knight, 1980) . In monoecious lines, if the temperature is high, they tend to produce more male flowers and in dioecious lines (?,?) sometimes, the male will be converted to female line. Conversion male flowers into female flowers occurs if the temperature is low below the critical level. These kinds of males are called as *teratalogical male.* Papaya is susceptible to ring spot virus, (PRSV)), which is a devastating disease of papaya and this disease is serious in north-western and northeastern regions but incidence of the disease is rarely noticed in tropical region.

Guava (*Psidium guajava*)

Guava is widely distributed with highest productivity in Gujarat although best quality fruits are produced in Uttar Pradesh. Among the tropical and subtropical fruits, this crop is more tolerant to environmental stress (Marler, 1994). The optimum temperature lies between 23°C and 28°C (Samson,1980). However, the season of availability of fruit is markedly governed by agroclimatic conditions.

Litchi (*Litchi sinensis*)

Litchi is grown in Bihar, Tripura, Assam, West Bengal, Uttar Pradesh, parts of Punjab and Haryana. Of the total area, 75% is produced from Bihar. Attempts have been made to introduce this fruit in Maharashtra, Gujarat, Karnataka and Tamil Nadu but without success. Prevailing temperature, rainfall pattern and soil types largely determine its success. Subtropical climate with dry period preceding flowering and rain free months at the time of maturity are essential. The maturity period varies with the region; the fruits mature in eastern region 30-45 days earlier than the western region (Ripening is early in Tripura than Bihar and Uttar Pradesh).

Temperate Fruits

Temperate zone favours the cultivation of apple, pear, peach, plum, apricot, stone fruits but cherry fruits are adapted to humid temperate region. Among temperate fruits, apple is the most popular and it is largely grown in Kashmir, Himachal Pradesh and hills. Area under apple in Arunachal Pradesh, Nagaland and Sikkim is negligible. A significant breakthrough in temperate fruits was made through the development of low chilling cultivars of Apple- (var. Parlins beauty; Peach- Shan-e- Punjab and Florida Sun). Many low-chilling stone fruits are grown in *tarai* region of northwestern India of Uttar Pradesh. Chilling requirement is the important criteria which decide the flowering and fruiting in temperate fruits. The chilling requirement can be defined as the duration (hours) of temperature below 7.2°C, which is necessary for the buds to have proper foliation and flowering. The chilling requirement for temperate fruits are given in Table 2.

Table 2. Chilling requirement for temperate fruits

S.No	Crop	Temperature (°C)	Hours	Remarks
1	Apple	Below 7.2	1200-1500	Low chilling cultivars are available
2	Pear	Below 7.0	1200	Patharnakh variety of fruits even at 150 hrs chilling
3	Peach	Below 7.2	800-1000	Shan-e- Punjab and Florida Sun varieties are low chilling peaches
4	Plum	Below 7.2	700-1000	European plums require less chilling hours

Arid Zone Fruits

Arid zone fruits are having certain characters to withstand the extreme climatic conditions. The characters are (i) deep root system to harness water from deeper areas (ii) leaf shedding in summer to conserve the moisture (iii) wax coating on the leaves to reduce the transpiration (iv) sunken stomata in the leaves to minimize loss of water from the leaves and (v) the synchronization of flowering and fruiting period with moisture available period (Ber, pomegranate, aonla, annona, jamun and tamarind).

VEGETABLES

Vegetable crops are important components in Indian Horticulture mainly due to their role in food and nutritional security. During 1991-92, total production of vegetables in the country was 58.54 million tonnes from 5.59 million ha of land with productivity of 10.5 t/ha. Now, India has emerged as the second largest producer of vegetables with a total estimated production (2003-04) of 93.16 million tonnes with the productivity of 14.76 t/ha. In 2003-04, Chattisgarh was credited with maximum productivity (25.43 t/ha), followed by Tamilnadu (24.86 t/ha) and Uttar Pradesh (18.14t/ha).

Table 3. Area and production of vegetable crops in India (2003-04)

S. No.	State	Area ('000 ha)	Production ('000 MT)	Productivity (t/ha)
1.	Andaman & Nicobar Islands	4.00	23.30	5.82
2.	Andhra Pradesh	248.00	2882.30	11.62
3.	Arunachal Pradesh	20.20	80.90	4.04
4.	Assam	198.80	1958.90	9.84
5.	Bihar	815.00	13296.0	16.31
6.	Chandigarh	0.1	1.70	17.00
7.	Chattisgarh	61.1	1554.1	25.43
8.	Dadra & Nagar Haveli	1.5	13.50	9.00
9.	Daman & Diu	0.0	0.0	0.0
10.	Delhi	43.7	626.0	14.32
11.	Goa	7.8	74.70	9.57
12.	Gujarat	248.30	3515.2	14.17
13.	Haryana	203.90	2703.3	13.25
14.	Himachal Pradesh	59.30	877.20	14.81
15.	Jammu & Kashmir	34.70	462.0	13.31

Contd...

16.	Jharkhand	110.60	1197.20	10.82
17.	Karnataka	363.30	4176.90	11.49
18.	Kerala	99.20	2602.00	13.5
19.	Lakshdweep	0.20	0.20	1.00
20.	Madhya Pradesh	164.70	2377.0	14.43
21.	Maharashtra	370.00	4132.0	11.17
22.	Manipur	13.40	86.00	6.41
23.	Meghalaya	32.70	270.50	8.27
24.	Mizoram	5.70	24.0	4.21
25.	Nagaland	11.90	88.10	7.40
26.	Orissa	655.30	8030.90	12.25
27.	Pondicherry	4.4	71.90	16.34
28.	Punjab	153.0	2588.00	16.91
29.	Rajasthan	116.50	527.60	4.52
30.	Sikkim	16.10	75.00	4.65
31.	Tamilnadu	187.10	4672.70	24.86
32.	Tripura	32.0	352.20	11.00
33.	Uttar Pradesh	819.5	14862.0	18.14
34.	Uttaranchal	41.0	447.3	10.83
35.	West Bengal	1165.9	18510.30	15.87
Total		**6308.9**	**93165.0**	**14.76**

Source: National Horticulture Board, 2003-04

Most of the vegetables are grown widely in the country through adjustment in season and supplemented irrigation. Cole crops, confined to temperate and subtropical regions, have also been extended to tropical regions with cooler night temperature, through the development of heat tolerant cultivars. However, quality is still superior in subtropical regions. Cucurbits, okra and chilli are widely grown across the country. However, muskmelon, watermelon etc. have specific climatic requirements. With supplemental irrigation and adjustment in season, most of the vegetables can be grown in majority of agroclimatic conditions. However, as rainfed crop, it has to be limited to the zone, which has better growth period. At the same time, in humid region, disease load has also to be accounted. Vegetables are generally grouped into three categories viz., tropical, subtropical and temperate/ cole crops based on their adaptability in the specific region. There is no curd formation in cauliflower or head formation in cabbage at the temperature below the optimum. Thermo sensitiveness of these crops impedes its adoption in the agroclimatic conditions not having required temperature regimes. However, through widening the genetic base, it has been possible to get curd formation even at little higher temperature. Since the most of vegetables are of short duration, through the adjustment in season of planting optimum potential can be achieved. However, muskmelon, watermelon, capsicum and onion are restricted. In onion, the short day condition favours the bulbing and long day condition hastened the formation. Maharashtra, Gujarat, Uttar Pradesh, Karnataka, Orissa and Bihar are major onion-producing states, accounting for more than 75% of production. Productivity is the highest in Gujarat followed by Punjab and Haryana. Onion is grown as irrigated as well as rainfed crop depending upon season by selection of suitable cultivars. *Kharif* onion has been successful in plains of Bihar, Uttar Pradesh and Rajasthan. The variability has enhanced the adaptability of onion in a wide range of agroclimatic regions. In potato, tuberization does not take place if difference in night and day temperature is not wide. For sprouting and initial vegetative growth, a temperature range of 22-34°C is the best. However, for better tuberization, 18-20°C will be optimum and short day condition favours the tuberization.There has been unprecedented growth in production of potato and other tuber crops in the last few decades owing to better cultivars and production technology. Major potato-growing states are Bihar, Uttar Pradesh, West Bengal, Gujarat and parts of north-eastern states. However, its production in Karnataka and Maharashtra has been successful. *Kharif* potato has also become popular. Its productivity is the highest in Gujarat and the lowest in Assam. Gujarat, Bihar, West Bengal and Uttar Pradesh account for more than 80% of potato area in the country. Cassava and sweet potato, being important tropical tuber crops, are grown widely. Major cassava-growing states are Tamil Nadu, Kerala, Karnataka, Orissa and Andhra Pradesh. Sweet potato is largely grown in Orissa, Uttar Pradesh, Bihar, Maharashtra, Tamil Nadu, Madhya Pradesh and Karnataka. Among the sweet potato-growing states, Bihar, Orissa and Uttar Pradesh account for 80% of area. Its productivity is 1.8–22.4 tonnes/ha, with the highest being in Gujarat and the lowest in Assam.

PLANTATION CROPS

Plantation crops are largely confined to coastal areas, which are characterized by little variation in temperature with high humidity. However, with development of cultivars and production technologies, cultivation of plantation crops has been extended to new areas. Major coconut-producing states are Kerala, Coastal Karnataka, Tamil Nadu, parts of Maharashtra, Gujarat, West Bengal and Orissa. Area under coconut has been extended to Bihar and its plateaus having good irrigation resources. Introduction of oil palm to the country has been successful. The potential states for growing oil palm are Andhra Pradesh, Karnataka, Kerala, Tamil Nadu and Maharashtra. Arecanut is confined to Kerala, Karnataka, coastal Maharashtra, north West Bengal and northeastern states. The agroclimatic conditions play a vital role in optimization of the yield of coconut, arecanut and cashew is confined to humid tropics as flowering is adversely affected at lower temperatures as well as in dry conditions.Cashew, an important export-earning crop, is grown largely in Kerala, Karnataka, Tamil Nadu, Orissa and Andhra Pradesh. It has a potential for cultivation in Bihar, Madhya Pradesh and northeastern region.

SPICES

The varied agro-climatic condition prevailing in India is highly suitable for the almost all spices (production of tropical black pepper to temperate saffron). Major spices are mainly grown in humid tropics like Western Ghats and NE states. Out of 21 agro-ecological zones, spices are grown mainly in 14 different agro-eco sub regions in India. Rajasthan, Madhya Pradesh, Kerala, Karnataka, Tamilnadu and Northeastern states are important spice growing states.Among spices, black pepper, large and small cardamom, ginger, turmeric and chilli are important. Of which, pepper and cardamom are confined to Kerala and Karnataka, while ginger, turmeric and chilli are widely cultivated across the country. There is a scope for extending area under black pepper in Tamil Nadu, Andhra Pradesh and northeastern states.

Interactive knowledge of agroclimatic regions added with information on climatic response of fruit and vegetable crops and varieties can be gainfully utilized to set higher economic returns. Efforts made through the All-India Coordinated Research Project have demonstrated that with little modification in production technologies, many horticultural and plantation crops can be adopted to a wide range of agro-ecological zones and the weakness can be converted into opportunity for extended harvest. Similarly, pepper and cardamom have poor growth at an average temperature below 15°C. However, the best productivity in large cardamom is achieved in humid subtropics.

Table 4. State wise area and production of spices in India (2002-03)

S. No.	State	Area ('000 ha)	Production ('000 ha)	Productivity (t ha)
1.	Andhra Pradesh	354.90	887.9	2.50
2.	Rajasthan	762.6	616.40	0.80
3.	Gujarat	473.5	505.30	1.06
4.	Karnataka	284.7	315.70	1.10
5.	Tamilnadu	159.3	250.40	1.57
6.	Orissa	253.8	224.00	0.88
7.	Madhya Pradesh	171.5	152.50	0.88
8.	Kerala	288.10	151.60	0.53
9.	Maharashtra	121.30	104.30	0.86
10.	Uttar Pradesh	42.4	100.60	2.37
11.	West Bengal	86.40	95.00	1.09
12.	Others	225.40	362.30	1.60
Total		**3224.0**	**3766.00**	**1.16**

Source: National Horticulture Board, 2002-03.

FLOWER

Floriculture has emerged as a viable agribusiness. The commercial cultivation of flowers in open field condition is much popular than the protected cultivation. But now-a- days, the protected cultivation opens up newer avenues for quality production and export to earn valuable foreign exchange. The growing environment plays profound influence on quality of cut flowers. Three major factors that regulate the quality include temperature, light and $Co_{2.}$

Table 5. Cultivation of horticultural crops in different agro ecological zones of India

S No	Agro ecological Zone (Region)	Active growing period	Climate	Soil type	Suitable Crops
1	Western Himalayas covering Ladakh and Gilgit districts	< 90 days	Cold arid	Shallow Skeletal soil	This region has high potential for growing almond, walnut, pecan nut, Chinese ber and sour pomegranate (Russian type). During summer period tomato, cole crops (Cabbage, Cauliflower, Sprouting broccoli and Brussels sprout) and brinjal can be cultivated
2	Western parts of Rajasthan, southern parts of Haryana and Punjab, the Kutch peninsula and northern parts of Kathiawar peninsula	< 90 days.	Hot arid	desert and saline soils	Date palm, citrus, grape, ber and aonla are potential crops for this area with supplemental irrigation. Cucurbits and cole crops can also be cultivated.
3	Deccan plateau including Raichur and Bellary in Karnataka and Anantapur in Andhra Pradesh	< 90	Hot arid	red and black soils	Deccan plateau including Raichur and Bellary in Karnataka and Anantapur in Andhra Pradesh
4	Northern plains, central highlands and Gujarat plains,	90–150	Hot and dry summer and mild winter	Alluvium derived soils	Citrus fruits, guava, mango and low-chilling stone fruits are grown. Banana and sapota find favour in plains of Gujarat. Besides cole crops, all vegetable crops have potential with a modification in production technology and cultivars.
5	Western parts of Madhya Pradesh, eastern parts of Rajasthan and Gujarat	90–150	Hot and dry summer and mild winter	Medium and deep black soils	Ber, pomegranate, aonla, tamarind, guava, citrus fruits, mango and sapota are grown. Banana is also grown with supplemental irrigation. Among the plantation crops, coconut, oil palm, cashew and black pepper are potential. All the vegetable except cole crops can be grown.
6	Parts of Maharashtra, northern parts of Karnataka and Andhra Pradesh	90–150	Hot and dry summer and mild winter	Shallow and medium (with inclusion of deep) black soils	This region has excelled in production of several fruit crops resultantly revolution in fruit production. Most of the fruit crops in the region are grown with supplemental irrigation. Grape, citrus fruits, ber, pomegranate, sapota, banana, mango and custard apple are major fruit crops grown in the region. This region has less problems of diseases, the major inputs

Contd...

Table 5. Contd ..

					being water and nutrition. Coconut, arecanut and oil palm and potato and onion are also potential in the region.
7	Deccan plateau and Eastern Ghat covering major part of Andhra Pradesh	90–150	Hot dry summer and mild winters	Black and red soils	This is highly potential for growing fruit crops with supplemental irrigation. The crops like sapota, custard-apple, mango, banana, citrus fruits, guava and pomegranate are largely growing in this region. Grape occupies a significant position. Cashew nut, oil palm, coconut and solanaceous vegetables and cucurbits can be grown in the region.
8	The region comprising Deccan plateau, Tamil Nadu uplands and western part of Karnataka	90–150	Hot dry summer and mild winters	red and black soils	Mango, sapota, guava, ber, aonla, grape, banana, tamarind and citrus fruits are major crops of this region. However, on hills low-chilling pear, plum and peaches may be grown. There is lot of variability in climatic conditions in this region, making different fruit crops suitable. Coconut, arecanut, cashew and all vegetables except peas are grown in this region.
9	Northern plain	150–180	Hot humid	Alluvial	Mango, guava, citrus fruits (limes and lemon), papaya and low-chilling stone fruits are grown in this region. All vegetables depending upon the season can be grown.
10	Central highlands (Malwa and Bundelkhand)	90–150	Hot sub-humid	medium and deep black soils	Ber, pomegranate, aonla, bael and mango are potential fruit crops for the region under protective irrigation. All vegetables have potential depending upon the season.
11	Deccan plateau and central highlands (Bundelkhand)	150–180	Hot sub-humid	red and black soils	Besides high quality mandarins, it is potential growing region for guava, sapota, mango, ber and pomegranate. Cashew and seed spices may also be grown in this region.
12	Eastern plateau (Chhatisgarh region)	150–180	Hot sub-humid	red and yellow soils	Mango, citrus fruits, cashew nut, guava, papaya and pineapple are potential fruit crops of the region. On high hills, pear can also be grown successfully. Banana has also proved successful in the region. Cashew is also potential crop of this region.

Contd...

Table 5. Contd...

13	Eastern Chhotanagpur plateau and Eastern Ghats,	150–180	Hot sub-humid	Red loamy soil	Mango, guava, banana, papaya, jackfruit, pineapple, cashew nut and several minor fruits are grown in this region. Aonla, ber and pomegranate have potential for commercial exploitation. Banana and litchi can also be grown especially in plateau of Bihar. Cashew nut is also a potential vegetable which can be grown all round-the-year.
14	Eastern plain	180–210	Hot sub-humid	Alluvium derived soils	Mango, jackfruit, guava, bael, aonla, acid lime and banana are predominant fruit crops in this region. Coconut and arecanut are potential crops in this region. The region can grow all kinds of vegetables.
15	Western Himalayas	180–210	Warm sub-humid	Laterite soils	The region is suitable for growing apple, pear, peach, plum, apricot, and almond. Apple cultivation has been successfully exploited in the region. All temperate vegetables have potential in the zone.
16	Assam and West Bengal plains	< 210 days.	Hot humid to sub humid	Alluvial soil	Banana, pineapple, mango and sapota are predominant fruit crops although litchi is also grown in sub-zones of the region. Citrus fruits are also grown especially in Assam valley. Guava, aonla and bael have also potential in the region. Cashew nut, coconut, arecanut and most of the vegetable crops except peas have potential in the region. Assam tea and Darjeeling tea are the world famous.
17	Eastern Himalayas (Northern tip of the West Bengal, northern parts of Arunachal Pradesh and Sikkim)	< 210 days.	Warm humid	Hilly soil	Apple, pear and plum are potential crops in the region. Mandarin is successfully grown in valley areas. In the warm season, all kinds of vegetables can be grown. Large cardamom, turmeric and ginger have potential.
18	North-eastern hill (The region represents hilly states of Nagaland, Meghalaya, Manipur, Mizoram and south Tripura)	< 210 days	Warm humid	Red and laterite soils	Citrus fruits, banana, passion fruits and pineapple are predominant fruit crops. Region has potential for growing low-chilling stone fruits and passion fruit. Cashew and potato are most potential crop.

Contd...

Table 5. Contd...

19	Eastern coastal plain	150–210 days.	Hot sub-humid	Alluvial soils	The region has predominance of mango and banana, although pineapple is also potential in the region. In recent past, area under sapota and guava has also been extended. Pomegranate has problem of flowering in the region. Coconut, arecanut, cashew, black pepper and seed spices are grown in this region. All the vegetables except cole crops are grown.
20	Western Ghats and coastal plains	> 210 days.	Hot humid / humid	Red, lateritic and alluvium-derived soils	In western coastal Maharashtra, mango, sapota and cashew nut are predominant while in coastal Karnataka banana, pineapple, cashew nut and mango are largely grown. In coastal Kerala, banana and pineapple are predominant fruit crops. Mango is grown to a limited extent only. Coconut, spices and cashew nut and a number of vegetables except cole crops are grown in the region.
21	Island of Andaman Nicobar and Lakshadweep	< 210 days.	Hot humid	Red loamy and sandy soils	Coconut is predominant crop along with banana. Pineapple, mangosteen, sapota and mango are potential crops in this zone. Cashew, pepper, seed spices and vegetables also have potential in this zone.

(Anonymous, 2002)

Temperature

It is an important factor, which plays to determine the quality of flowers. The excess temperature during cultivation decreases the quality of cut flowers due to faster utilization of carbohydrate accumulation in tissues and rapid water loss from the entire plant. The diurnal variations in temperature play a significant role in determining the quality of cut flowers.

Light

Light intensity directly influences the efficiency of photosynthesis, which determines the carbohydrate content of flowers. Flowers containing relatively high amount of carbohydrates especially mobile sugars are of high quality. Light intensity also influences the colour of petals. Excessive shading of rose buds during the formation of anthocyanins in petals reduces the formation of blue colouration in petal. Higher intensity of light causes red colouration of tissue, leaf spot, browning and leaf drop.

Table 6. Optimum diurnal temperatures for important cut flower production

S. No.	Crop	Temperature	
		Day (°C)	Night (°C)
1.	Rose	28	18
2.	Carnation	18	13
3.	Chrysanthemum	16	9
4.	Gerbera	16	12
5.	Anthurium	24	18
6.	Tropical orchids (*Vanda, Dendrobium*)	27	16
7.	Sub-tropical orchids (Cattelya)	18	13
8.	Temperate orchids (*Cymbidium, Papiopedilum*)	13	10
9.	Lilium	25	10

CO_2 Enrichment

The plants of tropical origin have a capacity to utilize high CO_2 concentrations. Elevated CO_2 concentrations more than the ambient concentrations accelerate the vegetative growth, water use efficiency and photosynthetic capacity of flowers leading to production of quality, blemish-free blooms. Elevated CO_2 also plays a major role in shortening the culture time, increasing the productivity, reducing the no. of blind shoots, early flowering and better lateral branching in flowers. The ideal concentrations for CO_2 enrichment in different flowers are given in Table7.

Table 7. Ideal concentration of CO_2 for quality flower production.

S.No.	Crop	CO_2 concentration (ppm)	Toxic CO_2 concentration (ppm)
1.	Rose	1000-1500	2500-3000
2.	Carnation	1000-1500	2000-250
3.	Chrysanthemum	700-900	1500-3000
4.	Tulip	1000-1500	2000-2500
5.	Gerbera	300-1200	1600-2200

Rose

Queen of flowers "Rose" is grown in temperate, tropical and sub-tropical climates. Bright sunshine and free ventilation is essential for quality flowers of rose. Optimum temperature for rose flowering is 28^{0}C during day and 18^{0}C at night. Temperatures below 10^{0}C especially during night affect flower production and quality adversely. Light intensity is one of the most important factors influencing the growth and flowering of roses. The optimum light intensity ranged between 6000-8000 foot candles for rose flowering. The formation of red pigment in rose is very specific for temperature and light intensity. CO_2 content of the air is a limiting factor in photosynthesis. Plants were significantly large and more flowers were produced when grown in CO_2 enriched atmosphere. The concentration of CO_2 at 1000-1500 ppm inside the poly-house/ glass-house results in reduced number of blind shoots, higher yield, and also longer / stronger stems. A moderate level of 50-60 per cent relative humidity should be maintained in protected cultivation.

Carnation

It is a quantitative long day plant and providing long days promote early flowering. Carnation requires a day temperature of 18^{0}C and night temperature of 10-12^{0}C in winter and 12-15^{0}C in summer. Higher day and night temperatures during flowering induce abnormal flower opening and calyx splitting. The initial stage of flower bud development is influenced by light intensities. Apart from this, concentration of CO_2 plays an important role in flower production by 15-20 per cent. The best quality flowers were produced when CO_2 concentration in the green house was maintained at 500 ppm along with a day temperature of 14.0 to 15.4^{0}C.

Chrysanthemum

Light and temperature are the two important factors influencing the growth and flowering. It is a short day plant but requires high light intensity for vegetative growth. Most of the cultivars require warm nights at the time of flower bud formation. Light affects the growth of chrysanthemum quantitatively by influencing the photosynthetic activity and qualitatively due to phenomenon of photoperiodism. The optimum soil and air temperature were found to be 18-21^{0}C and 10^{0}C, respectively for vegetative growth. However, temperature between 15-20^{0}C is ideal for flowering. In controlled condition, an enrichment of CO_2 from 500 to 1000 ppm enhanced the flower production and quality.

Gerbera

In tropical and sub-tropical climate, gerberas are grown in the open, but in temperate climate, they are protected from frost and cultivated in green house/ protected condition. However, all commercial varieties and growers are cultivating Gerbera only in protected condition viz. green house/ poly-house/ shade-net house to meet the quality standards for export. The photoperiod and temperatures have been found to influence the growth and flower production. For profuse flowering, cooler green houses are the best with a day temperature of 16^0C and 12^0C at night. Flower induction is adversely affected below 10^0C and above 35^0C. Optimum humidity inside the polyhouse must be below 70% during day and below 85% at night. There should be minimum temperature difference between air and soil temperature to get higher flower production per unit area.

Anthurium

Commercial cultivation of Anthurium is going on in poly-house, shade-net house or even high-tech green house. The day and night temperature, light intensity and relative humidity play a vital role in growth and flowering of Anthurium. The optimum temperature for growth is 18-21^0C and minimum should not be less than 10-13^0C for a short period. The relative humidity should be around 80 % and it can be maintained by misting/fogging. Bright sunshine hour and filtered light is essential for profuse and quality flowers. The optimum light intensity should be between 20,000 and 25,000 Lux and it is maintained by 50 % shade net on the top and 25 % shade below it.

Dahlia

Dahlia plants are responsive to day and night temperatures and also to day length. Several cultivars of Dahlia grown under protected condition at different seasons initiate flower buds at all temperatures in a range of 10-21^0C at night and 15.6-26.7^0C during day. The delay in flowering varies with the cultivars. Tuberose root formation of dahlia enhances at lower temperature.

Gladiolus

It requires full exposure to sunlight for better growth, otherwise blasting may occur or plant may remain blind. The long day conditions of 12 to 14 hours photoperiod increase number of florets, spike length and percentage of flowering. Higher temperature during growth period i.e., from 20-30^{0}C enhances the flowering. Gladiolus prefers high light intensities with controlled temperature. Temperatures below 5^{0}C and any type of shade pose adverse effect on flowering and flower quality. Corm storage at low temperature (3 to 7^{0}C) is conducive for better growth and flower production.

Tuberose

Temperature is the most important factor affecting growth and initiation and subsequent flower development of bulbous plants. Optimum temperature for growth and development of tuberose flowers ranges between 20-30^{0}C. Temperatures above 38^{0}C as well as below 10^{0}C reduce spike length, weight and quality of flowers. Bulbs remain dormant during winter season when the temperature is below 10^{0}C. Although tuberose is not strictly photo sensitive, long day exposure promotes vegetative growth as well as early emergence of the first flower spike. A day length of about 16 hours promotes growth and flowering. Low light intensity adversely affects flowering. Therefore, in short day conditions during winter, flower production and its quality are reduced upto 80% in case of single varieties. However, double varieties of tuberose produce flower even with low intensity of light and short photoperiod with little bit reduction in comparison to normal season. Bending of spike is also a problem due to absence of light intensity especially during rainy season and continuous clouds (<72 hours).

Marigold

Growth and flowering in marigold is usually governed by day length and temperature. African marigold produces flower under long day at temperature of 12 to 23^{0}C. The species of Marigold shows the behavior of both types viz., long day and short day types. The flower quality and its yield was higher under 24^{0}C temperature. The various hybrids developed through intraspecific crosses had shown flowering in all seasons with a variation in flower quality and yield.

Orchid

A wide range of variation in respect of growing environment for orchid species and genera has been observed. Environmental conditions play a key role in growth and flower production and these can hardly be separated from nutrients and growing media. Light intensity is essential for growth and development of orchids. The requirement of light intensity varied with genotypes and species. The optimum requirement in general varies from 11,500 to 2,400-foot candles. For growing temperate orchids, the typical climatic conditions demand the availability of cool and humid micro-climate coupled with plenty of aeration. Such conditions are possible in shade-net houses. The double layer shade net of 50% and 24% is preferred for commercial cultivation. However, tropical orchids prefer warm and humid atmosphere. Most of the orchids are day neutral.

CONCLUSION

The agro ecological zones and their suitability for production of various horticultural crops would certainly increase the productivity and quality. The influence of climate on horticultural crops is also essential for forecasting of pests and disease, which in turn will be useful to the planners and policy makers. However, the data on this aspect is not sufficient to increase the area on cultivation of horticultural crops. Since, there is a significant variability within the zone, micro-level planning based on field performance would be required for gainful utilization of agro ecological zones. The horticultural crop possesses the characters like wider adaptability, climatic acclimatization, various production system and wide genetic base of horticultural crops is advantageous for extending the area under different zones for harnessing the maximum potential. Govt.of India has taken effort to promote the export of various horticultural crops grown under organic mode by declaring the Agri export zones (Tripura- pineapple, Assam- tea and ginger, Maharashtra- Grapes and pomegranate, Andhra Pradesh- Mango, Himachal Pradesh-Apple and etc.,). Horticulture is highly remunerative and can bring prosperity to the growers. The need of the hour is to select the best performing breed suitable for a particular climatic condition in order to boost the productivity.

REFERENCES

Anonymous, (2002). *In Book: Handbook of Horticulture,* Published by DIPA, ICAR, New Delhi.

Knight, R.J. (1980). In: *Tropical and sub tropical fruits: Composition, properties and uses* (Eds.S. Nagy and P.E.Shaw). AVI Publishing Inc., Westport, CT, chap.1.

Marler, T.K.(1994). In: *Handbook of Environmental Physiology of fruit Crops:* Vol.II,Subtropical and tropical crops(Eds.B.Schaffer and P.C.Andersen), CRC Press, Inc., Boca Raton, FL, pp.213-216

National Horticulture Board database, 2003, Govt.of India, Gurgoan, Haryana

National Horticulture Board database, 2004, Govt.of India, Gurgoan, Haryana

Samson, J.A (1980).Tropical fruits, Longman, London.

Climate Change and Food Security, 2008
Edited by: M.Datta, N.P.Singh & Er.D.Daschaudhuri, pp.267-274
New India Publishing Agency, New Delhi (India)
E-mail : newindiapublishingagency@gmail.com
Web: www.bookfactoryindia.com

18

Influence of Environment in Cultivating Edible Mushrooms

S. BISWAS AND N.P. SINGH

INTRODUCTION

Edible mushrooms are higher fungi occurring seasonally all over the world in various habitats varying from sandy plains to thick forests. About 2000 fungal species in the world have been identified as edible, of which, only a few (20) are cultivated commercially (Chang and Miles, 1991). These mushrooms are generally considered as the good source of protein, vitamins and minerals in addition to their flavour and condiment value. About 3.5-4% protein with 71-90% digestibility of fresh mushrooms is nearly twice as that in most of the vegetables (Sohi, 1988). Vitamins, especially those of B complex group like thiamine (B1), riboflavin (B2), niacin and biotin, and minerals like potassium, phosphorus, sodium, magnesium and calcium are present in adequate quantities (Crisan and Sands, 1978; Rai, 1995). Apart from this, they have certain important medicinal values, such as, the effects of antitumor (with *Lentinus edodes* = *Lentinula edodes*), hypolipidemic and lypocholesterolemic (*Pleurotus sajor-caju; L. edodes*), antibacterial (*Agaricus bisporus*), antifungal (*L. edodes*) and antiviral (*L. edodes*) (Cochran, 1978; Rai, 1995). The mushrooms can survive in a wide range of climatic conditions, but, for satisfactory commercial production they must be given conditions fairly near to the ideal at each successive stage of growth. Failure to do this will result in weaken or slower growth. A large number of climatic factors are involved in the growth and development of mushrooms. In the current treatise, a few of the major parameters such as temperature, humidity, rainfall, light and CO_2 contents in air have been discussed.

TEMPERATURE

Mushrooms grow in a wide range of temperature. Some show preference for low temperature (*Agaricus bisporus; Lentinula edodes, Pleurotus* spp.) while the others require higher temperatures (*Calocybe indica; Volvariella volvacea*) for their growth and development. Therefore, the optimum temperature requirement of any given mushroom should be ascertained before it is selected for commercial cultivation in a particular place. Further, it is necessary to keep in mind that different strains of the same species occasionally differ markedly in their response to temperature. Some of the important cultivated mushrooms and their optimal temperature requirements are presented in Table1. Kim (1978) has reported that spring and autumn seasons of Korea having a temperature range of 10-20^{0}C are the best natural conditions for growing button mushroom (*A. bisporus*). However, in Indian conditions, 16-18^{0}C is considered as the most suitable temperature for fructification of the mushroom, although its mycelial growth and spawn run require to some extent higher temperature, 22-25^{0}C (Sohi, 1988; Bahukhandi and Sharma, 2002)). Apart from this, a high temperature-tolerant button mushroom (*A. bitroquis*) has also been introduced for cultivation under the tropical and sub-tropical climatic conditions in India due to infeasibility of the common button mushroom (*A. bisporus*) for cultivation. As regards the milky/*dudh* mushroom, it grows well during summer and rainy season in Tripura (Biswas and Singh, 2005-2007, unpublished). This mushroom requires higher temperature (30-32^{0}C) for fructification than that required for mycelial growth which is in contrast to other edible mushrooms (Sharma *et al.*, 2006). Fluctuation of temperature is very much essential to induce mycelial differentiation in Shiitake mushroom (Komatsu, 1961). Thus, in certain cases, chilling treatment with ice is in practice for fruit body initiation. Ito (1978) has suggested maintaining 12-20^{0}C for fructification of shiitake mushroom. Oyster mushrooms (*Pleurotus* spp.) are mostly cultivated within the temperature range 20-30^{0}C. The growing season of oyster mushrooms in north India is October to April, while in north east hills during September to mid November and in Tripura through out the year (Bahukhandi and Sharma, 2002; Chandra *et al.*, 1998; Biswas and Singh, 2007).

Table 1. Temperature requirement for optimal growth of different cultivated mushrooms

Mushroom	Growth stage	Temperature requirement (^{0}C)	Source
Agaricus bisporus (Button mushroom)	Mycelial growth	26.7	Dugger (1901)
		25	Lambert (1932)
		24	Treschow (1944)
		24-25	Hayes (1978)
		22-25	Bahukhandi and Sharma (2002)
	Spawn growth	20-25	Edwards (1978)
	Fructification	20-22	Edwards (1978)
		19.44 (67^{0}F) -21.66 (71^{0}F)	Samp and Phelps (1986)
		16-18	Sohi (1988)
A. bitroquis (Button mushroom)	Spawn growth	29-31	Chadha and Sharma (1995)
	Fructification	25-26	Chadha and Sharma (1995)
Calocybe indica (*Dudh/Milky* mushroom)	Mycelial growth	25	Sharma *et al.* (2006)
	Spawn growth	27-28	Sharma *et al.* (2006)
	Fructification	30-32	Sharma *et al.* (2006)
Lentinula edodes (Shiikate mushroom)	Mycelial growth	25	Tokimoto and Komatsu (1978); Sharma *et al.* (2006)
		24-28	Ito (1978)
	Fructification	12-20	Ito (1978)
		23-25	Sharma *et al.* (2006)
Pleurotus eous (Oyster mushroom)	Fructification	21-35	Purkayastha and Chandra (1985)
P. eryngie (Oyster mushroom)	Mycelial growth	25	Zadrazil (1978)
P. flabellatus (oyster mushroom)	Fructification	20-30	Suman and Sharma (1999)
P. florida (Oyster mushroom)	Mycelial growth	30	Zadrazil (1978)
P. ostreatus (Oyster mushroom)	Mycelial growth	30	Zadrazil (1978)
P. sajor-caju (Oyster mushroom)	Mycelial growth	22-28	Bahukhandi and Sharma (2002)
	Fructification	22-30	Charkarvarty and Sarkar (1982)
		22-28	Bahukhandi and Sharma (2002)
		23.6-28.5	Biswas and Singh (2007)
Volvariella volvacea (Paddy straw mushroom)	Mycelial growth	35	Chang and Chu (1969); Biswas (1988)
		30	Chandra and Purkayastha (1977)
	Spawn growth	34-35	Chang (1978)
	Fructification	28-32	Chang (1978)
		30-43.2	Purkayastha *et al* (1981b)
		30-42	Suman and Sharma (1999)

The range of maximum temperature 20-28^{0}C and the minimum 12.2-20.5^{0}C has been determined as the most suitable period for *P. sajor-caju* (Chandra *et al.*, 1998). Biswas and Singh (2007) have reported that the maximum temperature of atmosphere has greatly influenced on- and that correlates negatively with- the yield. The spores of *V. volvacea* germinate well at 40^{0}C, but, the initial hyphal growth requires 35^{0}C (Chang and Chu, 1969). The optimum temperature for the growth of mycelia is 30-35^{0}C and the growth rate drops markedly at 20 and 40^{0}C (Chang, 1978; Biswas, 1988). Further, it is to note that the production of paddy straw mushroom is quite sensitive to environmental factors. Purkayastha *et al.* (1980, 1981a, b) have found that June and July in West Bengal with temperature range 22.3 - 43.2^{0}C are the most favourable months for paddy straw mushroom production, although, the air temperature of 28-32^{0}C is the most suitable condition for fruit body formation and further increase in temperature though enhances the fruit body development but reduces the yield and quality of the mushroom (Chang, 1978).

RELATIVE HUMIDITY

High relative humidity is favourable for growing all types of mushrooms. It also prevents drying up of compost and substrate surfaces. Sohi (1988) has suggested maintaining 80-90% relative humidity with moisture content of 62-63% (in plastic bag method) and 65-67% (in tray method) in compost for button mushroom cultivation. Moisture content of straw as well as relative humidity has been found to play certain important roles in *V. volvacea* cultivation. Under natural conditions bed moisture 65-70% and the relative humidity of 80-90% around the beds should be maintained to cultivate the mushroom (Chang, 1965; Yau and Chang, 1972). High moisture content (90-95%) is also required for *C. indica* cultivation (Purkayastha and Chandra, 1985). In case of oyster mushroom, high humidity (80%) is maintained by hanging wet gunny cloths inside the mushroom house for initiation of fruit bodies and their developments (Chandra *et al.*, 1995).

RAINFALL

So far, rainfall has no direct impact in mushroom cultivation under in house condition, except that it causes rising of relative humidity to a great extent which favours fruit body formation and its growth. On the other hand, in nature its effect is enormous. It moistens substrates and creates the conditions favourable for the mushrooms' growth.

Practically, all the cultivated mushrooms of India grow on damp soil rich in organic matter or on rotten wood of trees, etc. after rain or in rainy season (Purkayastha and Chandra, 1985).

LIGHT

Light is also an important factor for initiation of fruit bodies. In *Pleurotus* spp., exposure to light (artificial or sunlight) is necessary for initiation of fruit body primordia and thus light is given for at least 15 minutes per day (Zadrazil, 1978). There is, also, a positive phototropism in *Pleurotus* species during the development of fruit bodies (Block *et al.*, 1959; Gyurko, 1972). Stimulation and retardation of mycelial growth in light (2000 lux) are also evident with *V. volvacea* and *Macrolepiota procera* (parasol mushroom), respectively (Biswas, 1988).

CARBON DIOXIDE

It is commonly said that mushroom needs no fresh air during spawn growth. That is not true because accumulation of CO_2 only up to 4% appears to be beneficial to the mycelial growth of *Agaricus bisporus* (Edwards, 1978). The concentration of CO_2 must be under 0.08% for fruit body initiation of button mushroom and that even less (0.05%) in certain strains. Low CO_2 (not beyond 0.15%) has been found as critical factor for obtaining good yield (Sohi, 1988). Further, it is evident that 0.3-0.5% CO_2 in the air above the casing soil results elongation of stipes and inhibition of sporophore initiation. Good aeration to minimize CO_2 concentration is also advocated for *C. indica* fruit body formation (Purkayastha and Chandra, 1985). However, on the contrary, mycelial growth of *Pleurotus* spp. is stimulated by high CO_2 concentrations in air (Schanel, 1970; Zadrazil, 1975). It has been observed that CO_2 concentration in air up to 28% (of volume) stimulates the growth of *P. ostreatus* and *P. florida,* while, that up to about 22% stimulates the growth of *P. eryngie.*

CONCLUSION

Pertinent information on the influence of environment in cultivating edible mushrooms has been generated during the last few decades. These environmental factors play very important role in all stages of mushroom growth. So, much emphasis is needed to establish commercial mushroom farms in respect of regularisation of atmospheric temperature, humidity, light intensity and carbon dioxide concentration in air. Proper thickness of wall in mushroom house, ventilation with exhauster, using of insulating materials and arrangement of misting apparatus and light systems are required for preparation of mushroom house. Any faulty designing in mushroom house preparation will cause a great failure in successful cultivation. Further, it is ,worth while, to mention that the choosing of suitable mushroom species for cultivation under a given agro-climatic condition is an important part of mushroom improvement programme, particularly, for that place where mushroom cultivation is likely to be done under natural condition.

REFERENCES

Bahukhandi, D. and R. K. Sharma. (2002). Interspecific hybridization in *Pleurotus* species. *Indian Phytopath.* 55: 61-66.

Biswas, S. (1988). Physiological studies of some edible gill fungi. Ph. D thesis, University of Calcutta, West Bengal.

Biswas, S. and Singh, N. P. (2007). Effects of climatic conditions and substrates on the production of *Pleurotus sajor-caju in* Tripura. *Environment and Ecology,* 25: 168-172.

Block, S. S., Trao, G. and Han, L. (1959). Experiments in the cultivation of *Pleurotus ostreatus*. *Mushroom Sci.* 4: 309-325.

Chandra, A. and Purkayastha, R. P. (1977). Physiological studies on Indian mushrooms. *Trans. Br. Mycol. Soc.*, 69: 63.

Chadha, K. L. and Sharma, S. R. (1995). *Advances in Horticulture,* Vol. 13. Malhotra Publishing House, New Delhi, India. P-649.

Chandra, S., Singh, A. K., Bhat, M. V. and Kumar, S. (1998). Yield performance of Pleurotus sajor-caju on some selected substrates in North-Eastern region of India. *Mushroom Res.* 7: 79-80.

Chandra, S., Singh, A. K. and Kumar, S. (1995). *Cultivation of oyster mushroom.* Technical bulletin, ICAR Research Complex for NEH Region, Umroi Road, Barapani, Meghalaya.

Chang, S. T. (1965). Cultivation of the straw mushroom in S. E. China. *World Crops,* 17: 47-49.

Chang, S. T. (1978). *Volvariella volvacea,* pp573-603. In S. T. Chang and W. A. Hayes (eds), *The biology and cultivation of edible mushrooms.* Academic Press, New York, P-819.

Chang, S. T. and Chu, S. S. (1969). Factors affecting spore germination of *Volvariella volvacea. Physiol. Plant.* 22: 734-741.

Chang, S. T. and Miles, P. G. (1991). Recent trends in world production of cultivated edible mushrooms. *Mushroom J.*, 504: 15-17.

Cochran, K. W. (1978). Medicinal effects, pp. 169-187. . In S. T. Chang and W. A. Hayes (eds). *The biology and cultivation of edible mushrooms.* Academic Press, New York, P-819.

Crisan, E. V. and Sands, A. (1978). Nutritional value of edible mushrooms, pp.137-168. In S. T. Chang and W. A. Hayes (eds). *The biology and cultivation of edible mushrooms.* Academic Press, New York, P-819.

Dugger, B. M. (1901). Physiological studies with reference to the germination of certain fungus spores. *Bot. Gaz.*, 31: 38.

Edwards, R. L. (1978). Cultivation in Western Countries: Growing in house, pp299-336. In S. T. Chang and W. A. Hayes (eds), *The biology and cultivation of edible mushrooms.* Academic Press, New York, P-819.

Gyurko, P. (1972). *Die Rolle der Belichlung be idem Anbau des Austernseillings (Pleurotus ostreatus). Mushroom Sci.* 8: 461-469.

Hayes, W. A. (1978). Biological Nature, pp 191-217. In S. T. Chang and W. A. Hayes (eds), *The biology and cultivation of edible mushrooms.* Academic Press, New York, P-819.

Ito, T. (1978). Cultivation of *Lentinus edodes*, pp461-473. In S. T. Chang and W. A. Hayes (eds), *The biology and cultivation of edible mushrooms.* Academic Press, New York, P-819.

Komatsu, M. (1961). Morphological characters of the hyphae of Lentinus edodes (Berk) Sing., grown under the fluctuated temperatures and those during fruiting. *Rep. Tottori Mycol. Inst. (Jpn.)* 1:45-59.

Kim, D. S. (1978). Cultivation in Asian Countries: Growing in temperate zones, pp345-363. In S. T. Chang and W. A. Hayes (eds), *The biology and cultivation of edible mushrooms.* Academic Press, New York, P-819.

Lambert, E. B. (1932). Mushroom growing in the United States. U. S. Dept. Agr., Washington Circ. 251.

Purkayastha, R. P., Biswas, S. and Das, A. K. (1980). Some experimental observations on the cultivation of *Volvariella volvacea* in the plains of West Bengal. *Indian J. Mushrooms*, 6: 1-9.

Purkayastha, R. P., Biswas, S. and Das, A. K. (1981a). Factors affecting production of paddy straw mushroom *(Volvariella volvacea). Indian J. Mushrooms,* 7: 26-30.

Purkayastha R. P. and Chandra A. (1985). *Indian Edible Mushrooms.* Today & Tomorrow's Printers and Publishers, New Delhi, India, P.267.

Purkayastha, R. P., Das, A. K. and Biswas, S. (1981b). Cultural practices and seasonal conditions affecting production of paddy straw mushroom. *Taiwan Mushrooms*, 5: 10-16.

Rai, R. D. Nutritional and medicinal values of mushrooms, pp. 537-551. In K. L. Chadha and S. R. Sharma (eds). *Advances in Horticulture*, Vol. 13 (Mushroom), Malhotra Publishing House, New Delhi, India, P-649.

Samp, R. J. and Phelps, J. K. (1986). Observations of the effects of growing temperature on first break mushrooms, pp265-270. In P. J. West, D. J. Royse and R. B. Beelman (eds). *Cultivating Edible Fungi*. Elsevier, P-677.

Schanel, L. (1970). Effect of carbon dioxide on the growth of wood-decaying fungi. *Folia Biol.* (Prague) 11: 67-72.

Sharma, S. R., Kumar, S. and Sharma, V. P. (2006). Physiological requirement for cultivation of Malaysian strain of shiitake, *Lentinula edodes*. *J. Mycol. Pl. Pathol.* 36: 149-152.

Sharma, V. P., Sharma, S. R. and Kumar, S. (2006). Comparative studies on substrate treatment methods for cultivation of *Calocybe indica*. *J. Mycol. Pl. Pathol.* 36: 145-148.

Sohi, H. S. (1988). Mushroom culture in India – Recent research findings. *Indian Phytopath.* 41: 313-326.

Suman, Y. S. and Sharma, B. K. (1999). Cultivation of *Volvariella volvacea* and *Pleurotus flabellatus* in submontane and low hills of Himachal Pradesh. *J. Mycol. Pl. Pathol.*, 29: 250-251.

Tokimoto, K. and Komatsu, M. (1978). Biological nature of *Lentinus edodes*, pp.445-459. In S.T. Chang and W.A. Hayes (eds), *The biology and cultivation of edible mushrooms*. Academic Press, New York, pp.819.

Treschow, C. (1944). Nutrition of the cultivated mushroom. *Dansk. Bot. Ark.*, 11:1.

Yau, C. K. and Chang, S.T. (1972). Cotton waste for indoor cultivation of straw mushroom. *World Crops*, 24: 302-303.

Zadrazil, F. (1975). Influence of CO_2 concentration on the mycelium growth of three *Pleurotus* species. *Eur. J. Appl. Microbiol.* 1: 327-335.

Zadrazil, F. (1978). Cultivation *Pleurotus*, pp.521-557. In S.T. Chang and W.A. Hayes (eds), *The biology and cultivation of edible mushrooms*. Academic Press, New York, P-819.

Climate Change and Food Security, 2008
Edited by: M.Datta, N.P.Singh & Er.D.Daschaudhuri, pp.275-291
New India Publishing Agency, New Delhi (India)
E-mail : newindiapublishingagency@gmail.com
Web: www.bookfactoryindia.com

19

Climate & Livestock Production

CHANDER DATT AND N.P. SINGH

INTRODUCTION

Livestock plays a very important role in the mixed farming system of the country contributing about 25 per cent of Gross Domestic Products (GDP) from agricultural sector. India has the largest population of cattle and buffaloes and the highest producer of milk in the world. India has also the second largest population of goats and one of the leading countries in poultry sector. There has been a tremendous increase in milk, meat and egg production since independence. However, owing to various reasons viz., low genetic potential, scarcity of feeds and fodders, diseases, poor management practices etc., the productivity is quite low. Climatic conditions of a place have also an important bearing on livestock production and productivity. Climatic factors like ambient temperature, humidity, solar radiation and wind velocity are the important ones which make up the physical environment of the animal having direct and indirect influences on animal's responses towards climate. Temperature and humidity are the most crucial with the former being the most researched one. Extremes of hot and cold weather are creating stress to any animal and the effects are more severe under hot and humid climate. The varied effects of undesirable climate on animals are pertaining to feed and water intake, physiological, biochemical and endocrinological aspects, reproduction, health and productivity etc. Thus, the positive and negative effects of the environment on the livestock have been discussed in a way to undertake some protective measures.

LIVESTOCK POPULATION OF INDIA

India has the largest population of cattle and buffaloes comprising about 16.2 and 57 per cent of the total world population, respectively (Table 1). India ranks second in terms of goat population (16.7 %) and is also a leading country in terms of poultry population (5.2%).India is the largest producer of milk (92 million tones) comprising 15 per cent of the world's milk production. (Mangala Rai, 2006). The population of other domesticated species is also quite good.

Table 1. Livestock population (x 10^3) of India and the world (FAO,2001)

Species	India	World
Cattle	2,19,642	13,51,792
Buffalo	94,132	1,65,724
Goat	1,23,500	7,38,246
Sheep	58,200	10,56,184
Pigs	17,500	9,22,929
Camel	1,030	19,334
Asses	1,000	42,822
Horses	800	58,244
Mules	200	13,463

INTERACTION BETWEEN MACRO/MICROCLIMATE AND ANIMAL PRODUCTIVITY

It is evident from the diagram given below that the macroclimate (temperature, humidity, solar radiation, wind velocity etc.) of a particular place has a direct bearing on the microclimate of the animals which in turn affect their feed and water intake, various body functions thus affecting their overall productivity.

Not only physical environment is continually modified by variations in annual and perennial rainfall, temperature and solar insulation, it is also modified by human intervention. These and other processes not only change biomass production and species composition of vegetation but contribute to long term climate change.

CLIMATE AND SPECIES DISTRIBUTION

In many places of the world, livestock remains highly vulnerable to climate change because they are not kept in protected environment and not given supplemental feeds. As the air temperature rises (say >2O⁰C), the environment is congenial to rearing of goats / sheep but not to cattle. Across the globe, beef /milk cattle and sheep decline in numbers as precipitation increases. Humid and wet condition increase the probability of diseases. In contrast, goat and poultry birds may flourish in wet climate. In general, profitability from livestock tends to decrease, but the relative profitability from rearing of goat/ sheep / poultry birds increases with precipitation. In animal rearing, the impact of temperature is more than that of rainfall. In general, as the temperature rises, farmers will raise fewer animals per household. The net profitability will decline and there is less investment.

MACROCLIMATE OF A PLACE

(Ambient temperature, humidity, solar radiation, wind speed)

↓

Microclimate of the animal

Physiological responses	Feed and water intake	Milk production and composition	Blood metabolites	Endocrine glands
Body temperature		Milk yield	Glucose	Thyroid
Respiration rate		Total solids	NEFA	Adrenal
Pulse rate		Fat		
		Protein		
		Lactose		
		NEFA		
		Urea		

Source: Aggarwal and Singh, 2006

EFFECT OF HEAT STRESS ON FEED AND WATER INTAKE

Thermal stress results in decline of feed intake (Armstrong, 1994). It results in increased gut fill due to reduced rate of passage, increased water intake and respiration rate (McDowell, 1976). The effect of thermal stress on feed intake, physiological responses and milk yield is given in Table 2 (Lough *et al.,* 1990).

In lactating cows, dry matter intake begins to decline at mean environmental temperatures of 25^0-27^0C. (NRC, 1981). Other climatic factors like wind velocity, humidity and radiation are interrelated with temperature in affecting feed intake (Beede *et al.,* 1985). Water intake during thermal stress may increase by 0.2 to 2 folds compared to non-stress condition. Feed consumption and time spent for rumination decreased while water consumption increased when air temperature was above 24^{0}C (Korsun, 1993).

Table 2. DM intake, milk yield and physiological response in dairy cattle (Lough *et. al.,* 1990)

Measurement	Thermal comfort zone	Thermal stress zone
Rectal temperature (^{0}C)	39	39.9
Respiration rate (breath / min.)	49	83
Heart rate (beats / min.)	81	77
DM intake (Kg/ d)	17.8	14.8
Milk yield (Kg/ d)	23.9	21.6
Mammary blood flow during 4 min.	5.1	4.5

Singh and Bhattacharjee (1991) studied ME intake (Table 3) in different genetic groups of cattle exposed to different temperatures for 8 in 18 days. There was significant decline in feed intake at 32^0 or 37^{0}C as compared to 17^{0}C in all the groups, however, the magnitude of decline was more in *zebu* crosses compared to *zebu (Bos indicus)* cattle (Table 3). A combination of high ambient temperature, high humidity and lack of ventilation has more adverse effects. *Zebu* cattle are more tolerant to such conditions because they have pendulous sheath and dewlap which act as good radiators of heat and also that they have silvery coat and reflect the insulations well thus indicating that they absorb less of it. Therefore, under such conditions, *B. indicus* could perform better than *B. taurus.*

Table 3. Meatbolisable energy intake (KJ/Kg W$^{0.75}$/day) at different temperatures in various genetic groups

Exposure time(days)	Temp. (^{0}C)	Genetic group			
		Har.	Jersey x Har.	B.S.x Har.	H.F.x Har.
8	17	902	875	871	873
	32	791	749	716	708
	37	687	665	653	649
18	17	923	906	874	913
	32	780	757	711	694
	37	686	674	665	653

Har. = Haryana ; B.S.= Brown Swiss ; H.F.=Holstein Friesian

In climate chambers (18° vs 30°C), there was an increase in water intake by 39 per cent, decrease in faecal water loss by 33 per cent and loss of water via urine, skin surface and respiratory evaporation by an increase of 15, 59 and 50 per cent, respectively (McDowell, 1972).

EFFECT ON PRODUCTION PERFORMANCE

There is a reduction in milk yield as the temperature rises above thermal neutral zone largely due to decline in feed intake. High yielders are the most affected animals and reduction is more in early lactation. Fat and protein contents are the lowest during hotter weather and highest during cooler months. These changes are partially due to difference in availability of forages and their quality. Milk yield remains relatively constant up to 24°C and then declines (Revagnolo *et al.*, 2000) .At temperature - humidity index (THI) more than 72, the drop averages 0.2 kg per unit of THI for milk and 0.009 kg/unit of THI for fat. Somatic cell count (SCC), an index of inflammatory condition of udder, is negatively correlated to fat, lactose and milk production (Singh and Ludri, 2001) and SCC are higher under hot humid conditions. Mehta (1985) reported the milk yield production data taken over a period of 18 fortnights (Table 4). Milk yield was the highest in winter and lowest in hot humid season. Sastry (1988) concluded that buffaloes in mild lactation yielded significantly less milk in hot humid period (July-Oct.) than those in similar stage of lactation in cold season (Nov-Feb.). The differences in milk yield could be mainly due to direct and indirect effects of climatic variables.

Table 4. Effect of season on milk production in buffaloes (Mehta, 1985)

Season	Temp.(°C)	RH(%)	Av.milk yield(kg/d)	Drop in yield (%)
Winter (Jan.-Feb.)	12.3-14.7	65-73	9.48	
Spring (Mar.-Apr.)	20.4-29.4	42.66	8.48	9.5
Hot & dry (May-June)	28.1-33.0	36.58	6.26	26.2
Hot & humid (July-Aug.)	29.1-30.5	64-74	5.21	16.0

Pigs are exposed to the impact of a number of environmental factors among which temperature is the key factor. When the temperature is below lower critical temperature (LCT), pigs must increase heat production through shivering and other metabolic processes to maintain body temperature. On the other hand, when temperature is higher than its effective critical temperature (ECT), the evaporative heat loss of pigs begins to rise particularly by increased respiration. The zone of thermal comfort (temperature between LCT and ECT) has been reported to be 12-22^{0}C for sow and 30-37^{0}C for piglets (Black *et al.*,1993). The effect of temperature on the production parameters in pigs has been shown in Table 5 (Stansbury *et al.*,1987).

Table 5. Performance of pigs in different farrowing house temperatures (Stansbury *et al.*, 1987)

Variable	Temperature (^{0}C) 18	25	30
Litter weaning weight (kg)	63	61	52
Weaning No.	8.1	8.9	8.3
Pig weaning weight(kg)	7.8	6.8	6.4
Sow feed intake(kg/d)	6.5	6.2	4.2
Sow wt.loss(kg/lactation)	3.1	7.9	24.2

For each 1^{0}C increase in ambient temperature above 16^{0}C, daily voluntary energy and feed intake of lactating sows decreased by 2.4 MJ DE and 0.17 kg, respectively. Mullan *et al.,* (1992) reported that feed intake was depressed by 25 per cent and milk yield by 15 per cent for sows housed at 30^{0}C compared to those housed at 20^{0}C and the high temperature had a direct effect on milk yield. An ambient temperature above ECT of lactating sows leads to a reduction in feed intake and milk yield.

CLIMATE AND REPRODUCTION

High ambient temperature suppresses oestrus symptoms resulting in long service periods and interference in ovulation. Other reproductive parameters like age at puberty, oestrous cycle length, duration of oestrus, ova abnormalities, embryonic mortality, gestation length and faetal size, incidence of oestrus with weak signs, percentage of silent heat, ovulation failure, service period and number of services per conception may increase (Hafez,1968).Summer stress has negative effect on fertility in lactating cows mainly due to high ambient temperature (Wolfenson *et.al.*,2000). Heat stress reduces the length and intensity of oestrus. Heat stress can cause increased cortisol secretion and ACTH (Adreno corticotrophin hormone) has been found to block estradiol induced behaviour. The effects could be due to hypothalamus- pituitary- adrenal axis. There is a significant decline in conception rate. Effect of high temperature on early embryonic development are more than in the later stages of pregnancy. Acute thermal stress increases blood concentration of prolactin (Wetterman and Tucker, 1974), anti dieuretic hormone (El-Nauty *et.al.,* 1980), aldosterone in nonruminants (Lipsett *et.al.,*1961) and catecholamines and glucocorticoids in cattle (Alvarez and Johnson ,1973) .

Table 6. Seasonal incidence of oestrus / inseminations and conception rate of buffaloes (Mehta, 1985)

	Seasons				Country	Source
	Winter	Spring	Summer	Autumn		
Occurrence of oestrus	46.36	15.25	7.22-9.50	21.67	India	Kumar *et al.,* (1986)
Incidence of inseminations (%)	30.65	18.71	20.31 40.95 (Apr-June)	30.27	Pakistan India	Pasha *et al.,*(1986) Reddy *et al.,* (1987)
	31.77		12.34 (Rainy)		India	Tailor and Jain (1987)
	37.45	10.46	15.21	36.81	India	Srivastava and Kharche (1986)
Conception (%)	45.91	28.75	24.00-26.31 34.71 (June) 10.09 (June)	39.47 56.11 (Oct.) 33.95 (Oct.)	India Pakistan India	Kumar *et.al.,*(1986) Pasha *et al.,* (1986) Reddy *et al.,* (1987)

There is a seasonality in buffaloes with regard to heat and conception rate. Singh and Singh (1985) reported data on calvings observed over a period of one year at one organised farm. Post – partum oestrus interval and service period was the longest (175.4 and 248.9 days) among animals that calved in January (cold) and the shortest (61.9 and 114.3 days) among animals that calved in September (warm humid season).

It appears that by the time uterine involution is complete among buffaloes calving in winter, the weather starts warming up with the result that animals shy off from breeding .But those animals calving in hot and humid months pass off into winter favorable for conceiving and their reproductive set up is ready for next conception. The climate, particularly temperature has a significant effect on reproductive behaviour of buffalo owing to its effect on animal reproductive systems as well as effect on availability of nutrients to the animals. Sastry and Georgie (1978) analysed the meteorological data and reproductive parameters over a period of 30 years. The correlation between conceptions and temperature, RH (Relative humidity) or rainfall was significant at a few centres while the correlation between rainfall in the preceding 3 month and conception was significant at all centres except one. Khurana and Madan (1985) reported the highest thyroxin secretion rate and plasma thyroxin level during cold season (T=7.4^0-20.3^{0}C. RH=76-100 %) as compared with hot dry season (T=25.2-38.5^{0}C, RH=51-97%).Values in hot humid season were intermediate. Showering, splashing etc. have positive influence on reproductive efficiency through increased estrus and conception rate .Changes in housing systems also gave good results (confined vs open).

Table 7. Incidence of calving, post-partum oestrus interval, and post-partum service period (Dhanani *et al.*, 1985; Singh and Singh, 1985)

Month	Calvings	Post-partum oestrus interval (days)	Post-partum service period (days)
January	3	175.4	248.9
February	0	123.6	229.0
March	0	167.0	192.8
April	8	162.0	235.0
May	7	160.3	244.3
Jun	18	143.0	234.0
July	38	115.4	209.6
August	40	95.4	155.7
September	26	61.9	114.3
October	27	74.4	168.7
November	12	117.0	246.8
December	7	189.0	170.1

Pigs are not able to sweat and are more sensitive to heat than cold conditions (Ingram, 1965). An ambient temperature above ECT of lactating sows leads to a reduction in feed intake, milk yield, reproductive performance and growth of piglets. High ambient temperature causes delay in rebreeding after mating (Barb *et al.*, 1991). To reduce heat load wallows, spray or water drips can be used.

Decreasing the protein level in diet and increasing energy density helps in alleviating the heat effect of high temperature.

Table 8. Correlation between conception per cent and different climatic factors at ten Centres (Sastry and Georgie, 1978)

Centre	Maximum temperature	Minimum temperature	Relative humidity (08.30 hrs)	Relative humidity (17.30 hrs)	Rainfall	Rainfall in the preceding 3 months
Bangalore	-0.0038	-0.8169**	-0.2516	-0.2362	-0.4888	0.6777*
Hyderabad	-0.4918	-0.7757**	-0.0011	-0.2868	-0.0391	0.5619*
Vishakhapatnam	-0.8577**	0.8978**	0.1667	-0.5908*	0.0186	0.7692**
Hisar	-0.4951	-0.4523	0.5006	0.9371**	-0.1373	0.7997**
Lucknow	-0.5458*	-0.6272*	0.5014	0.1819	-0.4479	0.9381**
Jabalpur	-0.5684*	-0.4340	0.3710	0.1227	-0.3629	0.9266**
Bombay	-0.2318	-0.5590	-0.1773	-0.3264	-0.5557*	0.6141*
Khanapara	-0.4314	-0.5425	0.5472	0.6137*	-0.7724**	0.5773
Gudlavalleru	-0.6431*	-0.7206**	0.2074	-0.2342	-0.4185	0.7983**
Bareilly	-0.4472	-0.5592*	0.2582	0.0701	-0.4259	0.9399**
Overall	-0.1904*	-0.3245**	0.7423**	0.0170	-0.3724**	0.9250**

* $P<0.05$, ** $P<0.01$

COLD STRESS AND ANIMAL PRODUCTION

The adaptability to cold is well developed in most of the domestic animals. Cold weather is not such a stress as hot humid weather. Young ones of any species viz., chicks, calves, lambs, pigs etc. are more sensitive to cold damp weather compared to adults. It is because more body heat is lost as they have larger surface area per unit weight. The extent of response to cold weather depends on the effective temperature of the environment relative to LCT of the animal and duration of exposure. Exposure to temperature below LCT induces an increase in metabolic rate. Prolonged exposure induces a process of acclimation to cold involving changes in hormonal and sympathetic nervous activity and increase in resting metabolic heat. An extreme cold causes decline in feed efficiency and animal production (Young, 1981) which could be due to increased energy expenditure (Webster, 1976) and in part due to decrease in extent of digestion (Christopherson, 1976). In general, there is an increase in dry matter intake during cold season. The effect on digestibility largely depends in the over all composition of the diet.

When feed intake is limited in cold adapted animals, there is a decline in body protein synthesis The percentage of protein in animal ration may be decreased because animal's energy needs an increase while that of protein remains almost constant during cold stress.

EFFECT OF DROUGHT ON ANIMAL PRODUCTION SYSTEM

Animal production is correlated with fodder supply, the production of which becomes restricted due to drought condition. Droughts of 1980's and 1990's resulted in reduced livestock production and maize yields including a 50 per cent crop failure in one region of South Africa. The impact of the drought in Vryburg district during 1960 showed ,on an average, 80 - 82 % decline in above ground biomass production of major grass species. While drought effects are often measured in terms of economic losses such as crop failure and live stock losses, drought conditions are also usually associated with decreased levels of employment, income, nutrition and wealth. Under drought conditions, animals often consume poisonous plants causing toxicity. Due to water restriction, there is a reduction in feed intake, feed efficiency and weight gain. Low prices of livestock have been observed during drought.

CLIMATE AND ANIMAL HEALTH

Environmental effects on animal health involve complex interaction between environmental and animal factors. Out of the factors, the former one includes temperature, humidity, wind speed and solar radiation (Appleman and Bath, 1983). In those countries where the year is naturally divided into 'dry season' and 'the rains', the diseases in animals are seen to be different in 2 seasons. However, in temperate climates where rain may occur any time of the year and its distribution in space as well as time is varied, it is difficult to see any particular pattern of disease. Humidity affects the disease incidence in many ways. Some species of pathogens survive best at low humidity but others like wet climate. There are exceptions like *E. Coli* which survives at both high and low humidity but beyond RH of 84 % , it succumbs to. It is apparent that dry atmosphere of fine sunny summer after noon may contain more moisture than the air of cold dampy winter morning hours and the former situation is putting more negative effect on animal production system.

Tropical dermatophilosis of cattle is also a condition of rainy season. The liver dystrophy which results from photosensitization of facial eczema of sheep in New Zealand caused by fungus, is governed by the prevailing temperature and humidity. The incidence of hemorrhagic septicemia in cattle in south Asia occurs particularly after onset of monsoon. It has been noticed that hypo magnesaemia is higher where weather is rough with high wind and rain and it could be due to lower feed intake. Chances of occurrence of aflatoxicosis are more during hot and humid weather.

In north eastern region, special care should be taken for coccidiosis from May to September and for mite infection from October to December in Pigs (Bujarbaruah,2006).Monthly deworming for GI (Gastro intestinal) parasites during monsoon period(April -September is necessary in goats. Chances of coccidiosis are more from May to September.In rabbits, monthly preventive drug against coccidiosis is always helpful. Care should also be taken for mite infection and GI parasites from October to December. In poultry, it is advisable to take control measures for coccidiosis during April to September, for GI tapeworms from February to November. Special care should be taken for coccidial infection in cattle during monsoon.

Cold weather causes animals to go for shelter they find. Even a slight wind will diminish the insulating properties of an animal coat. If shelter is left ,the sheep are likely to suffer from pregnancy toxemia. Snow lying is the added factor to cause this disease. Sway back in sheep may be due to copper deficiency but low temperature may aggravate the incidence. Frost is known to have at least two effects on animal health. Heavily frosted grass appears to cause devitalisation of mucous membrane of the GI tract of sheep. In such cases, tissues can be easily invaded by *Clostridiun septicum* and toxemia so caused may kill the sheep.

Altitude has also got effects on animal health. At greater heights, low oxygen content of atmosphere may cause chronic hypoxia with difficult breathing and heart failure eventually. Breed differences matter in that regard as well as in case of heat tolerance.

Pigs in outdoor environment have been reported to have better health than indoor herds especially fewer respiratory problems and a lower incidence of enteric diseases (Thornton, 1990, Tubbs *et al.*, 1993). However, mortality due to urogenital diseases, heart failure and locomotor's problems have been reported to be higher in outdoor system (Karg and Bikei, 2002). Outdoor pigs have higher parasitic burden which increases the nutrient requirement for maintenance and reduces feed utilization efficiency. Occurrence of parasites is affected by season mainly due to variation in temperature. Many eggs deposited during summer may die rapidly due to high temperatures. Some eggs deposited in colder months by foraging pigs can not survive through lower temperatures, more moisture and higher sequestration of eggs in soil by rain and earthworms. Roepstorff and Murrell (1997) reported that both *O. dentatum* and *H. rubidus* are very sensitive to environmental factors and significant transmission occurred only under the most favorable conditions. Transmission was reduced during low temperature in winter.

Basic Management Strategies to Overcome Climatic Stress

(i) Physical modification of the environment

(ii) Genetic improvement of less sensitive breeds

(iii) Nutritional management

EFFECT OF LIVESTOCK ON ENVIRONMENT

Livestock contribute positively to the environment in many ways (Dixit, 2005) :(i) Saving of natural resources as dung is used as fuel in place of fuel wood (ii) Saving of chemical fertilizers as dung is used as manure and thus green house gas emission is prevented in the process of fertilizer manufacture. (iii) Saving land due to recycling of agricultural byproducts as animal feed (iv) Prevention of emission of greenhouse grasses specially due to draught power used in agriculture. However livestock also contribute negatively particularly in form of CH_4 emission which has several times (40-60) global warming potential as compared to CO_2 .Global warming and ozone depletion due to increased emission of green house gases(IPCC, 1996) in atmosphere has drawn worldwide attention. Carbon dioxide, methane and nitrous oxide are the main gases emitted from agricultural system including livestock. CH_4 is the second contributor to global warming after CO_2. Enteric fermentation of the feeds in the ruminants and emission from dung produced and its managemental aspects are 2 major sources of CH_4 production from livestock. Indian ruminants contribute about 13.2 per cent of total CH_4 emitted by world livestock (WRI, 1990). Annual CH_4 produced by Indian livestock is estimated to be 10.0 Tg (WRI, 1990), 10.4 Tg (EPA,1993) 8.9 Tg (Khan, 1996) , 7.26 Tg (Garg and Sukla, 2002) and 10.0g Tg (Singhal and Mohini, 2002).

Besides environmental pollution, CH_4 emission from ruminants represents a loss of 6-8 per cent of GE intake. Khan (1996) reported that cattle population in developed/developing countries led to 58 % and 42 % in CH_4 production . Indian cattle comprising about 16 per cent of world cattle population emit less that 10 per cent of total CH_4 produced by world cattle population .Loss of energy due to CH_4 production by ruminants is in the order of cattle > buffalo >goat > sheep. India is losing energy worth Rs. 2.2 crores a day in CH_4. Though CH_4 production can not be totally eliminated, however, if reduced by 20 per cent, Rs.45.38 lakhs could be saved (Singh, 2001). Global warming can have physical, environmental and socioeconomic consequences. Climatic changes could cause change in precipitation pattern, ocean circulations, marine system, soil moisture, water availability and rise in sea level. With rising temperature, alteration in vector-borne disease, global population would be more vulnerable to health hazards causing disruption in settlement patterns and large scale migration (Haque, 2001).

There are various factors which affect methane emission from ruminants like quality, type of diet consumed by the animals, feeding frequency, ruminant species, production level of the animals, forage type, use of ionophores, processing of feeds, use of halogenated compounds etc. So, a strategy involving these factors needs to be evolved so as to reduce CH_4 production by the ruminants.

CONCLUSION

Among the different climatic variables ,high temperature and humidity are prominent ones in influencing an animal's response with regard to feed and water intake, growth/milk production and composition, biochemical and endocrinological changes, reproduction, health and hence over all productivity. The interaction of these variables is complex and it is most difficult to alleviate the cumulative effects of these factors. Management strategies involving physical modification of the environment, selection or development of less sensitive animals / breeds and nutritional management if followed, can help in alleviating the ill effects of unfavorable climatic conditions. The animals also have both negative and positive effects on environment .The negative effects being chiefly due to methane emission which has several times global warming potential as compared to carbon dioxide. Methane emission from ruminants also reduces the efficiency of energy utilization. So, a strategy involving modification of the factors affecting methane production *viz.*, quality and type of diet consumed by the animals, feeding frequency, production level, feed processing methods, use of ionophores / halogenated compounds should be followed so as to minimize the production of methane by the ruminant animals and its adverse effect on the climate.

REFERENCES

Alvarez, A. and Johnson, M.B.(1973). Environmental heat exposure on cattle plasma catecholamine and glucocoricoids. *J. Dairy Sci.* 56:189.

Aggarwal, A. and Singh, M. (2006).Impact of microclimate modification on production of dairy animals during summer. *Indian Dairyman*.58:49-59.

Appleman, R.D. and Bath,D.L. (1983).Facilty and equipment needs of dry cows.4 *ASAE SP*-4-83:273-80.

Armstrong, D.V. (1994). Heat stress interaction with shade and cooling. *J.Dairy Sci.*77:2044-2050.

Barb, C.R., Estienne, R.R., Kraeling, M.J., Marple, D.N., Rampace G.B., Rahe ,C.H. andSartin,J.L. (1991).Endocrine changes in sows exposed to elevated ambient temperature during lactation. Domest. *Anim. Endo.*8:117-127.

Beede, D.K., Collier,R.J., Wilcox,C.J. and Thatcher ,W.W. (1985).Effect of warm climates on milk yield and composition : short term effects. In: *Milk Production in Developing Countries*(Ed.Smith,A.J.).

Black,J.L., Mullan B.P., Lorschy, M.L. and Giles,L.R. (1993)Lactation in the sows during heat stress. *Livestock Production Science*.35:153-170.

Bujarbaruah,K.M. (2006). *Steps towards increasing livestock production in north eastern region*.ICAR Research Complex for NEH Region , Meghalaya-793103.*Res. Bulletin* No.45.

Christopherson, R.J. (1976).Effects of prolonged cold and outdoor winter environment on apparent digestibility in sheep and cattle. *Canadian J. Animal Sci.*56:201-121.

Dixit,A.K. (2005).Some interrelationships between livestock and environment in India. *Ind.J.Anim.Sci.*75:1444-1447.

Dhanani, J.S., Khangharani, M.U. and Seikh, Z.A.(1985).Reproductive performance of Kundi buffalo at Kundi Buffalo Farm at Rohri.*IBID*. p.952.

El-Nouty, F.B., Albana, I.M., Davis , T.P. and Johnson, H.D.(1980). Aldosterone and ADH response to heat and dehydration in cattle. *J.Appl.Physiol. Environ.Exercise Physiol.*48:249.

EPA. (1993). *International Anthropogenic Methane Production*. EPA-230-R-93-010 Washington, D.C.

FAO. (2001).FAO Production Year Book.FAO,Rome.Vol.55.

Garg A.and Sukla, P.R. (2002). *Emission Inventory of India.* Tata McGraw hill Publication Co. Ltd.p.04.

Hafez, E.S.E.(1968). *Adaptation of Domestic Animals.* Lea and Febiger. Philadelphia,U.S.A.

Haque, N. (2001).Environmental implications of methane production: diet and rumen ecology. In:"*Nutrition and dietetics under clinico therapeutic conditions of pet and farm animals*" (Short course) held at IVRI, Izatnagar from March 20 to April 18,2001.pp355-361.

Ingram, D.L.(1965). Eavporative cooling in the pig.*Nature.*207:415-416.

Karg, H and G. Bilkei. (2002). Causes of sow mortality in Hungarian outdoor and indoor pig production units.*Berliner and Munchenr Tierarztliche Wochenschrift.*15:366-368.

Khan, M.Y. (1996).Feeding strategies for management of gaseous pollutants produced by farm animals.In: *Monitoring effects of environmental pollutants on animal health* (Short Course) held at IVRI, Izatnagar.pp116-118.

Khurana, M.L. and Madan, M.L. (1985).Thyroxin secretion rate in buffaloes during hot dry, hot humid and cold seasons. *In Proc: First World Buffalo Congress,*Cairo.p-1165.

Korsun, B.A. (1993). Feeding cows under shaded awnings during summer. *MolchnoM'Yasne Skotarstvo*, 82: 52-54.

Kumar,S.,Purbey,L.N. and Aggarwal.(1986).Effect of season on conception rate in a sample of non - descript buffaloes. *Indian Vet. Med. J.*10:7.

Lipsett , M.B. , Swartz, T.L. and Thon , N.A.(1961). Hormonal control of sodium , potassium, chloride and water metabolism. In. *Minearl Metabolism,* Vol-IB(Eds.:C.L.Comar and F.Bommar),Academic Press , New York.

Lough, D.S., Beede , D.L. and Wilcox C.J.(1990). Effects of feed intake and thermal on mammary blood flow and other physiological measurements in lactating dairy cows.*J. Dairy Sci.*73:325.

McDowell, R.E. (1972). *Improvement of Livestock Production in Warm Climate.* W.H. Freeman and Co., San Francisco, CA.

McDowell, R.E., Hooven. N.W. and Camoens J. K.(1976). Effects of climate on Holsteins in first lactation. *J.Dairy Sci.*59:965-976.

Mehta, N.N.(1985). Seasonal variation in milk, fat per cent and lactose in buffaloes(*Bubalus bubalis).Livestock Advisor.*10:5.

Mullan, B.P., Brown, W. and Kerr, M. (1992). The response of the lactating sows to ambient temperature. *Proc. Nutr. Soc.Aust.*17:215 (abstract).

NRC.(1981). Effect of environment on nutrient requirements of domestic animals. National Academy Press. Washington,D.C.

Pasha T.N., Chaudhary, M.A., and Jabbar, M.A.(1986).Effect of month and season on hereditary behaviour of Nili Ravi buffaloes in rural conditions.*Buffalo Journal*.2:135.

Ravagnolo,O, Misztal, I. and Hoogenboom, G. (2000).Genetic component of heat stress in dairy cattle and development of heat index function. *J. Dairy Sci*.83:2120-25.

Reddy, A.O., Tripathi, V.N. and Raina,V.S. (1987). Effect of climate on the incidence of oestrous and conception rate in Murrah buffaloes. *Indian J.Animal Sci*.57:204.

Roepstroeff, K. and Murrell, K.D.(1997).Transmission dynamics of helminth parasites of pigs under continuous pastures: *Ascaris suum* and *Trichuris suis.Inernational J.Parasitology*.27:563-572.

Sastry, N.S.R. and Georgie, G.C.(1978). An appraisal of factors contributing to a seasonal breeding tendency in Indian water buffaloes. *In Proc.:Indo-Soviet Symp. Buff. Breeding*.NDDB,Anand, India.p-137.

Singh, M. and Ludri, R.S. (2001).Somatic cell count in Murrah buffaloes during different stages of lactation, parity and season. *Asian Aust. J. Anim.Sci*.14:189-192.

Singh, G.P. (2001). Livestock production and environmental protection. *In Proc.: "Animal Nutrition Conference"* held at NDRI, Karnal, Haryana from Nov.7-9,2001.

Singhal , K.K. and Madhu Mohini. (2002).Report on *"Uncertainty in methane and nitrous oxide from livestock in India"*. National Dairy Research Institute, Karnal-132001, Haryana.

Singh, K. and Bhattacharjee, N.K.(1981). Thermosensitivity of *Bos indicus* cattle and their F1 crosses with three breeds of *Bos Taurus.Anim. Prod*.52:57.

Singh,G and Singh,G.B. (1985).Effect of month of calving on post partum oestrous interval and service period in Murrah buffaloes. *In Proc: First World Buffalo Congress*,Cairo.p-960.

Srivastava , H.K. and Kharche, K.G.(1986). Studies on incidence of normal and abnormal cycling, post partum oestrous interval frequency of oestrous and calvings in buffaloes. *Livestock Advisor*.11:23.

Stansbury,W.F., McGlone, J. and Tribble, L.F.(1987). Effects of season, floor type, air temperature and snout coolers on sow and litter performance. *J. Anim. Sci*. 65:1507-1513.

Tailor, S.P. and Jain, L.S.(1987).Reproductive performance of medium sized buffaloes. *Indian J. Dairy Science*.11:179.

Thornton, K.(1990).*Outdoor pig production.* Farming Press Limited. Ipswich, UK.

Tubbs, R.C., Hurd, S., Dargatz , D. and Hill,G.(1993). Pre weaning morbidity and mortality in the United States swine herd. *Swine Heath and Production.* 1:21-28.

Webster, A.J.F. (1976). In:*Principles of Cattle Production* (Eds.:Swan, M. and Broster,W.H.),Butterworths,London.pp103-120.

Wetterman, R.P. and Tucker ,H.A. (1974).Relationship of ambient temperature to serum prolacin in heifers.*Proc.Soc.Expt,Biol.Med.*146:908.

Wolfenson, D., Roth, Z. and Meidan. (2000).Impaired reproduction in heat stressed cattle-basic and applied aspects. *Animal Reproduction Science.*34:89-101.

Young, B.A.(1981).Cold stress and its effects on animal production .*J. Anim. Sci.* 61: 154:163.

Climate Change and Food Security, 2008
Edited by: M.Datta, N.P.Singh & Er.D.Daschaudhuri, pp.293-298
New India Publishing Agency, New Delhi (India)
E-mail : newindiapublishingagency@gmail.com
Web: www.bookfactoryindia.com

20

Incidence of Strongyle Infection in Cattle and Pig with Relevance to Weather Parameters

S. BANDYOPADHYAY

INTRODUCTION

Prediction of helminthic infections is impractical. The aim of any control programme must therefore, ensure that parasite populations do not exceed levels compatible with economic production. The parasitological monitoring (e.g. faecal egg counts or pasture sampling) at intervals, forecasting on the basis of meteorological data and computer simulation may provide an alternate approach to control parasitic infection (Brunsdon, R.V., 1980) in a given geographical area.The most common gastrointestinal parasite prevalent throughout the year in Meghalaya, India is Strongyle infection. This is because of the high rainfall and humidity prevalent in the North-Eastern region.

As the prevalence of this infection is mainly dependent on rainfall and humidity, study was being initiated to identify the relationship between rainfall/temperature and strongyle infection. The study would be also helpful in predicting/ forecasting the disease occurrence based on the prevailing pattern of rainfall in a particular location.

MATERIALS AND METHODS

A total of 303 cattle and 253 pig faecal samples were collected from Govt. livestock farms located at Kyrdemkulai, Upper Shillong and Jowai in Meghalaya during the year 2001 and 2002. Meteorological data were collected from Govt. Meteorological department of Shillong. Samples were collected in the early morning and were processed and examined using standard parasitological procedures viz. sedimentation technique and self floatation technique. The egg per gram of faeces (epg) was counted using stoll egg counting method.

RESULTS AND DISCUSSIONS

The incidence of strongyle infection in different areas of Meghalaya and the Meteorological parameters are presented in Table 1.

Table 1. Monthly Egg per gram of faeces (epg) of stronogyle infection and meteorological parameters in Govt. pig and cattle Farms in Meghalaya.

Areas	Animals	Moths	Strongyle sp.	Meteorological data		
				Temp. (°C) (Max)	Temp. (°C) (Min)	Temp. (°C) (mm)
Jowai	Pig	Apr-Jun	240	26.72	18.9	341.4
		Jul-Sept.	426.66	26.67	20.09	693.4
		Oct- Dec.	330	21.59	14.2	91.5
		Jan-Mar	140	17.8	10.36	28
Kyrdemkulai		Apr.-Jun	175	27.12	20.81	116
		Jul-Sept.	270	29.55	23.63	417.7
		Oct.-Dec.	260	23.65	17.89	63.3
		Jan-Mar	140	20.42	12.89	73.7
Jowai	Cattle	Apr-Jun	142.85	26.72	18.9	341.4
		Jul-Sept.	287.5	26.67	20.09	693.4
		Oct-Dec.	133.33	21.59	14.2	91.5
		Jan-Mar	100	17.8	10.36	28
Kyrdemkulai		Apr. -Jun	142.85	27.12	20.81	166
		Jul-Sept.	350	29.55	23.63	417.7
		Oct-Dec.	150	23.65	17.89	63.3
		Jan-Mar	160	20.42	12.89	73.7
Upper Shillogn		Apr-Jun	133.33	20.61	12.63	117.4
		Jul-Sept.	314.28	21.5	15.68	284.6
		Oct.-Dec.	233.33	18.34	8.85	54.8
		Jan-Mar	133.33	13.96	3.4	16.63

Table 2. Statistical analysis of parasitic infection and meteorological parameters

Parasite	Temperature (max)	Temperature (min)	Rainfall
Strongyle sp.	r = 0.528319	R=0.531665	R=0.713168
	r^2 =0.279121	R^2 = 0.282668	R^2 =0.508609
	B= 0.023432 ± 0.0088	B = 0.028716 ± 0.0107	B = 1.546692 ± 0.6900
	Significant at 10% level	Significant at 5% level	Significant at 1% level

Rainfall and egg per gram of faeces (epg) of strongly infection has shown a linear and positive relationship. Rainfall contributed maximum effect on parasitic infection as compared to maximum and minimum Temperature (Figs. 1,2 and 3). Rainfall contributed more than 50 per cent for the occurrence of the parasitic infection in cattle and pig but maximum and minimum environmental temperature contributed above 25 per cent for the occurrence of strongyle infection in animal (Table 2). Regression analysis between strongyle infection and rainfall showed that 1 per cent increase in rainfall could predict 0.03 per cent increase in strongyle infection. Pitois and Leimbacher (1973) observed a close relationship between the frequency of spring rainfall and severity of *Fasciola hepatica* infection in sheep and cattle.

The predicted strongyle infection was calculated using the equation depicted in the regression analysis (Fig 1) which showed a higher strongyle infection than the observed infection (Fig 2). This might be due to anthelmintic treatment and other control measures taken by the Govt. farm for preventing the parasitic infection. This might also be due to the fact that 50 per cent of strongyle infection is dependent on rainfall. Onyiah (1985) also predicted the environmental temperature and development of parasites on pasture using Stochastic Development Fraction Model (SDFM).

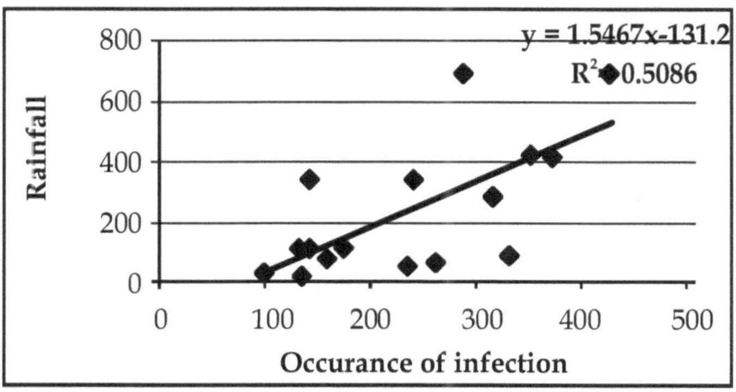

Fig. 1. Relationship between rainfall and occurrence of Strongyle infection in Meghalaya during 2001-02.

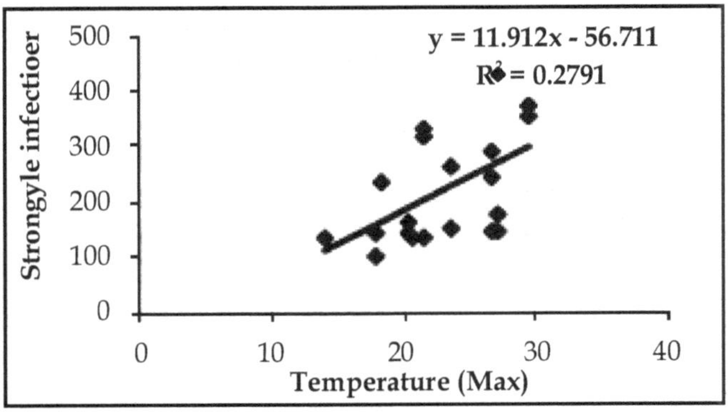

Fig. 2. Relationship between maximum temperature and occurrence of Strongyle Infection.

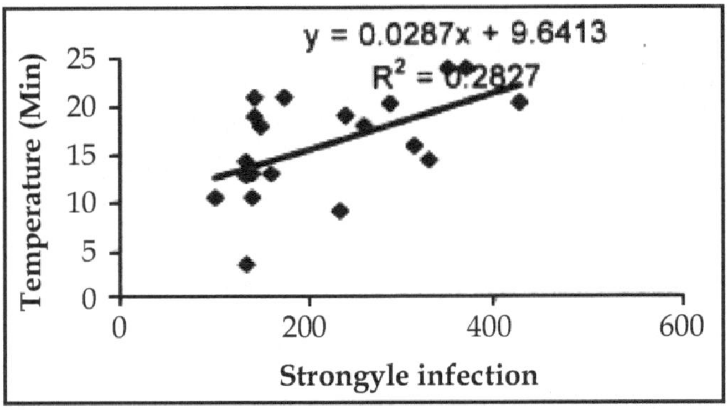

Fig. 3. Relationship between minimum Temperature and occurrence of Strongyle infection.

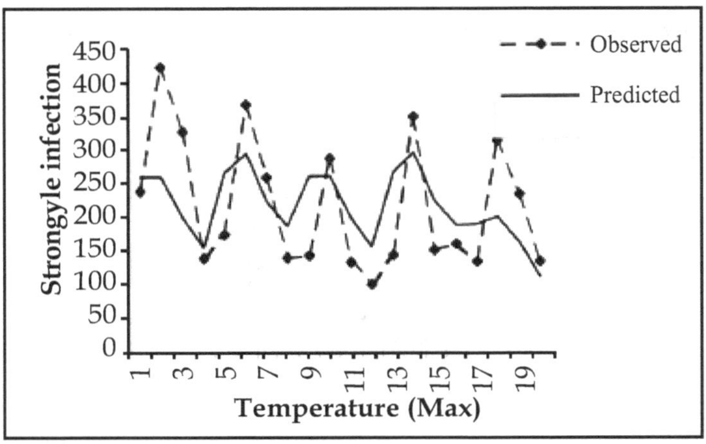

Fig. 4. Predicted Strongyle infection depending on maximum temperature

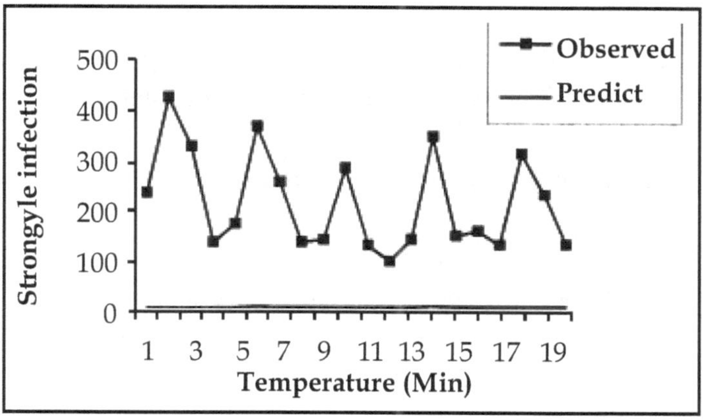

Fig. 5. Predicted and observed Strongyle infection in relation to minimum temperature.

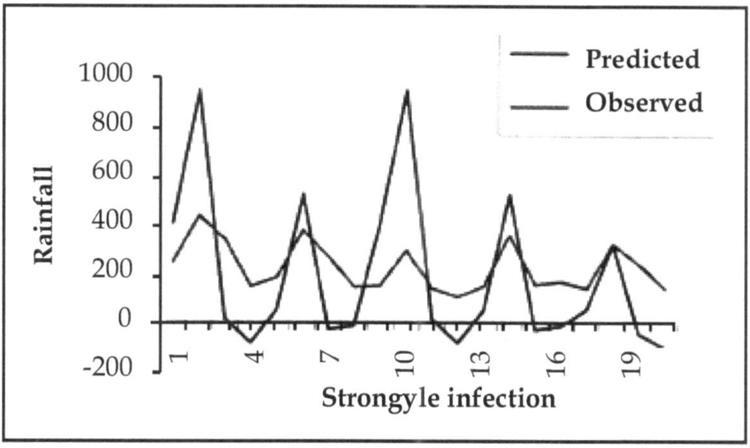

Fig 6. Predicted and observed Strongyle infection in relation to rainfall

Finally the multiple regression of disease infected with all the above mentioned parameters were analysed. The coefficient of multiple determination (R^2) explained more when we include temperature (max, min) and rainfall together as compared to single multiple regression of individual factor like rainfall, maximum temperature and minimum temperature. Interestingly, except rainfall, all other factors are statistically insignificant both at 5 per cent and 10 per cent probability level. But, the coefficient of correlation with rainfall is significant at 1 per cent probability level. From the above discussion, it may be concluded that the occurrence of the strongyle infection can mainly be predicted through rainfall instead of temperature.

REFERENCES

Anonymous, (1986). Manual of Veterinary Parasitological laboratory techniques. Bulletin No. 418, Ministry of Agriculture, Fisheries and Food. Her Majesty's Stationery Officer, London. pp 9-11.

Brunsdon R. V. (1980). Principles of helminh control. *Veterinary Parasitology* 6 : 185-15 Pitois M, Leimbacher F. 1973. Forecasting fascioliasis. *Recueil-de-Medecine-Veterinaire-del'Ecole-d'Alfort* 149 : 1293-02.

Onyiah L. C. (1985). A stochastic development fraction model for predicting the development of the nematode parasite of sheep, Haemonchus contortus from egg to infective larvae under the influence of tempeaature. *Journal of Thermal Biology*. 10 : 191-97.

Climate Change and Food Security, 2008
Edited by: M.Datta, N.P.Singh & Er.D.Daschaudhuri, pp.299-317
New India Publishing Agency, New Delhi (India)
E-mail : newindiapublishingagency@gmail.com
Web: www.bookfactoryindia.com

21

Climate & Freshwater Fisheries

B. SANTHOSH, M. DATTA AND N. P. SINGH

INTRODUCTION

One of the most important changes in the global environment which affects the existence of all living things , is the climate change. The climate change globally and locally also changes the aquatic habitat which, in turn directly and indirectly affects the environment. Anyway, the changes in climate are certain to occur and may be inevitable. The only possible way is to analyze the situation and find out the reasons for such changes and take immediate steps to minimize the impact of these climate changes. Global warming has eventually become synonymous with climate change because almost all the variables of the ecosystem are influenced directly or indirectly at different levels.

Aquatic environment is extremely diverse and influenced by the climate changes to a large extent both locally and globally. The changes in the ocean and its life are well studied at different levels with all possible details compared to the freshwater habitats. Unlike agriculture and animal husbandry, the vast majority of the fish production still comes from the natural habitat or the wild collections. Agriculture and animal husbandry usually focus on few important species but for fisheries, all the species of fishes including most of the aquatic living organisms are commercially important. The media (water) in which the aquatic organisms dwell, is much more important to all living organisms of the earth and so the influence by climate changes on water becomes very much relevant in the present situation. Freshwater habitats are limited and are much more complex to investigate in respect of climate changes due to the variability and diverse micro and macro level interactions. A holistic approach to deal with all these problems has not got much attention both globally and locally even though , it is the need of the hour to address the issues of impact of climate changes in the aquatic environment.

Policy decisions and research results could not successfully regulate, control or modify present day climatic changes, even though most of the major causes leading to climatic deterioration could have been regulated or controlled by concerted efforts from different angles. It is usually noticed that due to sheer ignorance and lack of foresightedness of people, they end up doing irreversible damages to the environment only for petty gains. It should be understood that it is these small actions of individuals and nations that make up the global environment.

From the beginning of the last century, there have been extreme changes in the surface temperature of the earth, increased evaporation, increased surface runoffs, severe floods, cyclones and seawater influx into fresh water areas thus affecting the aquatic environment. The global carbon dioxide level in the atmosphere has increased from 280 parts per million by volume (ppmv) to about 369 ppmv. Meanwhile, there was an increase in temperature by 0.6°C, mean sea level has risen by 10-20cm and there was 40% decline in Artic sea ice thickness. There has been a notable increase in the frequency of floods in all the large rivers and annual water discharge from all the six large eurasian rivers to Arctic Ocean has increased by 7% (Mall *et al.*, 2006).

In India ,the annual mean temperature has risen by 0.4°C between 1901 and 1982 which resulted in warmer nights in the northern region of the country. (Kalshian, 1999). Decrease in snow cover in Himalayan-Karakoram is bound to make serious consequences in the hydrology of Ganga and Brahmaputra (Hasnain, 1999).

Fisheries in broader sense include fishes, shell fishes and all other living organisms including the aquatic plants in the aquatic environment. As the fishes are cold blooded vertebrates, they are always prone to the ill effects of climate changes. Each group of fish is locally adapted with the hydrological condition of the microhabitat and any shift in the climate can make significant change in their physiology and reproductive success. So global warming can affect the physiology of the fishes directly because the climate changes directly affect the water quality parameters and limnology which , in turn has direct impact in fish populations. The changes in the wind pattern and rainfall also have a direct impact on fishes. Melting of polar ice and rise in sea level cause direct influx of seawater into the freshwater which can change the current succession pattern of aquatic life. Fisheries are an integral part of human existence not only from the nutritional and economic angle but in the recreational point of view also. Thus climate changes can bring far reaching consequences to this sector (Ficke *et al.,* 2005).

The changes in air temperature can cause a proportional change in water temperature which directly influences fisheries right from individual fishes to the ecosystem in which they live. This is because fishes are poikilothermic and often show behavioral thermoregulation by moving to a congenial temperature zone (Neilsen *et al.,* 1994; Nevermann and Wurtsbaugh, 1994; Brio, 1998). So, temperature is one of the most important physical factors which affect the physiology, reproduction and life history of fishes and aquatic life as a whole.

Apart from temperature, hydrology is also another important factor affecting the fisheries and fishes at different levels. The changes in the precipitation and evaporation can bring about extreme changes in the life of fishes and their ecosystem as a whole.

So, it can be generally stated that the effect of climatic changes on inland fishes needs to be analysed and studied in a detailed manner to arrive at reasonable corrective measures. This is limited by the fact that most of the models in all aspects of inland fisheries are developed based on regional climatic effects only without giving due importance to individual fish species or groups or the ecosystem including the aquaculture system. So the study remains incomplete.

CLIMATE CHANGE IN AQUATIC ENVIRONMENT

Approximately, 6% of the total earth land surface area is wetlands. High diversity and immense productivity are the special features of wet lands. Table 1 shows the distribution of wet lands in India. Low water levels and rich fauna / flora make this one of the most vulnerable ecosystem for the climate changes.

The predicted consequences of climate changes in the coming years are the rise in temperature, shift of rainfall pattern from traditional areas to non traditional areas, increase in evaporation and precipitation, soils becoming drier, drop in water level in lakes, decrease in the river discharge, reduction in the water availability for agriculture, domestic and industrial use etc. The most important and widely discussed impact is the rising sea level, coastal flooding and the influx of seawater into freshwater areas. (Sugunan and Maurye, 2003).

Table 1. Distribution of Floodplain wetlands in India. *(Source:* Sugunan and Sinha, 2001)

State	Distribution (by district)	River basin	Local name	Area (ha)
Arunachal Pradesh	East Kameng, Lower Subansiri, East Siang, Dobang valley, Lohit, Changlang & Tirap Dibang, Lohit Dihing & Tirap	Kameng, Subansiri,	*Beel*	2,500
Assam	Brahmaputra & Barak valley districts	Brahmaputra & Barak	*Beel*	10,000
Bihar	Saran, Champaran, Saharsa, Muzaffarpur, Darbhanga, Monghyr	Gandak & Kosi	Maun, Chaur & Dhar	40,000
Manipur	Imphal, Thoubal & Bishnupur	Iral, Imphal & Thoubal	Pat	16,050
Meghalaya	West Khasi hills and West Garo hills	Someshwari & Jinjiram	*Beel*	213
Tripura	North, South & West Tripura districts	Gumti	*Beel*	500
West Bengal	24-Parganas North & South, Hooghly Nadia, Murshidabad, Maldah, Cooch Behar Burdwan, North & South Dinajpore and Midnapore	Hooghly, Ichhamati Bhagirathi, Chumi, Kalindi,Dharub, Dharala, Pagla, Jalangi, Behula, Torsa and Mahananda	*Beel Charha* & *Baor*	42,500
	Total			**2,02,213**

The changes in the atmospheric temperature affect the water temperature. The breeding cycle of many temperate fishes depend upon the water temperature. The climate changes also bring changes in rainfall pattern and the anticipated range of rainfall variation for Asia is -15% to + 18%. (Mc Lean *et al.,* 1998). A greater frequency of heavy rainfall is expected in this region (Bhaskaran *et al.,* 1995; Bhaskaran and Mitchell, 1998) Reibsame *et al.,*(1995) predicted a greater seasonal difference in the water flow and a delay in the flood peak in the Mekong river. Late start of monsoon is also predicted by Bhaskaran *et al.,* (1995). The climate changes can alter the hydrography / hydrology of the rivers and can have a strong impact on the fish biodiversity of the lakes and rivers. Table 2 shows the open water fish resources in

India and the possible impact of global warming. The rise in water temperature and reduction in water flow in rivers and lakes will increase the concentration of pollutants and deteriorate the water quality.

Table 2. Open water fishery resources of India, mode of fishery management and possible impact of global warming on management mode. (Sugunan and Praveen Maurye, 2003)

Resources I	Resource size	Management mode	Likely impact of GW	Reason
River(km)	29,000	Capture	Decrease	Flow decreases
Mangroves(ha)	3,56,000	Subsistence	Increase	Sea level rises
Estuaries (ha)	3,00,000	Capture	Increase / Decrease	Unstable Gradient Zone
Freshwater ponds (ha)	23,80,000	Aquaculture	Decrease	Evaporation/Alg al bloom
Brackish water ponds(ha)	14,37,000	Aquaculture	Decrease	Overflow/salinit y rise
Estuarine wetlands (bheries) (ha)	39,600	Aquaculture	Decrease	Evaporation loss
Lagoons (ha)	190,500	Capture	Increase	Sea level rises
Large & medium reservoirs (ha)	16,67,809	Enhancement	Decrease	Low flow from rivers
Small reservoirs(ha)	14,85,557	Cultural-based	Decrease	Desiccation
Floodplains wetlands(ha)	2,02,213	Cultural-based	Decrease	River connection cutoff
Upland lakes(ha)	7,20,000	Not known	Decrease	Low water availability

TEMPERATURE CHANGES AND ITS EFFECTS IN INLAND FISHERIES

Rise of temperature can create serious consequences in the Inland fisheries. Fishes can tolerate a certain level of temperature fluctuation depending upon their adaptability. The distribution of fishes is very limited in too hot /cold waters. All the freshwater fishes are poikilothermic and observe behavioral thermoregulation. They move from a thermal microhabitat to a region with optimum temperature region where there are least disturbances in their physiology (Neilsen *et al.*, 1994; Nevermann and Wurtsbaugh, 1994; Brio, 1998). The smaller rivers and lakes generally do not have such type of wide temperature regimes which could have created unfavorable situations. Anadromus fishes like hilsa and salmon can tolerate saline water influx and a wide range of salinity. But most of the freshwater fishes are very sensitive to saline waters (Sugunan and Maurye, 2003).

The changes in the temperature due to the climate changes will affect the fresh water species both at individual level and ecosystem level. Certain fishes are well adapted for changes in temperature (eurythermic) and others are stenothermic. The temperature effect is different in fishes of different region such as tropical, temperate etc. Tropical fishes like Tilapia and Rohu are more tolerant to higher temperature and temperature fluctuations at the higher levels. A lot of studies have already been done on the physiological effect of temperature on various species (Kutty and Sukumaran, 1975; Welcomme, 1979; Val and Almeidia- Val, 1995; McKenzie et al., 1996; Hori, 2000; Dizikowski *et al.,* 2001; Glasser, 2003; Chatterjee *et al.,* 2004; Moyle and Cech, 2004). Slight increase in temperature can give better production in temperate species but beyond a limit, will reduce its growth, activities, reproduction and thus result in mortality (Ktchell *et al.,* 1977; Myrick and Cech, 2000; Hill and Magnuson, 1990: Kling *et al.* 2003; Langford, 1983; Gerdaux, 1998) The overall fishing in rivers of United States is in declining phase because of the sudden changes in the water temperature upto> 1°C. Seven warm water species of the rivers in US, had suffered the rise of temperature above its tolerance level (EPA, 1995). The impact of various meteorological factors such as temperature/sea level rise, changes on precipitation, wind direction/velocity and flood/droughts on fisheries and aqua ecosystems are presented in Table3.

EFFECT OF TEMPERATURE ON FISH PHYSIOLOGY

Fishes are cold blooded animals living in water medium which has high heat capacity and conductance and all their body activities are critically dependent on the water temperature. If the temperature moves towards the critical level for a particular fish species, the protein structure and function will be impaired significantly. For the survival and maintenance of the normalcy of the body, they develop heat shock proteins (Lindquist and Craig, 1988; Parsell and Lindquist, 1993). When there is a change in the temperature, the structural lipids of the biological membranes of fishes also changes. If the temperature is just above the tolerance level, hydrophobic domino of the phospholipids in the membrane expands and the molecular geometry changes towards hyper fluid state. As the temperature increases , the membranes start loosing its functions. The temperature above critical levels affects muscle activity, cardiovascular, reproductive and osmoregulatory systems (Sugunan and Maurye, 2003). The energy budget of yellow perch as a function of temperature is given in Fig. 1.

Table 3. Effect of various meteorological parameters on fisheries and aquaecosystems

Drivers	Biophysical Effects	Implications for fisheries and aquaculture
Temperature rise	Less dissolved oxygen; Algal blooms; Increased incidence of disease and parasites; Changes local ecosystems (microhabitat) Increase / decrease of competitors, predators and invasive species; Changes in plankton species and quantity.	Low survival and mortality, Migration and change of habitat, increased risk of predation or shift in population structure and composition. For aquaculture, changes in infrastructure and operating costs due to oxygen stress, algal blooms, diseases, pest, predators, chances for mass mortalities changes in the performance of species.
	Longer growing seasons; Less natural mortality in winter; Enhanced metabolic and growth rates.	Potential for increased production in some areas and for some species, especially for aquaculture.
	Enhanced primary productivity	Aquaculture and production and capture fisheries of some areas will increase but the species composition will be changed.
	Changes in pattern and success of migrations, reproductive abnormalities and abundance and sex ratios.	Shift in species composition, decline in catch, changes in reproductive pattern and seed availability for aquaculture.
	Variation in the species availability.	Aquaculture production and species changes. Potential species loss and altered species composition for capture fisheries.
	Macrophyte growth and eutrophication	Loss of certain species reduction / shift in recruitment of certain species.
Sea level rising	Loss of land	Reduction in area available for aquaculture. Loss for the freshwater fisheries.
	Changes in estuary systems	Shifts in species composition, distribution and abundance of fish stocks and aquaculture seed. Migration of freshwater species to interior.
	Salt water infusion into groundwater	Damages to freshwater culture and capture fisheries. Reduced freshwater availability for aquaculture. Water extraction for non fisheries purpose will destroy the habitat. Good for the culture of brackish water species.
Changes in the wind direction, velocity etc.	Changes in the mixing of water and variations in productivity.	Mortality or reduced growth or enhanced growth in some regions; increased operational costs.
El Nino	Alterations in the rainfall pattern, flood or drought	Loss of habitat and production, spread of certain species, damages to the river flow and riverine fisheries

Contd...

Table 3. Contd...

Increased incidence of storms	Loss of coastal ecosystems such as mangrove forests, mixing or influx of seawater, damages to the capture and culture fisheries; damages to culture fisheries	Reduced production through capture fisheries and aquaculture. Fast waves and storm surges make damages to inland aquaculture and fisheries. Aquaculture sector will be damaged by mixing of water from different ponds, from sea or from natural waters and escape of many fishes through sudden flood created by storms.
Higher inland water temperatures	Reduction in water quality parameters especially dissolved oxygen. Changes in thermal stratification and reduced mixing of water in lakes, reduction in primary productivity. Sometimes enhancement of productivity. Shift of location of certain temperature sensitive species. Increased metabolic rate, growth and in certain cases the reverse also.	Reductions in fish stocks. Increase in fish production both in culture and capture and in some areas decrease in production. Changes in the operational and capital cost in culture sector and operational cost in the capture Changes in the species suitable for culture, changes in the specie composition in capture
	Severe changes in timing and success of migrations, spawning and peak abundance.	Potential loss of species or shift in composition for capture fisheries; Changes on seed availability for aquaculture. Changes in the stocking and harvest seasons.
Changes in precipitation, intensity, frequency, seasonality and water availability	Changes in fish movement, migration and recruitment patterns and also in recruitment success.	Changes in the abundance and composition in capture fisheries Changes on seed availability for aquaculture.
	Changes in the water availability for aquaculture. Changes in the water quality causing more disease. Increased / decreased freshwater supplies with greater risk of drought.	Higher operational costs of maintaining pond water levels and from stock loss. Reduced production. Severe conflicts on water uses. Possible changes in the culture species.
	Drastic changes in lake and river levels and the overall movement patterns of surface water.	Changes in the distribution, composition and abundance of fish stocks. Increased operational costs and possible migration of the fishers.
Frequent droughts	Decrease in water quality and availability for aquaculture. Changes in the salinity Seawater infiltration due to excess ground water use.	Loss of wild and cultured stock. Increased operational cost for culture and pollution problems Loss of production and possible migration and social conflicts.

Sudden decline in growth was reported as the temperature increases beyond 25°C. So, the optimum temperature for maximum growth and metabolic activities is 23°C and beyond 25°C , starvation and mortality was observed (Ficke and Myrick, 2004). The rapid decline in growth is noted (Fig.1)as temperature increases beyond 25°C. The vertical dotted line at 23° C indicates the optimal temperature for maximum metabolic activity (Ficke and Myrick, 2004). Here, SDA is the energy allocated to specific dynamic action for food digestion and processing.

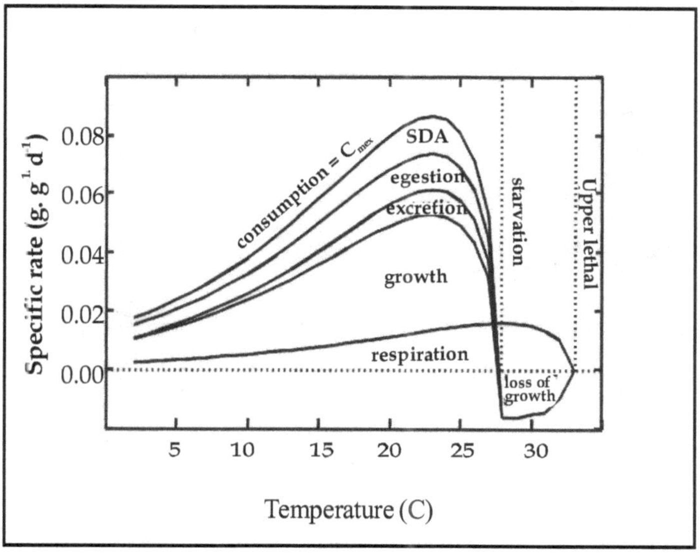

Fig.1. The energy budget of a yellow perch as a function of temperature

WATER QUALITY PARAMETERS

Most of the water quality parameters like dissolved oxygen, pH, nutrient concentration and even the toxic effect of pollutants depend on the temperature. Adequate dissolved oxygen is essential for almost all aquatic living organisms (Kalff 2000; Stickney 2000). Dissolved oxygen level above 4mg/l is ideal for fish. DO level below 2mg/l is critical and results in mortality. The effect of temperature on oxygen saturation is described in Fig.2. Increase in temperature lowers the oxygen saturation level and oxygen level is critically low in higher temperature (Ficke and Myrick, 2004). Increase in temperature leads to eutrophicaton in shallow water bodies and ultimately result in heavy growth of macrophytes. (Ficke and Myrick, 2004). Studies revealed that 2-3°C increase in temperature can shoot up the macrophyte growth in lakes by 300-500% level. (Kankaala *et al.,* 2002).

Even a slight variation in global temperature can induce a change in the stratification of lakes in temperate zone(Gaedke *et al.* 1998; Topping and Bond 1988). Thermal stratification can induce production of algal blooms especially the blue green algae (de Souza *et al.* 1998; George *et al.* 1990; Jones and Poplawski 1998). These are non edible to fishes and can cause fish mortality due to starvation or due to the low dissolved oxygen content or due to the toxins produced by these algae (George *et al.*, 1990; Kangur *et al.*, 2002). Some fishes may survive after consuming this toxic metabolites but are not recommended for human consumption (Banens and Davis 1998: de Magalhães *et al.*, 2000).

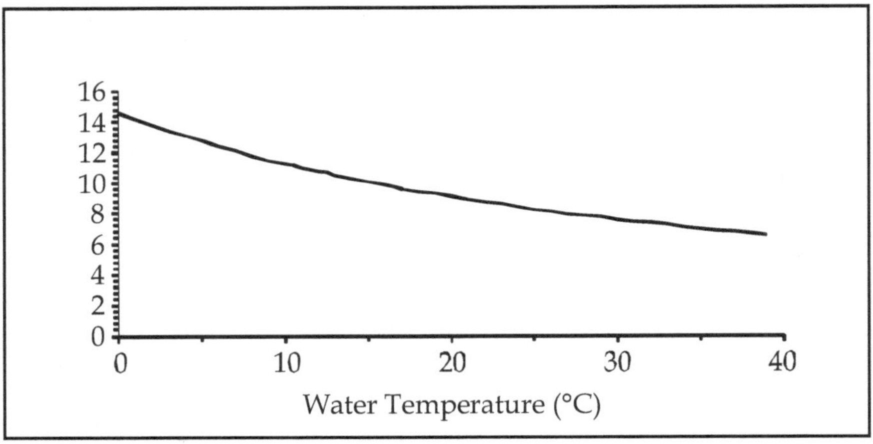

Fig.2. Effect of temperature on dissolved oxygen

The massive fish mortality in Great lakes during 1984 was due to the ultimate effect of a thunderstorm which changed the ecology of the Nyanza Gulf of Lake Victoria (Ochumba 1990; Verschuren *et al.*, 2002). The studies by Verburg *et al.*, (2003) indicated that the mixing of different layers in lakes was reduced due to the changes in climatic conditions thus affecting the fish production.

The increase in annual atmospheric temperature will change the current distribution pattern of the freshwater fishes of temperate and polar region. The compression or expansion of thermal habitats will affect the sensitive fishes (thermal refugees) to move towards a region congenial to them (Christie and Regier 1988; Gerdaux 1998; Ficke and Myrick, 2004). Historically, the distribution of many fish species is determined not by physical barriers but by temperature isoclines, both elevational and latitudinal (Moyle *et al.*, 1982; Baltz *et al.*, 1987; Ficke and Myrick, 2004). The effect of global warming on the distribution of freshwater fishes by northward expansion of warm water species in North America, Europe, and Asia, and the southward expansion of warm water species in Australia and South America was predicted by Ficke and Myrick, (2004). This kind of a migration of fishes to a new regime can make serious consequences in the existence of several species of fishes by predation or competition. The distribution of Mahaseers in Indian subcontinent is a typical example of the temperature barriers. There was a marked withdrawal of the availability of these fishes which move towards the cooler regions in a stretch of 100 years.

EFFECT OF CLIMATIC CHANGE ON PARASITES AND DISEASES

There is a host parasite system evolved through centuries existing in freshwater ecosystems. The changes in the climate can alter the equilibrium by increasing the reproductive ability of the parasite or by reducing the host resistance. High water temperature in winter may lead to higher survival of the parasite thereby increasing the infection load in the summer months of the year (Marcogliese 2001). On the other hand, the higher temperature hampers the immune response of the fishes in turn favouring the bacterial and protozoan infections (Hefer and Pruginin 1981; Wedemeyer 1996; Ozer and Erdem 1999).

The changes in the limnology associated with climate changes favour the causative agents of fish diseases to spread widely in more areas and also to more hosts. Fishes migrating from warmer regions also aid introduction of new parasites and diseases. (Schisler *et al.*, 1999; Hiner and Moffitt 2001; Marcogliese 2001).

EFFECT OF CLIMATIC CHANGE ON WATER BALANCE AND WATER CYCLE

Though there is a prediction of more rains in the coming years, there may be water stress especially in small streams, watersheds and lakes. The increase in solar radiation and temperature will increase evaporation. Shortage of rainy days and rainstorms will reduce the water availability (Magnuson *et al.* 1997; Kling *et al.*, 2003). It is also predicted that evaporation and evapotranspiration often exceeds precipitation which will have ultimately serious impacts in freshwater fisheries of tropical and subtropical water bodies (Welcomme 1979; Hulme 1994; Irion and Junk 1997). Jacobs (1992) reported 2-3°C increase in atmospheric temperature thus resulting 10- 15% increase in evaporation in Mekong river system. There will be a total change in the stream ecology and the migratory fishes will also be affected.

The biodiversity of freshwaters has declined much faster than both terrestrial and marine environment for the past 30 years (Jenkins, 2003; UNESCO, 2003). Postel and Richter (2003) indicated that the decline in river discharge (water availability) is the major cause and it is still increasing because of the climate changes. River discharge is one of the important parameters determining the ecological space and habitat heterogeneity which results in better survival, growth and reproduction of the fishes. Richness in fish Species increases with river discharge (Oberdorff *et al.,* 1995; Poff *et al.,* 2001). Increased and sustained river discharge promotes the existence of rich and diverse fish fauna. A combined effect of climate change and water withdrawal is going to affect the water availability in all the major rivers and this will cause extinction of many fish species by 2070 (Marguerite *et al.,* 2005).

SOCIOLOGICAL IMPLICATIONS OF CLIMATIC CHANGE

Majority of the fishing community (200 million approximately) and their dependants are poor and live in the areas vulnerable to vagaries of climate. Their livelihood resources are much affected by climate changes. This high level of vulnerability is a matter of great concern considering the contributions of fisheries in the nutritional security and poverty alleviation. Regional pattern of climatic changes are difficult to predict. Historically, these resource dependent communities in the developing countries are well adapted for climate variability by occupational or geographical mobility but in future the climate change will bring extreme events in their life since people will not be able to opt for geographical or occupational mobility. Moreover, increase in population, reduction in catches, emergence of large scale producers and diversion of the resources for domestic or industrial purpose coupled with the extreme events will make their life miserable. In 2001, the Intergovernmental Panel on Climate Change (IPCC) assessed the global warming to be initiated by 2050 and identified the potential vulnerable hotspot regions and countries, that will suffer seriously because of the climate change. According to IPCC, highest predicted climate changes will be in Asian land mass, Amazon basin and Western Sahara region. The level of projection of vulnerability or risk is dependent on the economical growth. Largest number of poor fishing communities are found in South and Southeast Asia but the worst affected will be those in African countries because of the extreme poverty. Ghana, Indonesia and Bangladesh are most nutritionally dependent countries on fishes. Most sensitive countries in Asia are China, Indonesia, India and Vietnam, in Africa, Mauritania and in Latin America, Peru. Considering several factors like risk of exposure, sensitivity, adaptive capacity, the countries are classified in two groups one which is vulnerable with rapid development and higher emission scenario and the other with local development and lower emission scenario is given in Table 4. The highly sensitive three big ecosystems are coral reefs, great lakes of Africa and river basins / flood plains of Bangladesh (Allisen *et al.*, 2005).

Table 4. Countries with high fish vulnerability indices to climate change under two scenarios (Allison *et al.*, 2005)

Rapid Development, High Emissions Scenario (IPCC A1F1)		Local Development, Lower Emissions Scenario (IPCC B2)	
Angola	81.97	Mauritania	83.10
Mauritania	81.18	Angola	82.15
Niger	79.24	Zimbabwe	79.32
Congo, Dem Rep	78.82	Niger	78.95
Mali	78.01	Congo, Dem Rep	76.03
Sierra Leone	77.09	Mali	75.92
Burkina Faso	76.01	Mozambique	75.13
Burundi	74.96	Russian Federation	74.33
Mozambique	74.86	Sierra Leone	73.61
Zimbabwe	74.55	Senegal	73.31
Senegal	73.70	Botswana	72.96
Guinea-Bissau	72.97	Zambia	71.78
Côste d'Ivoire	71.18	Burundi	71.68
Sudan	70.68	Burkina Faso	71.57
Russian Federation	70.57	Peru	70.98

CONCLUSION

The global climate changes coupled with the shift of land use patterns and the anthropogenic pressure on already shrinking freshwater resources will be subjected to serious consequences in the coming centuries than any other system. Wetlands and lakes are more seriously affected by pollution. Blockage of inflow of water, reclamation and constructions coupled with the climatic changes will have severe effects on freshwater fisheries. There will be more algal blooms, eutrophication, mass mortalities of fishes, and increased incidence of diseases and extinction of several species of fishes. This will reduce the 'vitally important protein supplement' of the poor people of the developing countries (Ficke and Myrick, 2004). A collective and serious effort is required to reduce the global warming and climate changes and only the anthropogenic pressure to the freshwater system, will help to conserve and sustain the freshwater fisheries of the world. The following problems will be thoroughly investigated and necessary action should be taken.

- Identify the major risk factors and vulnerabilities in fish communities and aquatic ecosystem with the climate change.
- Analyse these 'climatic impacts' which the planners and researchers are going to face. Investigate the actions that can be taken for producing a desired outcome.
- Delineate the most urgent priorities of the resources and resource management now and for the coming few decades.
- Find out the solutions available to minimize or to prevent the negative impacts of the climate change.
- Investigate the possibilities and identify the priorities for researchers, planners, administrators and fishermen to deal with this uncertainty around the climate change.

Necessary channels should be opened for the administrators to implement the policies in an effective manner . No doubt, the major risk is the light of already known facts like an increasing population and an adversely changing climate. Attempt may be necessary to render a concerted effort to nurture freshwater resources for their conservation and productivity enhancement.

REFERENCES

Allison, E. H., W. N. Adger, M. C. Badjeck, K. Brown, D. Conway, N. K. Dulvy, A. Halls, A. Perry and J. D. Reynolds, (2005). Effects of climate change on the sustainability of capture and enhancement fisheries important to the poor: analysis of the vulnerability and adaptability of fisher folk living in poverty. Project No. R4778J. Fisheries Management Science Programme Department for International Development 167pp.

Baltz, D. M., B. Vondracek, L. R. Brown, and P. B. Moyle. (1987). Influence of temperature on microhabitat choice by fishes in a California stream. *Transactions of the American Fisheries Society* 116:12-20.

Banens, R. J., and J. R. Davis. (1998). Comprehensive approaches to eutrophication management: the Australian example. *Water Science and Technology* 37: 217-225.

Bhaskaran, D., Mitchell, J.F.B., Lavery, J. R. and Lal, M. (1995). Climate response of the Indian subcontinent to doubled CO_2 Concentrations. *International Journal of Climatology*, 15: 873-892.

Bhaskaran, D. and Mitchell, J.F.B., (1998). Simulated changes in South East Asian precipitation resuming from anthropogenic emissions, *International Journal of Climatology*, 18: 1455-1462.

Brio, P.A., (1998). Staying cool: behavioural thermoregulation during summer by young of year brook trout in a lake. *Transactions of the American Fisheries Society*, 127: 212-222.

Chatterjee, N. A., K. Pal, S.M. Manush, T. Das and S.C. Mukharjee., (2004). Thermal tolerence & oxygen consumption of *Labeo rohita* & *Cyprinus carpio* early fingerlings acclimatised to three different temperatures. *Journal of Thermal Biology*, 29: 265-270.

Christie, G. C., and H. A. Regier. (1988). Measures of optimal thermal habitat and their relationship to yields for four commercial fish species. *Canadian Journal of Fisheries and Aquatic Science* 45:301-314.

De Magalhães, V. F., R. M. Soares, and S. M. F. O. Azevedo. (2000). Microscystin contamination in fish from the Jacarepaguà Lagoon (Rio de Janeiro, Brazil): ecological implication and human health risk. *Toxicon* 39:1077-1085.

De Souza, R. C. R., M. C. Carvalho, and A. C. Truzzi. (1998). Cylindrospermopsis raciborskii (Wolosz.) Seenaya and Subba Raju (Cyanophyceae) dominance and a contribution to the knowledge of Rio Pequeno Arm, Billings Reservoir, Brazil. *Environmental Toxicology and Water Quality* 13:73-81.

EPA (1995) *On-Site Meteorological Program Guidance for Regulatory Modeling Applications*. EPA-450/4-87-013. June. Triangle Park, North Carolina: U.S. Environmental Protection Agency. TIC 210292.

Ficke, A. and C. A. Myrick, (2004). *The Potential of Anthropgenic Climate Change on Freshwater Fisheries*. WWF publication, Gland, Switzerland. 93pp.

Gaedke, U., D. Ollinger, P. Kirner, and E. Bauerle. (1998). The influence of weather conditions on the seasonal plankton development in a large and deep lake (L. Constance) III: The impact of water column stability on spring algal development. Pages 71-84 *in* D. G. e. a. George, (ed.) *Management of Lakes and Reservoirs during Global Climate Change*. Kluwer Academic Publishers, The Netherlands.

George, D. G., D. P. Hewitt, J. W. G. Lund, and W. J. P. Smyly. (1990). The relative effects of enrichment and climate change on the long-term dynamics of *Daphnia* in Esthwaite Water, Cumbria. *Freshwater Biology* 23:55-70.

Gerdaux, D. (1998). Fluctuations in lake fisheries and global warming. Pages 263-272 *in* D. G. e. a. George, (ed.) *Management of Lakes and Reservoirs during Global Climate Change*. Kluwer Academic Publishers, The Netherlands.

Hasnain, S. I. (1999). *Himalayan Glaciers: Hydrology and Hydrochemistry*. Allied Publishers, New Delhi, India . 234pp.

Hill, D. K., and J. J. Magnuson. (1990). Potential effects of global climate warming on the growth and prey consumption of Great Lakes fish. *Transactions of the American Fisheries Society* 119:265-275.

Hiner, M., and C. M. Moffitt. (2001). Variations in infections of *Myxobolus cerebralis* in field-exposed cutthroat and rainbow trout in Idaho. *Journal of Aquatic Animal Health* 13:124-132.

Hori, H.(2000). *The Mekong: Environment and Development*. United Nations University, New York.

Hulme, M. (1994). Global climate change and the Nile Basin. P. Howell, and J. Allan, (eds.) *The Nile: Sharing a Scarce Resource*. Cambridge University Press.

Irion, G., and W. J. Junk. (1997). The large Central Amazonian River floodplains near Manaus. Pages 23-46 *in* W. J. Junk, (ed.) *The Central Amazon Floodplain: Ecology of a Pulsing System*. Springer-Verlag, Berlin, Heidelberg.

Jacobs, J. W. (1992). International River Basin Development and Climatic Change: The Lower Mekong of Southeast Asia. Doctoral. University of Colorado, Boulder, Co.

Jones, G. J., and W. Poplawski. (1998). Understanding and management of cyanobacterial blooms in sub-tropical reservoirs of Queensland, Australia. *Water Science and Technology* 37(2):161-168.

Jenkins M., (2003). Prospects for biodiversity. *Science*, 302: 1175- 1177.

Kalshian, R. (1999). On high flame. *Outlook*, 16V: 54-58.

Kalff, J. (2000). *Limnology*. Prentice Hall, Upper Saddle River, New Jersey.

Kangur, A., P. Kangur, and E. Pihu. (2002). Long-term trends in the fish communities of Lakes Peipsi and Vortsjarv (Estonia). *Aquatic Ecosystem Health and Management* 5:379-389.

Kankaala, P., A. Ojala, T. Tulonen, and L. Arvola. (2002). Changes in nutrient retention capacity of boreal aquatic ecosystems under climate warming: a simulation study. *Hydrobiologia* 469:67-76.

Kitchell, J. F., D. J. Stewart, and D. Weininger. (1977). Applications of a bioenergetics model to yellow perch (*Perca flavescens*) and walleye (*Stizostedion vitreum vitreum*). *Journal of the Fisheries Research Board of Canada* 34:1922-1935.

Kling, G. W., K. Hayhoe, L. B. Johnson, J. J. Magnuson, S. Polasky, S. K. Robinson, B. J. Shuter, M. M. Wander, D. J. Wuebbles, D. R. Zak, R. L. Lindroth, S. C. Moser, and M. L. Wilson. (2003). *Confronting Climate Change in the Great Lakes Region: Impacts on our Communities and Ecosystems.* Union of Concerned Scientists, Cambridge, Massachusetts, and Ecological Society of America, Washington, D.C.

Kutty, M. N., and N. Sukumaran. (1975). Influence of upper and lower temperature extremes in the swimming performance of *Tilapia mossambica. Transactions of the American Fisheries Society* 104:755-761.

Lindquest, S. and E. A. Craig, (1998). The heat -shock proteins. *Annual Review of Genetics*, 22: 631-677.

Langford, T. E. (1983). *Electricity Generation and the Ecology of Natural Waters.* Liverpool University Press, Liverpool.

Magnuson, J. J., and coauthors. (1997). Potential effects of climate changes on aquatic systems: Laurentian Great Lakes and Precambrian Shield region. *Hydrological Processes* 11:825-871.

Mall, R. K., Akhilesh Gupta, Ranjeet Singh, R. S. Singh and L. S. Rathore, (2006). Water resources and climate change: An Indian perspective. *Current Science,* 90: 1610-1626.

Marcogliese, D. J. (2001). Implications of climate change for parasitism of animals in the aquatic environment. *Canadian Journal of Zoology* 79:1331-1352.

Margueritea, X., M. L. Lodge, Joseph, A., Michel, M. R., S., C. Kerstein and V. V. Detlefp, (2005).

Scenarios of freshwater fish extinctions from climate change and water withdrawal. *Global Change Biology*, 11: 1557–1564.

McKenzie, D. J., G. Serrini, G. Piraccini, P. Bronzi, and C. L. Bolis. (1996). Effects of diet on responses to exhaustive exercise in Nile tilapia (*Oreochromis nilotica*) to three different temperatures. *Comparative Biochemistry and Physiology Part A* 114: 43-50.

Mc Lean, R. F., Sinha, S. K., Mirza, M. Q., and Lal, M. (1998). Tropical Asia, in The Regional Impact of Climate Change. An assessment of Vulnerability, R. T.

Watson, M. C. Zinypwera, R. H. Moss and D. J. Dokken (eds). Cambridge University Press, Cambridge, 381-407.

Moyle, P. B., and a. coauthors. (1982). *Distribution and ecology of stream fishes of the Sacramento-San Joaquin drainage system,* volume 115. University of California Press, Berkeley, California.

Myrick, C. A., and J. J. Cech. (2000). Temperature influences on California rainbow trout physiological performance. *Fish Physiology and Biochemistry* 22:245-254.

Moyle. and J. J. Cech. 2004. Fishes: an Introduction to Ichthyology, 5th edn. Prentice Hall, Englewood Cliffs, N.J.

Nevermann, D. and W.A. Wurtsbaugh, (1994). The thermoregulatory function of dial vertical migration for a juvanile fish Cuttus extensus. *Oecologia,* 98: 247-256

Nielsen, J.L., T.E. Lisle and V. Ozaki, (1994).Thermally stratifies pools and their use bye steelhead in northern California streams. Transations of the American Fisheries Society, 123: 613-626

Oberdorff T., J.F. Gue´gan and B. Hugueny, (1995). Global scale patterns of fish species richness in rivers. *Ecography,* 18; 345–352.

Poff, N. L., P. L, Angermeier and S. D. Cooper, (2001). Fish diversity in streams and rivers. *In*: Global Biodiversity in a Changing Environment: Scenarios for the 21st Century (eds. Sala, O.E., Chapin, F. and Huber-Sannwald, E.), pp. 315–349. Springer, New York.

Ochumba, P. B. O. (1990). Massive fish kills within the Nyanza Gulf of Lake Victoria, Kenya. *Hydrobiologia* 208: 93-99.

Ozer, A., and O. Erdem. (1999). The relationship between occurrence of ectoparasites, temperature and culture conditions: a comparison of farmed and wild common carp (*Cyprinus carpio* L., 1758) in the Sinop region of northern Turkey. *Journal of Natural History* 33:483-491.

Parsell, D. A. and S. Lindquest, (1993). The function of heat- shock protein in stress tolerence: degradation and reactivation of damaged proteins. *Annual Review of Genetics,* 27: 437-496.

Postel S. and B.Richter, (2003). *Rivers for Life: Managing Water for People and Nature.* Island Press, Washington.

Reibsame, W.E., Strzepek, K.M., Wescoat Jr., J.L., Perritt, R., Gaile, G.L., Jacobs, J., Leichenko, R., Magadza, C., Phien, H., Urbiztondo, B.J., Restrepo, P., Rose, W.R., Saleh, M., Ti, L.H., Tucci, C. & Yates, D. (1995). Complex River Basins. In Strzepek, K.M. & Smith, J.B. (eds), *As Climate Changes : International Impacts and Implications,* 57-91. Cambridge University Press.

Schisler, G. J., P. G. Walker, L. A. Chittum, and E. P. Bergersen. (1999). Gill ectoparasites of juvenile rainbow trout and brown trout in the Upper Colorado River. *Journal of Aquatic Animal Health* 11:170-174.

Stickney, R. R. (2000). *Encyclopedia of Aquaculture.* John Wiley and Sons, New York.

Sugunan, V.V. and Praveen Maurye, (2003). Global warming and climate change perspective on impact over inland fisheries. *Fishing Chimes,* 23: 27-36.

Sugunan, V.V. and M. Sinha, (2001). Sustainable capture and culture fisheries in freshwater of India. In: *Sustainable Indian Fisheries.* T.J. Pandian (ed.). National Academy of Agricultural Sciencs, New Delhi, 43-70.

Topping, J. C., Jr., and J. P. Bond. (1988). *The Potential Impact of Climate Change on Fisheries and Wildlife in North America.* The Climate Institute, U. S. Environmental Protection Agency.

UNESCO, (2003). The United Nations World Water Development Report: *Water for People, Water for Life.* UNESCO & Berghahn Books, Paris.

Val, A. L., and V. M. F. d. Almeida-Val. (1995). *Fishes of the Amazon and Their Environment: Physiological and Biochemical Aspect.* Springer-Verlag, Berlin, Heidelberg.

Verburg, P., R. E. Hecky, and H. Kling. (2003). Ecological consequences of a century of warming in Lake Tanganyika. *Science* 301:505-507.

Verschuren, D., and coauthors. (2002). History and timing of human impact on Lake Victoria, East Africa. *Proceedings of the Royal Society of London, Series B: Biological Sciences* 269: 289-294.

Wedemeyer, G. (1996). *Physiology of Fish in Intensive Culture Systems.* Chapman and Hall, New York.

Welcomme, R. (1979). *Fisheries Ecology of Floodplain Rivers.* Longman, Inc., New York.

Climate Change and Food Security, 2008
Edited by: M.Datta, N.P.Singh & Er.D.Daschaudhuri, pp.319-325
New India Publishing Agency, New Delhi (India)
E-mail : newindiapublishingagency@gmail.com
Web: www.bookfactoryindia.com

22

Impact of Climate on Fish Diseases

ASHUTOSH D DEO

INTRODUCTION

Aquaculture had undergone rapid expansion and intensification in the recent past and has taken the shape of a full-fledged diversified industry. But the risk involved in aquaculture activity has also increased significantly due to the outbreak of diseases which has a devastating effect on the aquaculture crops. Fish are aquatic and poikilothermic, their existence and performance are dominated by the quality of their environment .The aquatic environment is a complex system subject to constant physical and chemical changes brought about by natural and manmade processes. Therefore, choosing the proper site to develop aquaculture facility is paramount to its success.

A disease is an abnormal condition of an organism caused by infection, inherent weakness or environmental stress; it prevents the organism from functioning normally which ultimately leads to death (Avault, 1998). In an aquatic environment, there is a profound balancing of relationship between host (fish), pathogen and the environment existing and this is the fundamental concept to disease susceptibility of fish (Snieszko, 1973). A disease organism may be present in the environment or host, but will not cause disease as long as the host has high natural resistant or the physical condition of the host is strong. Once the imbalance is created to any of the component, opportunistic pathogen invades the host resulting to the outbreak of disease.

ENVIRONMENTAL STRESS

Stress is a condition in which an animal is unable to maintain a normal physiologic state because of various factors adversely affecting its well being (Plumb, 1999). Stress is caused by placing a fish in a situation which is beyond its normal level of tolerance .The list of stressors include (i) Physical stressor (temperature, light, dissolved gases *etc.*), (ii) Chemical stressors (poor water quality, pollution, metabolic wastes accumulation, diet composition), (iii) Biological stressors (population density, pathogen) and (iv) Procedural stressors (handling, shipping etc.)

CLIMATE AND FISH DISEASES

To know the impact of climate on fish diseases, one has to realize that in the aquatic environment, there is a great physical and chemical variability with the difference in the latitude than in the terrestrial environment. In a very large water body, the variation is minimal and the conditions are relatively stable while in small water bodies or in pond, variations are greater (Wedemeyer *et al.,* 1999). Climatic factors dictate the amount of rainfall, solar radiation, evaporation and temperature range from region to region thereby affecting the pond aquaculture (Kelly and Kohler, 1997). The combination of climate change and environmental degradation have created ideal conditions for the emergence, resurgence and spread of diseases in aquaculture systems . Generally, fishes are adapted to live within a relatively narrow range of physico-chemical parameter and the tolerance range to such physico-chemical parameter varies considerably from species to species (Wedemeyer *et al.,* 1999). Harvell *et al.,* (2002) opined that many pathogens of terrestrial and aquatic texa are sensitive to temperature, rainfall, and humidity, which creates synergism that affects the biodiversity.

CLIMATIC FACTORS AND ITS IMPACT ON FISH DISEASES

Temperature

Fresh water fishes are ectotherm that cannot regulate their body temperature through physiological means (Moyle and Cech, 2004).The temperature of the water is regulated mainly by the temperature of the surrounding air (Boyd and Tucker, 1998). Water temperature is one of the "master variable" in a fresh water system because of its wide spread effect on the life history, physiology, and behavior of most freshwater organism including fishes (Reynolds and Casterlin,1979).

Pathological spinal curvatures have been reported due to the effect of temperature variations either above or below the optimal level for the species (Polo *et al.*, 1991) and thermal shock (Liu *et al.*, 1994). Higher temperature can cause asynchronous development of organ systems in fish resulting in abnormalities (Brown and Nunez, 1998). Changes in temperature disrupts the structural integrity of fish, brings variations in gill ventilation rate and cardiac output (Reynolds and Casterlin, 1979). Following changes in temperature fish makes compensatory change in blood pH, osmotic and ionic balance for the maintenance of cellular metabolism (Crawshaw, 1979). Temperature has strong influence on dissolve oxygen (DO) content of water which is responsible for increased gill ventilation rate. The rapid temperature change brings about pathological changes in the gills and respiration is impaired by copious mucus secretion. The gills turn white, indicating the circulatory disorder as a result, death occurs. Carps fail to produce antibody at temperature below 12^{0}C (Kabata, 1985). Low temperature suppresses the ability of T cell to produce macrophage activating factor (MAF) which results in impaired cell mediated immunity (Roberts, 2001). At critical temperature, wound healing process is slow and the skin wounds remain open for longer time inviting the invading pathogen (Kabata, 1985). Increased climate temperature can increase pathogen development and survival rates, disease transmission, and host susceptibility (Harvell *et al.*, 2002). Transmission of parasite depends on host condition, the presence of intermediate hosts required for the completion of life cycle, water quality and temperature (Wootton, 1999; Marcogliese, 2001). Climate change can alter host-parasite dynamics by changing transmission opportunities and changing host susceptibility (Roessig *et al.*, 2004). Whirling disease caused by *Myxobolus cerebralis* is particularly detrimental to juvenile fish, causing mortality up to 89% within four weeks time and thus can severely limit recruitment in wild situation or production of fish under aquaculture condition has been reported to be linked with higher temperature (Gilbert and Granath, 2003). Hinter and Moffitt (2001) reported severe damaging impact of same pathogen on rainbow trout and cut throat trout in the United States when summer temperature was very high. The influence of temperature on the occurrence of parasitic diseases (Ich disease or white spot disease, *Chilodenella* infection, Costiasis all caused by protozoan, *Dactylogyrus* a monogenean fluke infection, Diplostomiasis caused by a digenean Trematode and crustacean disease by *Larnea* sp. infection), Viral diseases (Spring viraemia of carps, Swim Bladder Inflammation, Viral Hemorrhagic Septicemia, Lymphocystis disease), Bacterial diseases (eye diseases of carps, Columnaris disease and furunculosis), and fungal diseases (Branchyomyces infection and Epizootic Ulcerative Syndrome) are well documented by Kabata (1985), Lilley *et al.* (1998), Wedemeyer *et al.*, (1999), Chinabut and Roberts (1999), Roberts(2001), and Hoole *et al.*,(2001). The toxicity of common pollutants (eg. organophosphate and heavy metals) to fish has generally found to increase at higher temperature and the studies on bioaccumulation have shown a positive correlation between temperature and pollutant uptake (Mohopatra and Saha, 2000).

Solar Radiation

The seasonal variation in solar radiation is responsible for the seasonal variation in air temperature which generally influences water temperature (Kelly and Kohler, 1997). Solar radiation also varies with latitude. The photoperiod and light intensity are two separate factors which affects aquaculture. In aquaculture system, the quantity of light which falls on the water ,regulates the development of plankton and thus the availability of natural foods that saves the fish from nutritional deficiency disorders (Hoole *et al*, 2001).The evaporation of water is also a result of light intensity and photoperiod.

Precipitation

Precipitation occurs mainly as rain but may also occur as hail, sleet, snow, fog or dew. The amount of rainfall an area receives depends on the geographical location and the climate of that area. Epizootic Ulcerative Syndrome (EUS) that outbreaks in freshwater fish in Asia , has been reported with the rainfall events and associated low and / or decreasing water temperature, alkalinity, and hardness and chloride concentration (Lilley *et al.*, 1998).Floods are responsible for the spread of EUS infection by the spread of infected fish or the causal fungus as reported from Bangladesh which suffered a devastating flood in the year 1988 followed by the rapid spread of EUS (Lilley *et al*, 1998).

PREDICTING EARTH'S CLIMATE

Predicting trends in the earth's climate is not so easy but it can be done using a variety of techniques *viz.* Mathematical simulations of the earth's climate such as global circulation models (GCMs) which requires validation by comparing the predictions with observed climatic conditions (Rodo and Comin, 2003), forecasting by analogy (Gunn, 2002) and by the examination of the recent trends (Mann *et al.*, 1998). Each model has its advantage and disadvantage over others and the selection of model totally depends on the experience and expertise of the person involved in the prediction process.

CONCLUSION

The degree to which an individual system, a particular species, or even a single population will be affected by the climatic factors is difficult to predict. Fish being a poikilothermic animal, 90% of its activity is directly influenced by a single climatic factor,viz., temperature. Other factors like solar radiation, rainfall, evaporation, system of winds have indirect impact on the occurrence of diseases in fish. Fish is vitally important as a protein source in developing countries and provides numerous recreational opportunities in industrialized nation. Any such agro-meteorological prediction or information will help the farmers, entrepreneurs and the persons involved in aquaculture activity to take adequate measures to combat fish diseases thereby enhancing the fish production and ensuring fish to all.

REFERENCES

Avault J. W.(1998). Fundamentals of aquaculture. AVA Publishing Company, USA: 889pp.

Boyd C.E. and Tucker, G.S. (1998). Pond aquaculture water quality management Kluwer Academic Publisher, USA: 700pp.

Brown C.L. and Nunez J.M. (1998). Disorders of Development:1-17. *In: Fish Diseases and Disorders*, vol.2:Non-infectious Disorders, Leatherland J. F. and Woo P. T. K. (eds.) CABI Publishing UK: 386pp.

Chinabut S. and Roberts R.J.(1999). Pathology and Histopathology of Epizootic ulcerative Syndrome(EUS) The Aquatic Animal Health Research Institute. Bangkok: 33pp.

Crawshaw L.I. (1979). Responses to rapid temperature change in lower vertebrate ectotherms.*Am.Zool.*19:225-237.

Gilbert M. A. and Granath W.O.J. (2003) bv Whirling disease of salmonid fish:life cycle, biology, and disease. *Journal of Parasitology* 89 (4):658-667.

Gunn J. M. (2002). Impact of the (1998) El Nino event on the lake char ,*Salvelinus namaycush* population recovering from acidification. *Environmental Biology of Fishes*.64: 343-351.

Harvell C. D.,Mitchell C.E.,Ward J. R.,Alitzer s., Dobson A.P., Ostfeld R.S. ,Samuel MD (2002). Climate warming and disease risks for terrestrial and marine biota.*Science*, 296(5576):2158-2162.

Hinter M. and Moffitt C.M. (2001). Variations in infections of Myxobolus cerebralis in field exposed cutthroat and rainbow trout in Idaho. *Journal of Aquatic Animal Health* 13 :124 – 132.

Hoole D; Bucke D; Burgess P and Wellby I. (2001). Diseases of carp and other cyprinid fishes. Fishing News Books London :264 pp.

Kabata Z. (1985). Parasites and diseases of fish cultured in the tropics.Taylor and Francis, London,:318pp.

Kelly A.M. and Kohler C. C. (1997). Climate, site and pond design. *In: Dynamics of Pond Aquaculture*, Egna S.E. and Boyd C.E.(eds) CRC Press, New York:437pp.

Lilley J.H, Callinan S.; Chinabut S. ;Kanchananakhan I.M. Mac Rae; and Phillips M.J.(1998). Epizootic Ulcerative Syndrome (EUS) Technical Handbook. The Aquatic Animal Health Research Institute, Bangkok:88pp.

Liu H. W.,Stickney R.R .,Dichoff W.W. and Mc Caughran D.A.(1994). Effect of environmental factors on egg development and hatching of Pacific halibut, Hippoglossus Stenolepis. *Journal of the World Aquaculture Society* 25:317-321.

Mann M.E.R.; Bradley, R.S. and Hughes M.K. (1998). Global scale temperature pattern and climate forcing over the past six centuries. *Nature* 392:779-787.

Marcogliese D. E. (2001). Implication of climate change for parasitism of animals in the aquatic environment. *Canadian Journal of Zoology* 79:1331-1352.

Mohopatra B. C. and Saha C. (2000). Aquatic pollution and management. CIFA,Bhubneshwar :363pp.

Moyle P.B. and Cech J.J. (2004). Fishes: An introduction to Icthyology,5th edition, Prentice Hall, Engle Wood Cliff, NJ.:726pp.

Plumb J. A. (1999). Health maintenance and principal microbial disease of cultured fishes, Iowa State University Press USA: 328pp.

Polo A., Yufera M.and Pascual E.(1991). Effect of temperature on egg and larval development of *Sparus aurata* L. *Aquaculture* 92:367-375.

Reynolds W.W. and Casterlin M.E. (1979). The role of temperature in the environmental physiology of fishes. In: *Environmental Physiology of Fishes*, Ali .A.(ed.), Plenum Press, New York:723pp.

Roberts R.J. (2001). Fish Pathology . W.B. Saunders, London , :472pp.

Rodo X. and Comin F.A.(2002). Global climate : Current Research and Uncertainties in the climate System,Springer-Verlag,Berlin : 286pp.

Roessig J. M.; Woodley C.M.; Cech J.J. and Hansen L.H.(2004). Effect of global climate change on marine and estuarine fishes and fisheries. *Reviews in Fish Biology and Fisheries* 14:251-275.

Snieszko S.F. (1973). Recent advances of scientific knowledge and development Pertaining to diseases of fishes. *Advances in Veterinary Science and Comparative Medicine* 17: 291-314.

Wedemeyer G. A., Meyer F.P., and Smith L. (1999). Environmental stress and fish diseases. Narendra Publishing House, New Delhi: 192pp.

Wootton R. J. (1999). Ecology of teleost fishes, Kluwer Academic Publication, London.: 386pp.

Index

A

Abiotic factors 192
Acclimation 283
Accumulation 201, 202, 208, 210, 211, 259, 271, 320
Acid lime 247
Adequate rains 133
Adreno corticotrophin hormone 280
Adult emergence 177
African marigold 263
Age at puberty 280
Agri export zones 264
Agricultural 21, 22, 24, 25, 27, 28, 35, 37, 38, 39, 40, 41, 42, 43, 44, 46, 48, 49, 50, 53, 54, 55, 56, 57, 58, 59, 60, 62, 63, 64, 65, 66, 67, 73, 76, 77, 79, 80, 82, 86, 87, 99, 101, 103, 104, 105, 106, 107, 108, 109, 111, 112, 113,, 116, 117, 120, 122, 124, 126, 147, 161, 169, 170, 172, 180, 182, 183, 185, 186, 188, 193, 225, 238, 275, 286, 317
Agricultural community 169
Agricultural development potentials 104
Agricultural operations 39, 54, 56, 57, 77, 126, 169, 193
Agricultural planning 46, 48, 63, 213
Agriculture operations 192
Agri-horticultural crops 191
Agrimet department of IMD 188
Agri-portal 33
Agro-advisories 53
Agro advisory network programmes 54
Agro advisory service 54, 187, 216
Agro ecological sub regions 243
Agro meteorological advisory services 194
Agro-biodiversity 79
Agro-climatic zones 22, 24, 57, 80, 192
Agroclimatology 44
Agro-ecological regions 54, 83
Agro-ecozones 105
Agromet advisory bulletin 26
Agromet field unit 54
Agrometeorological 25, 38, 45, 50, 51, 66, 119, 149, 172
Agro-meteorological advisory service 64
Air pollutants 40
Air pollution 2, 98
Air temperature 11, 12, 15, 66, 103, 108, 114, 150, 154, 261, 270, 277, 278, 290, 322

Albinism 109, 111, 112
Algal blooms 308, 312
Alkalinity 322
Alphonso 246
Alpine zone 103, 104, 192
Alternate spells of rain and dry heat 114
Ambient temperature 275, 278, 280, 282, 288, 289, 291
Anaerobic soil conditions 44
Animal health 284, 285, 289, 314, 317, 323, 324
Animal husbandry 77, 80, 100, 299
Annual 9, 15, 16, 18, 41, 43, 49, 71, 72, 77, 92, 93, 96, 101, 109, 120, 121, 122, 123, 125, 126, 137, 149, 150, 154, 160, 193, 195, 200, 202, 203, 204, 205, 206, 207, 208, 212, 216, 222, 223, 225, 238, 276, 286
Annual average rainfall 216, 222
Annual drought probability 72, 77
Annual rainfall 15, 71, 72, 120, 125, 130, 136, 137, 141, 143, 150, 154, 160, 193, 195, 202, 203, 207, 212, 222, 223, 225
Anthelmintic treatment 295
Anthurium 116, 262
Anti dieuretic hormone 281
Apple 28, 250, 264
Apple scab 164
Apricot 250
Aquaculture system 301, 322
Aquatic ecosystems 172, 315
Aquatic habitat 299
Aquatic organisms 299
Aquatic texa 320
Arable land 21, 41, 129
Arctic ocean 300
Arid 38, 44, 46, 243, 246, 248, 250
Arid zone fruits 250
Aridity 31
Aromatic plants 243
Artificial fertilization 120
Asynchronous development 321
Atmospheric continuum 184
Atmospheric environments 165
Atmospheric factors 108
Atmospheric temperature 302, 308, 310
Automatic weather stations 67, 216
Available land 79
Average annual surface water flow 96

B

Bacterial diseases 322
Bacterial pathogens 174
Bamboo species 191
Banana 29, 38, 41, 114, 223, 245, 246
Banana (*musa* sp) 245
Behavioral thermoregulation 301, 303
Bioaccumulation 322
Biochemical 275, 287, 317
Biodiversity 79, 85, 98, 302, 310, 314, 316, 320
Biological observations 164
Biomagnification 172
Biophysical factors 21, 199
Biotechnological tools 100
Biotic agents 171
Biotic factors 192
Biotic stresses 55
Black pepper 243, 254
Blast in rice 226
Blast infestation 225
Bright sunshine 167, 168, 260, 262
Buffaloes 275, 276, 279, 281, 289, 290
Bulbous plants 263
Button mushroom 268, 270, 271

C

Cabbage 114, 182, 187, 223, 253
Canopy temperature 184

Capsicum 114, 253
Carbohydrate 259
Carbon dioxide 2, 4, 5, 13, 14, 17, 40, 271, 274, 286, 287
Cardamom 254, 255
Carnation 116, 261
Cashew 114, 254
Cattle 80, 97, 275, 276, 277, 278, 281, 285, 287, 288, 290, 291, 293, 294, 295
Cauliflower 28, 114, 223
Cell structure 40
Cellular metabolism 321
Central highlands 84
Cessation of rainy season 140
CH_4 production 286, 287
Chemical spray 182
Chemical stressors 320
Cherry fruits 250
Chickpea 47, 68, 129, 131, 178, 180
Chickpea and pigeon pea 129
Chilli 29, 252, 254
Chilling requirement 250
Chloride concentration 322
Citrus 114, 164, 167, 223, 247, 248
Citrus scab 164
Climate 1, 2, 5, 13, 18, 19, 21, 37, 53, 55, 65, 67, 79, 80, 85, 86, 100, 101, 103, 119, 120, 121, 122, 124, 126, 127, 129, 130, 149, 150, 163, 166, 167, 171, 172, 174, 184, 187, 191, 199, 215, 229, 243, 267, 275, 276, 277, 279, 280, 281, 284, 288, 289, 290, 293, 299, 300, 301, 302, 304, 309, 310, 311, 312, 313, 314, 315, 317, 319, 320, 321, 322, 323, 324, 325
Climate change 1, 2, 5, 13, 18, 19, 21, 37, 53, 65, 79, 85, 100, 101, 119, 121, 122, 126, 129, 149, 163, 171, 187, 191, 199, 215, 229, 243, 267, 275, 276, 293, 299, 301, 310, 311, 312, 313, 314, 315, 317, 319, 320, 321, 324, 325
Climate database 65
Climate sensitive sectors 85
Climate variability 310
Climatic changes 169, 229, 231, 233, 235, 236, 287, 300, 301, 310, 312
Climatic conditions 191, 231, 233, 236, 243, 250, 264, 275, 287, 308, 323
Climatic constraints 149
Climatic requirements 37, 150
Climatic water balance 137, 138
Cloud cover 6, 53, 57, 62, 185
Cloudy conditions 168
Cloudy weather 114, 115
CO_2 concentrations 4, 260, 271
Coarse cereals 88, 129
Coastal erosion 18, 233, 238
Coccidiosis 285
Coconut 28, 41, 106, 114, 115, 254
Coefficient of variability 136, 147
Cold blooded vertebrates 300
Cold injury 45, 50, 111, 112
Cold spells 37
Cold stress 111, 112, 283, 284, 291
Cold weather 45, 226, 275, 283, 283, 285
Cold weather forecast 226
Cold/frost 17, 38
Color infra red (cir) films 183
Computer based simulation model 176
Computing power 23
Conception 280, 281, 282, 283, 289, 290
Conducive ranges 150
Control measures 57, 163, 164, 165, 226, 285, 295
Convection 44
Conventional agriculture 98
Convergent world 14
Cool weather 247
Copious mucus secretion 321

Coral reefs 19, 311
Corm 263
Corn blight disease 184
Cortisol secretion 280
Cotton 29, 129, 178, 180, 183, 274
Critical agenda 103, 116
Critical analysis 82, 211
Critical temperature 45, 50, 279, 280, 321
Crop calendar 124, 125, 126
Crop contingency plans 63
Crop diversification 120, 227
Crop establishment operations 142
Crop failure 284
Crop growth 55, 103, 108, 109, 120, 134, 136, 142, 143, 166, 186, 192, 193, 194, 199
Crop insurance 67
Crop losses 41, 42, 100, 116, 182, 226
Crop microclimate 185
Crop microclimate changes 185
Crop morphology 66, 68
Crop phenology 174, 176, 178, 182
Crop planning 55, 100, 104, 124, 129, 140, 141, 199, 201
Crop production 32, 65, 78, 80, 82, 93, 109, 116, 130, 131, 143, 192
Crop productivity 42, 46, 82, 85, 95, 103, 192, 194
Crop stress 174, 183
Crop variety 165
Crop weather models 24, 25, 68
Cropping intensity 126
Cropping season 223
Cropping strategies 201
Cropping systems 89
Crysanthamum 116
Cucurbits 252
Curative 164
Cut flowers 255, 259
Cyclone 6, 17, 39, 41, 42, 47

D

Dahlia 262
Damping of disease 226
Damping off in nurseries 226
Dashehari 246
Day neutral 115, 264
Deccan plateau 84
Deficit of monsoon rainfall 179
Degradation 1, 45, 86, 91, 92, 93, 94, 95, 97, 98, 99, 101, 120, 248, 316
Degree day concept 174, 180, 181, 182
Degree days based model 188
Denitrification 44
Dependable rainfall 160
Desertification 18, 38, 43, 45, 86, 101
Deteriorating quality 120
Development 14, 18, 31, 35, 36, 37, 38, 39, 40, 41, 45, 46, 49, 50, 66, 68, 103, 104, 105, 106, 107, 108, 114, 117, 119, 120, 124, 130, 160, 163, 164, 165, 166, 168, 169, 172, 176, 177, 180, 182, 184, 185, 186, 187, 199, 213, 226, 229, 231, 232, 233, 234, 235, 240, 290, 295, 298, 321, 322, 323, 324, 325
Dew 108, 109, 175, 185, 322
Dew period 175
Diapause 179, 187
Digestion 283, 307
Digital elevation model 48
Dioecious lines 249
Disaster management 50, 232
Disasters 19, 36, 37, 39, 45, 46, 50, 51
Disease buildup 164
Disease development 41, 163, 164, 165, 166, 168, 169
Disease forecasts 163
Disease organism 319
Disease severity 168, 169, 170
Disease symptoms 165
Disease transmission 321

Dissolve oxygen 321
Dissolved 307, 308, 320
Distribution 2, 19, 36, 72, 95, 104, 133, 135, 137, 151, 152, 154, 160, 161, 166, 173, 193, 196, 203, 204, 205, 206, 207, 208, 212, 222, 223, 276, 284, 301, 302, 303, 308, 309, 316
Diurnal temperature 17
Diversification of the cropping pattern 100
Downpours 8
Drought 7, 8, 17, 27, 34, 37, 38, 39, 43, 46, 47, 51, 53, 55, 72, 73, 76, 77, 100, 226, 284
Drought stress 183
Dry and wet spells 55, 213
Dry desiccating winds 37
Dry hot weather 248
Dry matter 277, 283
Dry or wet spell 199
Dry rubber yield 149, 150
Dry spell 27, 38, 47, 133, 199, 200, 213
Dry weeks 151, 157
Dryland agriculture 66, 129, 130, 147
Dryland areas 130, 143
Dutch rule 175

E

Early and late blight of potato and tomato 167
Early maturing crop 134
Earthen dams 199
Eastern coastal plain 85
Eastern himalayas 85
Eastern plateau 84
Ecological balance 82
Ecological space 310
Economic criteria 171
Economic impact assessment 28
Economic injury 171
Economic studies 48
Edaphic factors 109
Edible mushrooms 267, 268, 271, 272, 273, 274
Egg 181, 182, 275, 324
Elevated co2 260
Elite breeds 100
Embryonic mortality 280
Encroachment 44, 97
Endocrinological 275, 287
Endset of rainy season 210
Enteric diseases 285
Enteric fermentation 286
Environmental factors 154, 270, 271, 279, 286, 324
Environmental planning 50
Environmental stress 249, 319, 320, 325
Epidemiology 166, 170
Epiphytotic conditions 167
Epizootic ulcerative syndrome 322, 323
Erosion 18, 39, 43, 44, 47, 49, 50, 92, 93, 95, 96, 98
Etl 178, 183, 188
Eutrophication 312, 313
Evaporation 39, 40, 46, 66, 67, 68, 69, 109, 110, 116, 140, 162, 279, 300, 301, 309, 310, 320, 322, 323
Evapotranspiration 38, 40, 55, 70, 109, 130, 140, 141, 162, 309
Ever growing population 122
Evergreen 200
Expenditure 284
Expert systems 177, 178
Export quality 246
Extensive efforts 169
Extreme events 7, 17, 18, 38, 39, 41, 50, 121

F

Farm animals 40, 289
Farm holdings 79
Farm inputs 28, 57

Farm planning 65
Farm production 169
Farming community 29, 56, 58, 61, 90, 151
Fauna 191, 301, 310
Feed efficiency 284
Feed intake 277, 278, 279, 280, 283, 284, 285, 290
Female flowers 249
Fertility 38, 42, 43, 281
Fertilization 120, 185, 247
Fertilization or irrigation 185
Fertilizer use 27, 88
Finger millet 129
Fire outbreaks 40
Fish diseases 309, 319
Fish mortality 308
Fish production 299, 308, 323
Fish species 2, 301, 304, 308, 310, 313, 316
Fisheries 41, 42, 43, 77, 85, 100, 121, 298, 299, 305, 325
Flood 8, 18, 34, 36, 38, 39, 44, 47, 48, 49, 53, 85, 86, 96, 121, 223, 226, 241, 300, 302, 304, 311, 322
Flood mitigation 49
Flood plains 311
Flora 191, 301
Floriculture 116, 212, 243, 255
Flower drop 114
Flowering of rice 111
Flowers 39, 116, 177, 243, 249, 255, 259, 260, 261, 262, 263
Fog 40, 109, 322
Foliage 116, 185
Food requirements 82
Food security 1, 21, 37, 53, 65, 79, 103, 119, 126, 129, 149, 163, 171, 191, 199, 215, 229, 243, 267, 275, 293, 299, 319
Foodgrains production 88
Food-security 96
Forages 279
Forecast 22, 24, 25, 26, 27, 30, 31, 32, 36, 65, 127, 163, 173, 175, 176, 177, 180, 182, 188, 216, 225, 226
Forecasting models 166, 176, 188
Foreign exchange 243, 259
Forest 34, 35, 38, 39, 79, 84, 85, 86, 91, 96, 97, 98, 105, 115, 120, 191, 200
Forest degradation 97
Forest management 79, 91
Forest vegetation 191
Forestry 41, 42, 51, 82, 96, 97, 98, 101
Forestry systems 98
Forward and backward accumulation 201, 202, 208, 210, 211
Frequency of floods 300
Freshwater fish 315, 322
Freshwater fisheries 299, 309, 312, 313
Freshwater fishes 303, 308, 309
Freshwater resources 79, 312
Frost 17, 27, 29, 37, 38, 40, 45, 50, 109, 114, 115, 116, 216, 226, 262, 285
Frost occurrences 226
Fruit characteristics 244
Fruit quality 114, 247
Fungal growth 114
Fungal species 267
Fungicidal spray 174, 182
Future weather 176, 177

G

Gastrointestinal parasite 293
Geographic information system (gis) 184
Geographical information system 48
Geographical mobility 310
Geographical region 151
Geological 230, 231, 232, 234, 240, 241

Geological survey of india 230, 231, 232, 234, 241
Geomorphic data 232
Geomorphic units 232, 235
Geomorphological map 232
Geomorphology 235, 239, 240
Gerbera 116, 262
Gestation length 280
GHGs 2
Gill ventilation rate 321
Ginger 28, 254, 264
GIS software 232
Glaciers 1, 6, 17, 19, 236, 314
Gladiolus 263
Glass-house 261
Global circulation models 323
Global climate change 120, 122, 311, 314, 325
Global environment 299, 300
Global sea level 6
Global warming 1, 2, 5, 187, 286, 287, 288, 299, 300, 303, 309, 310, 312, 314, 317
Goat population 276
Grain and fruit spoilage 44
Grain sterility 111, 112
Grape 248
Great lakes of africa 311
Greenhouse gases 287
Green revolution 97, 129
Greenhouse grasses 286
Gross domestic products 275
Ground water 62, 96, 110, 172, 207, 232
Groundnut 16, 28, 68, 73, 114, 129, 212
Groundnut and other oilseed crop 212
Guava 28, 115, 116, 249

H

Habitat heterogeneity 310
Hail 27, 40, 109, 194, 322
Hardness 322
Healthy plant 184
Heart failure 286
Heat and cold waves 18, 55
Heat index 17, 290
Heat stress 37, 277, 281, 288, 290
Heavy metals 322
Heavy rainfall 36, 38, 43, 44, 114, 302
Heavy rains 7, 8, 112, 114, 193
Helicoverpa 177, 178, 180, 181, 183, 186
Helminthic infections 293
Heterogeneous world 14
High density planting 246, 248
High humidity 114, 116, 194, 224, 246, 248, 254, 270, 278
High intensity rainfall events 120
High relative humidity 194, 223, 270
High winds 38
Higher temperature 268, 307, 321
Himalayan mountain range 121
Holocene 230, 232, 234, 235, 236, 239
Horticultural crop 264
Host resistance 309
Host susceptibility 321
Host-parasite dynamics 321
Hot winds 114
Hot-weather 158, 159, 160
HTML 71
Human existence 301
Human health 98, 313
Human life 40, 46, 99
Humid 38, 84, 85, 106, 130, 131, 149, 161, 167, 185, 191, 200, 226, 243, 246, 248, 249, 250, 253, 254, 255, 264, 275, 277, 279, 281, 282, 283, 285, 289
Humidity 53, 57, 66, 67, 68, 175, 176, 177, 179, 185, 194, 197, 223, 224, 226, 268, 270, 271, 275, 276, 277, 278, 279, 282, 284, 285, 287, 293, 320

Humus soil 45
Hurricane 8, 41, 42, 51
Hydrogen sulphide 40
Hydrogeomorphology 35

I

Ice storms 37
Incidence 38, 40, 41, 46, 163, 174, 175, 176, 194, 249, 281, 282, 293, 294, 312
Incidence of oestrus 280, 281
Increase in CO_2 concentration 121
Incubation period 169
India meteorological department 151, 173
Indian council of agricultural research (ICAR) 60
Indian ruminants 287
Information technology 60
Infrared thermometry 184
Infrastructure 24, 34, 35, 36, 40, 42
Inland fisheries 43, 301, 303, 317
Inoculum 166
Input use efficiencies 120
Insect or disease attack 175
Insect pest 172, 174, 187, 192
Insect pests and diseases 194
Insect resurgence 172
Integrated pest management (IPM) 100, 172
Inter society consortium for plant protection 171
Intercropping systems 131
Interdisciplinary approach 171
Intergovernmental panel on climate change 310
IPCC 5, 13, 287, 310
Irrigation potential 43, 129
Irrigation scheduling 124
Isotherms 126

J

Java 70
Jhum cultivators 119, 120
Jhum cycle 119
Jhum lands 120, 191, 194

K

Kharif crop 142, 193
Kinnow mandarins 248
Kisan goshthi 32
Kyoto protocol 2

L

Lactating sows 280, 283, 290
Lactation 279, 288, 290
Land and water resources management 82
Land degradation 86, 92, 93, 95, 101
Land preparation 130, 143, 192, 201
Land quality 43
Land use changes 48, 99, 103
Land use policy 89, 90, 91
Landslide 39
Landward shifting 237, 238
Late blight 167, 170, 186, 226
Late holocene shifting 234
Latent period 169
Latex production 153
Latex vessels 149
Leaf wetness 31, 169
Leaf yellowing 111, 112
Lemon 247
Lethal 186, 187
Life cycle 165, 181, 186, 187
Life-support system 99
Light intensity 109, 260, 261, 262, 263, 264, 271, 322
Light trap 68, 177

Lightning 40
Lime and lemon 247
Limnology 301, 309, 314
Linseed 131
Litchi 114, 115, 116, 249
Literacy 79
Livelihood 79, 85, 90, 21, 310
Livestock 26, 27, 28, 29, 38, 39, 41, 42, 46, 56, 64, 275, 276, 277, 284, 286, 287, 288, 289, 290, 291, 294
Local area network 71
Local solutions 13, 14
Locust 41
Long day 253, 261, 263
Longer growing season 133
Lord Buddha 239
Loss in production 92
Low chilling cultivars 250
Low germination 114
Low humidity 114, 116, 285
Low light intensity 263
Low pod formation 112
Low relative humidity 111
Low temperature 109, 114, 115, 116, 181, 246, 263, 285, 286, 321
Lower critical temperature 280
Lowlands 235, 236

M

Macroclimate 167, 276
Macromolecules 45
Mahaseers 309
Maize 28, 29, 89, 105, 120, 131, 167, 168, 169, 223, 224, 284
Maize cultivation 224
Major cereal crop 225
Male flowers 249
Malformation 109, 247
Mandarin 247
Mango 28, 187, 223, 246, 247, 264
Mangrove roots 232, 234, 235
Mangrove roots and marine shells 234
Mangroves 19, 85, 232, 235
Manipur 15, 32, 105, 107, 110, 115, 154, 215, 216, 217, 222, 225
Marigold 263
Mass media 32, 54, 57, 61
Maximum temperature 7, 17, 57, 66, 68, 122, 176, 194, 196, 216, 217, 218, 268, 270, 296, 297
Meat 28, 100, 275
Metabolisable energy intake 278
Mechanical damage 38, 42, 49, 164
Medicinal 243, 244, 267, 273, 274
Medicinal values 267, 274
Medium range 30, 54, 57, 62, 64, 163, 173, 177, 183, 188
Medium range weather forecast 22
Metabolic activity 307
Metabolic heat 284
Meteopathological forecasting 163
Meteorological data 66, 200, 216, 225, 282, 293, 294
Meteorological station 166
Meteorology 64, 77, 170, 172, 173, 174, 216
Methane 2, 4
Microclimate 165, 166, 174, 185, 276, 288
Microhabitat 300, 303, 313
Micromet data 177
Mild tropical hill zone 103, 192
Mild tropical plain zone 103, 192
Milk 28, 100, 275, 276, 277, 278, 279, 280, 283, 287, 288, 290
Milk yield 277, 278, 279, 280, 283, 288
Minerals 267
Minimum temperature 17, 57, 66, 68, 122, 175, 176, 179, 193, 219, 220, 262, 295, 296, 297
Mite infection 285

Mitigation plans 34
Mizoram 32, 105, 107, 110, 121, 191, 192, 193, 194, 195, 196, 215
Moderate 85, 92, 103, 120, 150, 155, 181, 194, 261
Moderate wind 103
Moist deciduous forest 200
Moisture availability index (MAI) 143
Moisture regimes 70
Moisture stress 49, 114, 134, 143, 186 , 207, 212
Monocrop paddy 223
Monocrop vegetables 223
Monocropping 129
Monoecious lines 249
Monsoonal rain 193
Monthly 66, 69, 70, 71, 72, 77, 141, 150, 151, 152, 159, 160, 161, 167, 178, 179, 180, 193, 194, 196, 197, 198, 200, 201, 219, 221, 222, 246, 285, 294
Multiple regression 168, 169, 175, 178, 179, 297
Muskmelon 252, 253
Mustard 28, 68, 114, 129, 131, 223
Mycelial growth 268, 271

N

Nagpur mandarins 248
National agro advisory network 58, 60, 64
National land use policy 89, 90, 91
National remote sensing agency 232, 238
Natural enemies 177
Natural resources 19, 54, 79, 85, 94, 98, 119, 130, 188, 286
Natural rubber 160, 161
NCA 129
NCMRWF 54, 57, 60, 61, 62, 63, 64, 66, 67, 183, 185, 188
Negative effects 39
Negative impacts 38, 312
Nematodes 171, 174, 185, 186
Nitrate 44
NLB epidemics 168
North east india 103, 119, 120, 123, 127, 149, 199
Northern leaf blight of maize 167
Nutrient concentration 307
Nutritional deficiency 322
NWP models 31

O

Occurrence of diseases 323
Oestrous cycle length 280
Oilseeds 88, 105, 106, 112, 114, 129, 131, 137
Okra 252
Onion 28, 29, 253
Onset and cessation of wet season 201
Onset and endset of wet period 202, 208
Onset of monsoon 136, 210, 223
Onset of sowing rains 140
Optimal 103, 155, 268, 269, 307, 313, 321
Orchid 264
Organic agriculture 100
Organic matter 120, 271
Organophosphate 322
Orographic structure 121
Osmoregulatory systems 304
Osmotic and ionic balance 321
Outbreak of disease 320
Ova abnormalities 280
Ovulation failure 280
Oyster mushrooms 268
Ozone depletion 99, 286

P

Paddy cum fish culture 227
Paddy straw mushroom 270, 273, 274
Paleo-environmental conditions 232
Pan evaporation 109, 110
Papaya 28, 223, 249
Parasite 183, 188, 293, 298, 309, 321
Parasite populations 293
Parasitic diseases 321
Parasitic infection 293, 295
Pathogen 165, 166, 185, 187, 319, 320, 321
Pathogen and host 165
Pathogen development 185
Pathogens 174, 175, 185, 186, 285, 320
Peach 115, 116
Pear 250
Pearl millet 68, 129, 131
Pendulous sheath 278
Penman equation 110
Per capita 2, 3, 14, 21, 79, 80, 88, 96, 97, 122
Percent disease index (PDI) 167
Perennial crop 41, 149
Pest and disease epidemics 109
Pest development 39, 40, 185
Pest forecasting 173, 176, 178, 185, 186, 188
Pest management 100, 171, 172, 173, 174, 178, 188
Pest phenology 172, 174, 178
Pest population 171, 175, 176, 177, 178, 179, 183, 187
Pest risk analysis 174, 186, 187
Pest scenario 174, 187
Pesticides 55, 63, 151, 172, 178
Pet 70, 109, 110, 116, 124, 127, 140, 141, 142, 240, 289
Phenology 66, 68
Pheromone trap 176, 178
Phonological stage 199
Photo period 175, 262, 263, 322
Photosensitization 285
Photosynthesis 109, 111, 260, 261
Physical environment 95, 165, 174, 238, 275, 276
Physical stressor 320
Physiological 169, 272, 273, 274, 275, 277, 278, 290, 304, 316, 317, 320
Physiological activity 169
Physiological responses 277
Phytotoxic 164
Pineapple 115, 223
Plankton 322
Planning commission 80, 100
Plant disease 163, 165, 174, 184, 189
Plant disease development 163
Plant disorders 109
Plant growth 38, 43, 109, 114, 165, 183
Plant nutrition 100, 165
Plant protection 171, 178, 187, 188, 227
Plant root zones 44
Plant vigour 44
Plasma thyroxin 282
Pleistocene 234, 235, 236
Pleistocene ice age 234
Plum 115, 250
Poikilothermic 301, 303, 319, 323
Polar ice 301
Pollination 39, 115, 247
Pollutants 40, 289, 303, 307, 322
Poly-house 261, 262
Pomegranate 114, 115, 251, 264
Pooled data 168
Population density 178, 179, 180, 320
Population pressure 119
Postharvest losses 227
Post monsoon 11, 71, 72, 77, 86, 120, 201, 206, 207, 218, 220, 222
Post rainy season 131
Potato blight 170, 175, 182

Potential evapotranspiration 70, 109, 130, 140, 141, 162
Poultry sector 275
Poverty 79, 99, 310, 311, 313
Poverty alleviation 99, 310
Powdery mildew 226
Precipitation 42, 43, 48, 96, 121, 149, 151, 154, 161, 176, 201, 235, 277, 287, 301, 304, 309, 313
Precipitation concentration index (PCI) 151
Predictions 123, 163, 323
Pregnancy 281, 285
Pre-holocene 232
Preventive 90, 164, 178, 285
Probability estimates 157
Probability of rainfall 132
Probability studies 151
Procedural stressors 320
Production index 123
Productivity 21, 56, 73, 82, 85, 86, 87, 88, 89, 90, 91, 92, 93, 95, 96, 97, 99, 100, 101, 103, 109, 112, 114, 116, 120, 124, 126, 130, 192, 194, 199, 223, 275, 276, 288
Projected population 88
Prolactin 281
Puddling 201
Pulse availability 86
Pulses 86, 88, 89, 106, 112, 114, 223
Pulses and oilseeds 137

Q

Quarantine 187
Quaternary period 229, 235

R

Rabi crop 142, 207, 209, 212
Rain 8, 42, 44, 48, 53, 67, 103, 108, 109, 114, 115, 185, 193, 200, 201, 203, 207, 208, 216, 223, 225, 248, 250, 271, 284, 285, 286, 322
Rain storms 44
Rainfall 9, 10, 14, 36, 38, 39, 43, 44, 46, 47, 48, 67, 68, 69, 71, 72, 76, 77, 96, 125, 130, 131, 132, 133, 134, 135, 136, 137, 140, 141, 142, 143, 146, 147, 149, 150, 151, 152, 154, 155, 157, 159, 160, 161, 162, 168, 173, 175, 176, 179, 180, 182, 188, 191, 193, 194, 195, 196, 197, 199, 200, 201, 202, 203, 204, 205, 206, 207, 208, 209, 210, 211, 212, 213, 216, 221, 222, 223, 225, 236, 293, 295, 297, 301, 302
Rainfall availability 151
Rainfall characteristics 131, 199, 202
Rainfall distribution 72, 134, 135, 151, 193, 196, 222
Rainfall index 123
Rainfall pattern 132, 133, 135, 137, 193, 213, 249, 301, 302
Rainfall probabilities 71, 162
Rainfall values 133
Rainfall variability 38, 43, 71, 72, 137, 143, 146, 151, 161
Rainfed agriculture 43, 129, 130
Rainfed areas 96
Rainfed crop 65
Rainstorms 309
Rainy season 96, 131, 133, 135, 136, 140, 141, 142, 201, 210, 211, 225, 248, 263, 268, 271
Range fires 38
Recreational opportunities 323
Regression analysis 295
Regression equations 152
Relative humidity 57, 66, 67, 68, 111, 117, 130, 162, 175, 176, 179, 194, 197, 223, 224, 226
Remote sensing 31, 232, 238
Remote sensing techniques 174, 177
Reproductive 40, 112, 113, 120, 130, 184, 280, 282, 283, 289, 291, 300, 304, 309

Reproductive ability 309
Reproductive behaviour 282
Reproductive phase 114, 130
Respiration rate 277, 278
Retrieval 66, 70, 71
Rice blast 167
Rice production 39
Ring spot virus 249
Risk assessments 48
River basins 311, 316
River discharge 237, 301, 310
Root functioning 120
Root zone 143
Rose 116, 260, 261, 316
Rubber based farming community 151
Rubber cultivation 149, 150, 151, 155, 160, 161
Rubber yield 149, 150
Ruminants 2, 96, 287, 288

S

Saffron 243, 254
Salinity 42, 46, 95, 303
Salinization 41, 86
Salt deposition 42
Sand storms 45
Sapota 249
Satellite images 35, 238
Saturated moisture 114
Scab of citrus 167
Scheduling insecticidal 174, 182
Sea level changes 229, 233, 241
Seasonal 12, 16, 66, 71, 72, 116, 121, 130, 135, 136, 137, 149, 151, 164, 167, 200, 201, 203, 204, 207, 208, 227, 233, 247, 248, 274, 281, 322
Seasonal-rainfall 135
Seawater influx 300
Sedimentation technique 294
Seed germination 120
Seed storage 114
Seedling mortality 111, 112
Self floatation technique 294
Sequential 131, 142
Shelterbelts 49, 50, 55
Shifting cultivation 92, 93, 119
Shoreline 230, 231, 232, 234, 237, 238, 239
Short range 163, 173
Shortage 1, 122, 309
Shorter daylength 142
Shorter growing season 133
Shrubs 39, 49, 191
Sigatoka leaf spot 246
Skewness 160
Snow 37, 40, 45, 47, 50, 108, 109, 285
Snow cover 6, 17, 18, 45, 50, 300
Soil 21, 31, 38, 39, 41, 42, 43, 44, 45, 46, 47, 48, 49, 51, 55, 57, 62, 66, 70, 77, 79, 80, 82, 91, 92, 93, 94, 95, 96, 98, 99, 101, 103, 109, 114, 116, 117, 119, 120, 124, 129, 130, 141, 142, 143, 153, 154, 162, 174, 176, 177, 184, 185, 186, 187, 192, 199, 201, 202, 207, 213, 229, 244, 246, 247, 249, 262
Soil and vegetation 119
Soil borne pathogens 174, 185
Soil conservation structures 199
Soil erosion 92, 99
Soil moisture 31, 38, 43, 44, 70, 77, 141, 142, 143, 167, 176, 184, 186, 207, 287
Soil moisture storage 153
Soil nutrients 44, 49
Soil particles 49
Soil resources 94
Soil temperature 44, 45, 103, 109, 114, 162, 246, 262
Soil water availability 143, 144

Solar radiation 130, 154, 175, 186, 309, 320, 322, 323
Sorghum 28, 29, 68, 77, 129, 131, 186
Southern oscillation index 123
Southwest monsoon 130, 135
Soybean 68, 129, 223
Spatial 11, 35, 36, 48, 53, 62, 184
Spatial heterogeneity 154
Spatial information 62
Spatial variability 151
Spatial weather information system 36
Spices 105, 243, 254, 255
Spore germination 272
Spraying operation 182, 188
Sprinkling water 226
Stagnation/decline 120
Stepwise regression analysis 169
Sterility 40, 55, 111, 112
Stone fruits 250
Stratosphere 5, 6
Stream flow 43
Stress 29, 37, 39, 49, 77, 111, 112, 114, 134, 143, 174, 183, 184, 186, 247, 249, 275, 277, 278, 281, 283, 284, 286, 288, 290, 291, 319, 320, 325
Strong winds 27, 38, 42, 44, 55, 111
Strongyle infection 293, 294, 295, 296, 297
Structural lipids 304
Stubble 49
Sub-humid climates 149
Subsistence farming 129
Subsistence living 98
Subtropical climate 6, 96, 192, 247, 250, 268, 314
Suitability of clones 160
Sunshine 66, 67, 68, 69, 73, 82, 103, 111, 162, 167, 168, 261, 262
Surface water runoff 110
Surplus 70, 110, 117, 138, 140, 141, 142, 143, 162
Sustainability 13, 94, 95, 97, 98, 99, 100, 119, 121, 312
Sustainable agricultural productivity 103, 116
Sustainable land management 99, 101
Sustainable strategies 46
Sweet orange 116, 247

T

Tamralipta 239, 240
Tectonic activities 236
Tectonic changes 235, 236
Temperate 37, 121, 192, 215, 243, 246, 248, 250, 253, 254, 261, 262, 264, 284, 302, 304, 308
Temperate crop 248
Temperate fruits 104, 115, 250
Temperate hilly regions 114
Temperature 1, 5, 6, 7, 11, 12, 15, 16, 17, 18, 26, 31, 57, 66, 68, 85, 103, 108, 109, 111, 114, 115, 116, 119, 120, 121, 122, 124, 125, 127, 130, 142, 150, 153, 154, 162, 166, 168, 169, 170, 174, 175, 176, 177, 179, 181, 182, 184, 185, 186, 187, 188, 193, 194, 196, 197, 216, 217, 218, 219, 220, 224, 235, 236, 237, 239, 244, 246, 247, 248, 249, 250, 252, 253, 254, 255, 259, 261, 262, 263, 268, 269, 270, 271, 274, 278, 293, 295, 296, 297, 307, 308
Temperature -humidity index 279
Temporal distribution 199
Temporal variability 184
Ten day rainfall distribution 135
Terrestrial 310, 320, 324
Thaws 50
The mahabharata 239
Thermal comfort 278, 280
Thermal radiation 184

Thermal stress 29, 277, 278, 281
Threshold value 37
Thunderstorms 38
Thyroxin secretion rate 282, 289
Tomato 68, 114, 223, 226
Total geographical area 35, 94, 191, 215
Toxic metabolites 308
Toxins 308
Traditional crops 129
Transgenic plants 172
Transpiration 38, 96, 251
Transplanting of rice 201
Tree belts 49
Tree planting 98
Tree species 191
Tribe 191
Tropical 6, 17, 18, 19, 38, 41, 43, 46, 86, 96, 121, 191, 192, 200, 215, 216, 223, 224, 233, 268, 285, 304, 309, 314, 315
Tropical cyclone intensity 6, 17
Tropical dermatophilosis 285
Tropical fishes 304
Tropical forest type 200
Tropical orchids 264
Tropical storm 43
Troposphere 5, 6
Tuberose 262, 263
Turcicum blight 167, 169
Turmeric 254

U

Unproductive tillers 111, 112
Upland rice 131
Urogenital diseases 286
Uterine involution 282

V

Vagaries of climate 310
Vagaries of erratic weather conditions 192
Variation 11, 12, 71, 72, 104, 109, 112, 146, 199, 200, 213, 217, 218, 219, 220, 221, 222, 224, 247, 248, 254, 263, 264, 286, 290, 302, 308, 320, 322
Varietal characteristics 166
Vegetable 212, 251, 254
Vegetation 44, 47, 48, 50, 119, 191, 200
Vegetative growth 112, 253, 260, 261, 262, 263
Vegetative stage of crops 119
Vernalization 40
Viral diseases 322
Virulent strain 165
Visual basic program 70
Vitamins 267

W

Warm equable climate 150
Water 1, 2, 4, 21, 22, 28, 31, 34, 35, 36, 38, 39, 40, 42, 43, 46, 47, 48, 50, 55, 62, 70, 77, 79, 80, 82, 85, 86, 91, 92, 93, 95, 96, 98, 99, 100, 101, 109, 110, 117, 119, 120, 124, 126, 129, 130, 138, 140, 141, 142, 143, 144, 146, 149, 153, 161, 162, 184, 185, 199, 207, 213, 226, 227, 232, 237, 238, 247, 250, 251, 259, 260, 275, 276, 277, 278, 279, 283, 284, 287, 289, 290, 299, 300, 301, 302, 303, 304, 307, 309, 310, 312, 313, 314, 315, 316, 317, 320, 321, 322, 323
Water availability 42, 138, 143, 144, 149, 287, 302, 309, 310
Water balance 55, 70, 138, 140, 309
Water body 320
Water budget 109, 110
Water deficiency 130
Water deficit 70, 110, 117, 138, 140, 141, 143

Water erosion 92, 93
Water harvesting ponds 199
Water harvesting structure 110, 207
Water holding capacity 130
Water intake 275, 276, 277, 278, 279, 287
Waterlogging 92, 93, 95, 114
Watermelon 252, 253
Water movement 184
Water pollution 40, 99
Water quality 99, 320, 321, 323
Water resources 21, 35, 80, 82, 85, 95, 96, 100, 120, 146, 315
Water storage 110
Water stress 77, 309
Water surplus 70, 117
Water temperature 320, 322
Weather 19, 21, 22, 23, 24, 25, 26, 27, 29, 31, 32, 36, 37, 38, 39, 41, 45, 46, 47, 48, 50, 53, 54, 55, 56, 57, 58, 62, 63, 64, 66, 67, 68, 70, 71, 75, 121, 127, 144, 150, 153, 158, 159, 160, 163, 164, 165, 166, 167, 168, 170, 171, 172, 173, 175, 176, 177, 178, 179, 182, 183, 184, 188, 192, 193, 194, 215, 216, 225, 226, 247, 248, 275, 279, 282, 283, 285, 293, 313
Weather conditions 55, 104, 110, 163, 164, 165, 174, 175, 192
Weather disaster 39, 41
Weather forecast 22, 24, 25, 27, 31, 36, 188, 216, 225, 226
Weather information 22, 26, 36, 164, 172, 174, 182, 188, 216
Weather parameters 21, 53, 56, 146, 153, 160, 168, 173, 175, 178, 179, 182, 184, 192, 193, 194, 293
Weather sensitivity 31, 32
Weather systems 103, 116
Weather vagaries 112, 116
Weather variable 167
Web-based dissemination 60
Weekly 54, 58, 66, 69, 70, 71, 77, 100, 124, 134, 138, 144, 146, 200, 201, 202, 207, 208, 209, 210, 211
Weekly distribution of rainfall 146
Weekly rainfall probability 207
Weight gain 284
Western Himalayas 83, 84
Wet condition 277
Wet spells 38, 96, 110, 152, 157, 158, 213
Wet weeks 151, 157, 158
Wetlands 1, 48, 235, 301, 302, 312
Wetness duration 169, 185
Wetness period 177
Wheat 17, 28, 68, 88, 89, 129, 175, 176
Wheat rusts 175
White mold disease 185
Wild collections 299
Windbreaks 49, 50
Wind erosion 43, 49, 50, 92, 93
Wind speed 26, 29, 49, 55, 57, 66, 67, 68, 69, 182, 183, 284
Winter resistance 50
World cattle population 287
World commission on environment 98
Wound healing process 321

Z

Zebu cattle 278
Zebu crosses 278

□□□